CRM Short Courses

T0171869

The volumes in the **CRM Short Courses** series have a primarily instructional aim, focusing on presenting topics of current interest to readers ranging from graduate students to experienced researchers in the mathematical sciences. Each text is aimed at bringing the reader to the forefront of research in a particular area or field, and can consist of one or several courses with a unified theme. The inclusion of exercises, while welcome, is not strictly required. Publications are largely but not exclusively, based on schools, instructional workshops and lecture series hosted by, or affiliated with, the *Centre de Recherches Mathématiques* (CRM). Special emphasis is given to the quality of exposition and pedagogical value of each text.

More information about this series at http://www.springer.com/series/15360

Marc-Hubert Nicole

Editor

Arithmetic Geometry of Logarithmic Pairs and Hyperbolicity of Moduli Spaces

Hyperbolicity in Montréal

 Springer

Editor
Marc-Hubert Nicole
Institut de mathématiques de
Marseille (I2M)
Université d' Aix-Marseille
Marseille, France

ISSN 2522-5200   ISSN 2522-5219   (electronic)
CRM Short Courses
ISBN 978-3-030-49866-5   ISBN 978-3-030-49864-1   (eBook)
https://doi.org/10.1007/978-3-030-49864-1

Mathematics Subject Classification: 32Q45, 11G35, 14G35, 14G05, 14G25

This Springer imprint is published by the registered company Springer Nature Switzerland AG
The registered company address is: Gewerbestrasse 11, 6330 Cham, Switzerland

Preface

This book aims at introducing a wide audience ranging from number theorists with a basic course in algebraic geometry under the belt to complex geometers to some of the exciting developments around the theme of hyperbolicity and the fundamental conjectures of Bombieri, Lang, Vojta, and others, beyond the well-known case of curves. Recall that Faltings's theorem (née Mordell conjecture) states that the number of rational points of a curve of genus greater or equal than two over any number field is finite. Analogues of the abovementioned conjectures exist with suitable modifications over function fields, and we have tried emphasizing this point in the book especially for function fields of characteristic zero. Therefore, some readers might be stimulated to investigate the more subtle, yet still tractable analogues of the works mentioned therein over function fields of positive characteristic in higher dimensions. As of 2020, the original conjectures are still very wide open over number fields. Classes of examples for which they are known are documented in chapters "The Lang–Vojta Conjectures on Projective Pseudo-Hyperbolic Varieties" and "Hyperbolicity of Varieties of Log General Type". In some special cases, much stronger and presumably much harder to prove statements are expected to be true. For example, the generalized Fermat equation

$$x^m + y^n = z^r,$$

where x, y, z, m, n, r are positive integers with $m, n, r > 2$ and x, y, z are pairwise coprime, is known by a theorem of Darmon–Granville of 1995 to have at most finitely many solutions for fixed m, n, r, and its proof relies on orbifold curves and a dévissage to Faltings's theorem, see chapter "Arithmetic Aspects of Orbifold Pairs" for a sketch of the proof. On the other hand, Beal's much harder conjecture states that the generalized Fermat equation should have no such solutions with coprime factors whatsoever.

To capture the original atmosphere of the delightful lectures in Montréal in 2018 and 2019, we give a very brief description of the chapters' contents in French in the following lines; for more details (in English), the reader is referred to the introductory section of the corresponding chapter.

$$* * *$$

Cet ouvrage comporte quatre chapitres.

Le premier chapitre intitulé *Lectures on the Ax–Schanuel Conjecture* par Benjamin Bakker et Jacob Tsimerman, explique les grandes lignes de la preuve de la conjecture d'Ax–Schanuel pour les variations des structures de Hodge reposant sur les techniques de géométrie o-minimale. Les résultats originaux expliqués ici ont paru en 2019 sous forme d'article de recherche.

Le second chapitre intitulé *Arithmetic Aspects of Orbifolds Pairs* par Frédéric Campana, est un exposé des conjectures de Campana visant un auditoire varié (comportant des théoriciens des nombres, des géomètres arithméticiens et des géomètres complexes), y compris sa notion éponyme de paires orbifoldes placée littéralement au 'cœur' de ses conjectures.

Le troisième chapitre intitulé *The Lang–Vojta Conjectures on Projective Pseudo-Hyperbolic Varieties* par Ariyan Javanpeykar, est une introduction au thème de l'hyperbolicité et des conjectures de Lang–Vojta dans le cas projectif, ainsi qu'à une bonne dose de résultats dus à l'auteur.

Le quatrième chapitre intitulé *Hyperbolicity of Varieties of Log General Type* par Kenneth Ascher et Amos Turchet, continue d'explorer le thème de l'hyperbolicité et des conjectures de Lang–Vojta dans le cadre plus général des variétés quasi-projectives. Il fournit en particulier une présentation de résultats par les deux auteurs et leur collaboratrice DeVleming.

Remerciements : chacun des chapitres de cet ouvrage est basé sur un mini-cours donné à l'Université du Québec à Montréal (UQÀM) lors des conférences et ateliers suivants : Variétés de Shimura et hyperbolicité des espaces de modules, 28 mai - 1 juin 2018; Géométrie arithmétique des orbifoldes, 11–13 décembre 2018; Approximation diophantienne et théorie de distribution des valeurs, 13–17 mai 2019.

Je me dois de remercier chaleureusement mes co-organisateurs sans qui toute cette florissante activité internationale (en particulier la coopération Québec-France) n'aurait pas pu prendre place : Erwan Rousseau et Steven Lu pour les deux premiers événements; et Carlo Gasbarri, Nathan Grieve, Aaron Levin, Steven Lu, Erwan Rousseau et Min Ru pour le troisième événement. Le premier événement fut crucialement financé par le fonds québécois CRM-UMI-FQRNT, par des fonds français (fonds propres de l'UMI, ANR Foliage (projet ANR-16-CE40-0008) et IUF) et une modeste participation du CICMA. Merci à Emmanuel Giroux alors à la gouverne de l'UMI du CRM de m'avoir bien conseillé et outillé pour la recherche de financement à Montréal. Le deuxième événement fut financé par le CIRGET

(UQÀM) ainsi que les fonds français UMI, ANR Foliage, IUF. Le troisième événement fut financé par le CIRGET, les fonds français UMI, ANR Foliage, IUF ainsi que la NSF pour les participants états-uniens.

Marseille, France   M.-H. Nicole

Contents

Lectures on the Ax–Schanuel conjecture

Benjamin Bakker and Jacob Tsimerman

MSC codes 14D07, 32G20, 03C64, 11J81

1 Introduction to Transcendence

1.1 Preliminaries

We begin with some very basic definitions. For details on transcendence theory we refer to [28, Chap. 8].

Definition 1.1.1. Let L/K be a field extension.

(1) For a finite subset $\{\alpha_1, \ldots, \alpha_n\} \subset L$, an *algebraic relation over* K satisfied by $\{\alpha_1, \ldots, \alpha_n\}$ is a polynomial $p \in K[z_1, \ldots, z_n]$ such that

$$p(\alpha_1, \ldots, \alpha_n) = 0.$$

(2) $\alpha \in L$ is said to be *algebraic over* K if $\{\alpha\}$ satisfies a nonzero algebraic relation over K.

We will often use "$\{\alpha_1, \ldots, \alpha_n\}$ satisfies an algebraic relation over K" interchangeably with "$\alpha_1, \ldots, \alpha_n$ satisfy an algebraic relation over K".

Lemma 1.1.2. *Let* L/K *be a field extension.*

B. Bakker (✉)
Dept. of Mathematics, University of Georgia, Athens, GA, USA
e-mail: bakker@math.uga.edu

J. Tsimerman
Dept. of Mathematics, University of Toronto, Toronto, ON, Canada
e-mail: jacobt@math.toronto.edu

© Springer Nature Switzerland AG 2020
M.-H. Nicole (ed.), *Arithmetic Geometry of Logarithmic Pairs and Hyperbolicity of Moduli Spaces*, CRM Short Courses,
https://doi.org/10.1007/978-3-030-49864-1_1

(1) $\alpha \in L$ is algebraic over K if and only if there is a finite-dimensional K-vector subspace $V \subset L$ with $\alpha V \subset V$.
(2) If $\beta_1, \ldots, \beta_n \in L$ are algebraic over K and $\alpha \in L$ such that $\{\alpha, \beta_1, \ldots, \beta_n\}$ satisfies an algebraic relation over K that is nonconstant in α, then α is algebraic over K.
(3) The set $F \subset L$ of elements which are algebraic over K is a subfield.

Definition 1.1.3. Let L/K be a field extension.

(1) A finite subset $\{\alpha_1, \ldots, \alpha_n\} \subset L$ is *algebraically independent over* K if it satisfies no nonzero algebraic relation over K. A subset $\Sigma \subset L$ is *algebraically independent over* K if every finite subset is algebraically independent.
(2) $\alpha \in L$ is *transcendental over* K if $\{\alpha\}$ is algebraically independent over K.
(3) A *transcendence basis* for L over K is a maximal subset of L which is algebraically independent over K.

We will often use "$\{\alpha_1, \ldots, \alpha_n\}$ is algebraically independent over K" interchangeably with "$\alpha_1, \ldots, \alpha_n$ are algebraically independent over K."

Example 1.1.4. $e \in \mathbb{R}$ is transcendental over $\mathbb{Q}$, as is $\pi \in \mathbb{R}$.

Example 1.1.5. It is conjectured but not known that $\{e, \pi\} \subset \mathbb{R}$ is algebraically independent over $\mathbb{Q}$.

Lemma 1.1.6. *Any two transcendence bases of L/K have the same cardinality.*

Definition 1.1.7. The *transcendence degree of L over K*, denoted $\mathrm{trdeg}_K L$, is the cardinality of a transcendence basis of L over K.

Example 1.1.8. For any field K, it is easy to see that any nonconstant $f \in K(t)$ is transcendental over K and moreover that $\{f\}$ is a transcendence basis of $K(t)$ over K. Thus, $\mathrm{trdeg}_K K(t) = 1$.

Example 1.1.9. The transcendence degree of $\mathbb{C}$ over $\mathbb{Q}$ is equal to the cardinality of $\mathbb{C}$.

1.2 Classical Transcendence of the Exponential Function

Arithmetic Transcendence Naively we think of the exponential function e^z as highly transcendental. By this we mean that given $\alpha_1, \ldots, \alpha_n \in \mathbb{C}$, we expect algebraic relations among the arguments α_i to rarely translate into algebraic relations among the values e^{α_i}, and vice versa. There is one notable exception: since the exponential function is a group homomorphism $\mathbb{C} \to \mathbb{C}^*$, where $\mathbb{C}^* := \mathbb{C}\backslash\{0\}$, any $\mathbb{Q}$-linear relation

$$0 = r_1\alpha_1 + \cdots + r_n\alpha_n \text{ for } r_i \in \mathbb{Q}$$

leads to a "trivial" algebraic relation

$$1 = \left(e^{\alpha_1}\right)^{r_1 b} \cdots \left(e^{\alpha_n}\right)^{r_n b}$$

where $r_i = a_i/b_i$ in lowest terms and $b = \text{lcm}(b_1, \ldots, b_n)$.

If we assume $\alpha_1, \ldots, \alpha_n \in \mathbb{C}$ satisfy no $\mathbb{Q}$-linear relations, we have the following longstanding conjecture:

Conjecture 1.2.1 (Schanuel Conjecture). Let $\alpha_1, \ldots, \alpha_n \in \mathbb{C}$ be $\mathbb{Q}$-linearly independent. Then

$$\text{trdeg}_{\mathbb{Q}} \mathbb{Q}(\alpha_1, \ldots, \alpha_n, e^{\alpha_1}, \ldots, e^{\alpha_n}) \geq n. \tag{1}$$

Note that the conjecture is only interesting when the α_i are algebraically dependent—it is a statement about how algebraic relations among the α_i interact with algebraic relations with the exponentials.

The Schanuel conjecture remains wide open; to give a sense of how strong it is, we have the following example.

Example 1.2.2. Take $\alpha_1 = 1$ and $\alpha_2 = \pi i$. Then the conjecture implies

$$\text{trdeg}_{\mathbb{Q}} \mathbb{Q}(1, \pi i, e, -1) = \text{trdeg}_{\mathbb{Q}} \mathbb{Q}(\pi, e) \geq 2$$

that is, that e and π are algebraically independent over $\mathbb{Q}$.

By taking $\alpha_i \in \overline{\mathbb{Q}}$, we see that the statement of Schanuel's conjecture is optimal, since

$$n \geq \text{trdeg}_{\mathbb{Q}} \mathbb{Q}(e^{\alpha_1}, \ldots, e^{\alpha_n}) = \text{trdeg}_{\mathbb{Q}} \mathbb{Q}(\alpha_1, \ldots, \alpha_n, e^{\alpha_1}, \ldots, e^{\alpha_n}).$$

Moreover, in this case the conjecture says that $e^{\alpha_1}, \ldots, e^{\alpha_n}$ are algebraically independent over $\mathbb{Q}$, and this has in fact been verified:

Theorem 1.2.3 (Lindemann–Weierstrass). *Let $\alpha_1, \ldots, \alpha_n \in \overline{\mathbb{Q}}$ be $\mathbb{Q}$-linearly independent. Then*

$$\text{trdeg}_{\mathbb{Q}} \mathbb{Q}(e^{\alpha_1}, \ldots, e^{\alpha_n}) = n.$$

Formal Functional Transcendence The exponential function is also defined on formal power series $f \in \mathbb{C}[[t_1, \ldots, t_m]]$, and we may try to obtain functional analogs of the above arithmetic statements by simply replacing the extension

$\mathbb{C}/\mathbb{Q}$ by $\mathbb{C}((t_1,\ldots,t_m))/\mathbb{C}$. Given $f_1,\ldots,f_n \in \mathbb{C}[[t_1,\ldots,t_m]]$, both sides of (1) have a clear analog, and we might guess that the correct statement is

$$\mathrm{trdeg}_{\mathbb{C}}\, \mathbb{C}(f_1,\ldots,f_n,e^{f_1},\ldots,e^{f_n}) \geq n.$$

There are however several new phenomena we must take into account. First, any relation of the form

$$\zeta = r_1 f_1 + \cdots + r_n f_n \ \text{ for } \ r_i \in \mathbb{Q} \text{ and } \zeta \in \mathbb{C}$$

leads to a "trivial" algebraic relation among the exponentials $e^{f_1},\ldots,e^{f_n}$, so we should assume the f_i are $\mathbb{Q}$-linearly independent *modulo constant terms*.

Second, the f_i may now satisfy a *formal* relation $p \in \mathbb{C}[[z_1,\ldots,z_n]]$, meaning that

$$0 = p(f_1,\ldots,f_n).$$

Example 1.2.4. Not surprisingly, the f_i may be algebraically independent over $\mathbb{C}$ and still satisfy a formal relation. Indeed, $f_1 = t$ and $f_2 = e^t$ satisfy a formal relation, namely

$$p(z_1,z_2) = e^{z_1} - z_2.$$

Formal relations are in fact much easier to detect. By the formal implicit function theorem, the number of independent formal relations is encoded by the dimension of the kernel of the Jacobian matrix $J(f_1,\ldots,f_n) := \left(\frac{\partial f_i}{\partial t_j}\right)$ over $\mathbb{C}((t_1,\ldots,t_m))$, and the *formal* transcendence degree can be reasonably defined to be the rank of $J(f_1,\ldots,f_n)$. The correct analog of Conjecture 1.2.1—which is a theorem due to Ax [2]—says roughly that the *algebraic* transcendence degree of $\mathbb{C}(f_1,\ldots,f_n,e^{f_1},\ldots,e^{f_n})$ over $\mathbb{C}$ is at least n *more* than the *formal* transcendence degree of the f_i:

Theorem 1.2.5 (Ax–Schanuel, Theorem 3 of [2]). *Let* $f_1,\ldots,f_n \in \mathbb{C}[[t_1,\ldots,t_m]]$ *be* $\mathbb{Q}$-*linearly independent modulo* $\mathbb{C}$. *Then*

$$\mathrm{trdeg}_{\mathbb{C}}\, \mathbb{C}(f_1,\ldots,f_n,e^{f_1},\ldots,e^{f_n}) \geq n + \mathrm{rk}\, J(f_1,\ldots,f_n). \tag{2}$$

Of course, we always have

$$\mathrm{trdeg}_{\mathbb{C}}\, \mathbb{C}(f_1,\ldots,f_n) + \mathrm{trdeg}_{\mathbb{C}}\, \mathbb{C}(e^{f_1},\ldots,e^{f_n}) \geq \mathrm{trdeg}_{\mathbb{C}}\, \mathbb{C}(f_1,\ldots,f_n,e^{f_1},\ldots,e^{f_n})$$

from which we deduce the following weaker version, which is often what's used in applications.

Corollary 1.2.6 (Weak Ax–Schanuel). *In the setup of Theorem 1.2.5, we have*

$$\operatorname{trdeg}_{\mathbb{C}} \mathbb{C}(f_1, \ldots, f_n) + \operatorname{trdeg}_{\mathbb{C}} \mathbb{C}(e^{f_1}, \ldots, e^{f_n}) \geq n + \operatorname{rk} J(f_1, \ldots, f_n). \tag{3}$$

As a further corollary, we can deduce an analog of the Lindemann–Weierstrass theorem:

Corollary 1.2.7 (Ax–Lindemann–Weierstrass). *In the setup of Theorem 1.2.5, further assume*

$$\operatorname{trdeg}_{\mathbb{C}} \mathbb{C}(f_1, \ldots, f_n) = \operatorname{rk} J(f_1, \ldots, f_n). \tag{4}$$

Then

$$\operatorname{trdeg}_{\mathbb{C}} \mathbb{C}(e^{f_1}, \ldots, e^{f_n}) = n.$$

Condition (4) has a clear geometric interpretation: if the f_i converge in some ball centered at the origin, it means the image of the germ $(f_1, \ldots, f_n)$: $\mathbb{C}^m \to \mathbb{C}^n$ is (the germ of) an algebraic variety. This observation naturally leads us to the geometric approach of the next section.

Geometric Functional Transcendence Often a more geometric interpretation of the results of the previous section admits clearer generalizations to other settings. The key point is that if we replace the field $\mathbb{C}((t_1, \ldots, t_m))$ from the previous section with the subfield $\mathbb{C}\langle\langle t_1, \ldots, t_m \rangle\rangle \subset \mathbb{C}((t_1, \ldots, t_m))$ of power series that converge in some ball around the origin, it does not affect the transcendence statements (see [46]).

Now, transcendence statements about the field of convergent power series can be phrased in terms of the analytic varieties they parametrize. For example, consider the flat uniformization

$$\pi : \mathbb{C}^n \to (\mathbb{C}^*)^n : (z_1, \ldots, z_n) \mapsto (e(z_1), \ldots, e(z_n))$$

where[1] $e(z) = e^{2\pi i z}$. Both $\mathbb{C}^n$ and $(\mathbb{C}^*)^n$ can be endowed with obvious structures as complex algebraic varieties, and it is then natural to ask what algebraic subvarieties $L \subset \mathbb{C}^n$ also have algebraic "image." To formulate this precisely, for a complex algebraic variety X and a subset $Y \subset X$, we denote by Y^{Zar} the Zariski closure of Y in X. We make the following definition:

Definition 1.2.8. We say an algebraic subvariety $L \subset \mathbb{C}^n$ is *bialgebraic* if

$$\dim L = \dim \pi(L)^{\text{Zar}}.$$

In this case we will sometimes abusively refer to $\pi(L)$ as being bialgebraic as well.

---

[1] We could formulate everything with $e(z) = e^z$ and the statements would be identical.

Example 1.2.9. Building on the "trivial" algebraic relations from the previous subsection, any $L \subset \mathbb{C}^n$ which is a $\mathbb{C}$-translate of a linear subspace of $\mathbb{C}^n$ defined over $\mathbb{Q}$ is bialgebraic. Said differently, every coset $M \subset (\mathbb{C}^*)^n$ of an algebraic subgroup of $(\mathbb{C}^*)^n$ is bialgebraic.

In fact, cosets of subtori are the *only* bialgebraic subvarieties, as we shall show in Corollary 4.1.2:

Proposition 1.2.10 (See Corollary 4.1.2). *Every closed bialgebraic $M \subset (\mathbb{C}^*)^n$ is a finite union of cosets of subtori.*

Now consider the following situation. Let $W \subset \mathbb{C}^n \times (\mathbb{C}^*)^n$ be the graph of π, and let $\mathrm{pr}_1 : \mathbb{C}^n \times (\mathbb{C}^*)^n \to \mathbb{C}^n$ be the first projection. Suppose we have an algebraic subvariety $V \subset \mathbb{C}^n \times (\mathbb{C}^*)^n$, as well as an analytic component U of the intersection $V \cap W$. Let $\Delta \subset \mathbb{C}$ be the unit disk. Taking a local holomorphic parametrization $f = (f_1, \ldots, f_n) : \Delta^m \to \mathrm{pr}_1(U) \subset \mathbb{C}^n$, we see that on the one hand

$$\mathrm{rk}\, J(f_1, \ldots, f_n) = \dim \mathrm{pr}_1(U) = \dim U$$

while on the other hand, if we consider the formal power series expansions at the origin $f_i \in \mathbb{C}[[t_1, \ldots, t_m]]$,

$$\mathrm{trdeg}_{\mathbb{C}}\, \mathbb{C}(f_1, \ldots, f_n, e(f_1), \ldots, e(f_n)) = \dim U^{\mathrm{Zar}} \leq \dim V.$$

Given Example 1.2.9, for $f_1, \ldots, f_n$ to be $\mathbb{Q}$-linearly independent modulo constant terms, we equivalently must have that $\mathrm{pr}_1(U)$ is not contained in any proper bialgebraic subvariety $L \subset \mathbb{C}^n$, in which case Theorem 1.2.5 says that we must have

$$\dim V \geq n + \dim U. \tag{5}$$

Conversely, suppose that for any algebraic $V \subset \mathbb{C}^n \times (\mathbb{C}^*)^n$ and any analytic component U of $V \cap W$ that is not contained in the graph of a proper bialgebraic subvariety we have (5). Then given a holomorphic function $f = (f_1, \ldots, f_n) : \Delta^m \to \mathbb{C}^n$ whose image is not contained in any bialgebraic subvariety, define

$$F = f \times (\pi \circ f) : \Delta^m \to \mathbb{C}^n \times (\mathbb{C}^*)^n$$

and take $V = F(\Delta^m)^{\mathrm{Zar}}$, so that

$$\mathrm{trdeg}_{\mathbb{C}}\, \mathbb{C}(f_1, \ldots, f_n, e(f_1), \ldots, e(f_n)) = \dim V.$$

Some analytic component U of the intersection $V \cap W$ must contain $F(\Delta^m)$, and U cannot be contained in the graph of a proper bialgebraic subvariety by the assumption on $f(\Delta^m)$, so (5) would imply

$$\dim V \geq n + \dim U \geq n + \dim F(\Delta^m) = n + \operatorname{rk} J(f_1, \ldots, f_n).$$

Rephrasing, we have therefore proven the following statement is equivalent to Theorem 1.2.5:

Theorem 1.2.11 (Ax–Schanuel). *Let $W \subset \mathbb{C}^n \times (\mathbb{C}^*)^n$ be the graph of π, and suppose there is an algebraic subvariety $V \subset \mathbb{C}^n \times (\mathbb{C}^*)^n$ such that there is an analytic component U of $V \cap W$ of unexpected codimension:*

$$\operatorname{codim}_{\mathbb{C}^n \times (\mathbb{C}^*)^n}(U) < \operatorname{codim}_{\mathbb{C}^n \times (\mathbb{C}^*)^n}(V) + \operatorname{codim}_{\mathbb{C}^n \times (\mathbb{C}^*)^n}(W).$$

Then U is contained in the graph of a proper bialgebraic $L \subset \mathbb{C}^n$.

The moral is that "atypical" intersections between algebraic subvarieties of $\mathbb{C}^n \times (\mathbb{C}^*)^n$ and the graph of π are controlled by bialgebraic subvarieties.

We of course also have geometric versions of Corollaries 1.2.6 and 1.2.7:

Corollary 1.2.12 (Weak Ax–Schanuel). *Suppose there are algebraic subvarieties $V_1 \subset \mathbb{C}^n$ and $V_2 \subset (\mathbb{C}^*)^n$ such that there is an analytic component U of $V_1 \cap \pi^{-1}(V_2)$ of unexpected codimension. Then U is contained in a proper bialgebraic $L \subset \mathbb{C}^n$.*

Proof. Take $V = V_1 \times V_2$.   □

Corollary 1.2.13 (Ax–Lindemann–Weierstrass). *Suppose there are algebraic subvarieties $V_1 \subset \mathbb{C}^n$ and $V_2 \subset (\mathbb{C}^*)^n$.*

(1) If $\pi(V_1) \subset V_2$, then there is a bialgebraic $M \subset (\mathbb{C}^*)^n$ with

$$\pi(V_1) \subset M \subset V_2;$$

(2) If $\pi(V_1) \supset V_2$, then there is a bialgebraic $M \subset (\mathbb{C}^*)^n$ with

$$\pi(V_1) \supset M \supset V_2.$$

Proof. For the first part, we have a containment $V_1 \subset \pi^{-1}(V_2)$ which is an intersection of unexpected codimension unless $V_2 = (\mathbb{C}^*)^n$. Thus, provided V_2 is a proper subvariety, by the previous corollary we obtain $L \subset \mathbb{C}^n$ bialgebraic containing V_1. Replacing $\mathbb{C}^n$ by L and V_2 by $\pi(L) \cap V_2$, we may continue until $\pi(L) \cap V_2 = \pi(L)$—that is, until $\pi(L) \subset V_2$.

We leave the second part as an exercise.

□

Corollary 1.2.13 can be equivalently formulated as the following:

Corollary 1.2.14.

(1) For $V \subset \mathbb{C}^n$ algebraic, $\pi(V)^{\mathrm{Zar}} \subset (\mathbb{C}^*)^n$ is a finite union of cosets of subtori.

(2) For $V \subset (\mathbb{C}^*)^n$ algebraic and any component V_0 of $\pi^{-1}(V)$, we have that $V_0^{\mathrm{Zar}} \subset \mathbb{C}^n$ is a finite union of $\mathbb{C}$-translates of linear subspaces defined over $\mathbb{Q}$.

Note that it is really the first part of Corollary 1.2.14 that is the analog of Corollary 1.2.7. It can also be stated as:

Corollary 1.2.15. *For any closed algebraic $V \subset (\mathbb{C}^*)^n$, a maximal irreducible algebraic subvariety of $\pi^{-1}(V)$ is a coset of a subtorus.*

We leave it to the reader to show that Corollary 1.2.6 (resp. 1.2.7) is equivalent to Corollary 1.2.12 (resp. 1.2.13).

1.2.16   Semiabelian Varieties

Let $Y = A$ be a semiabelian variety with identity $0 \in Y$. Let $X = V$ be its universal cover with its natural structure as a complex vector space, $\pi : V \to A$ the covering map, and $\Lambda = \pi^{-1}(0)$, which is a discrete subgroup of V. The universal covering map π is then identified with the quotient map $V \to V/\Lambda$. Note that if we had started with V and $\Lambda \subset V$ a discrete subgroup, V/Λ is not guaranteed to have the structure of an algebraic variety, and if it does it may not be unique.

The bialgebraic $M \subset A$ are then cosets of algebraic subgroups of A, and the Ax–Schanuel conjecture was proven by Ax [3].

In fact, more generally still, it makes sense to allow X, Y to be (euclidean) open subsets of algebraic subvarieties $\check{X}, \check{Y}$, in which case we proceed as above defining the "algebraic subvarieties" of X to be intersections $V \cap X$ for V an algebraic subvariety of $\check{X}$, and likewise for Y.

1.2.17   Shimura Varieties

A Shimura variety is a quotient of a bounded symmetric domain by an arithmetic lattice in a semisimple algebraic group $\mathbf{G}$. We will discuss this case more precisely in Lecture 4; for now we just give an example:

Example 1.2.18. The (coarse) moduli space of principally polarized abelian varieties A_g is a Shimura variety. In this case A_g admits a uniformization $\pi : \mathbb{H}_g \to A_g$ realizing A_g as the quotient of Siegel space

$$\mathbb{H}_g := \{Z \in \mathrm{Mat}_{g \times g}(\mathbb{C}) \mid Z^t = Z \text{ and } \mathrm{Im}\, Z > 0\}$$

by the action of $\mathrm{Sp}_{2g}(\mathbb{Z})$ via

$$(\begin{smallmatrix} A & B \\ C & D \end{smallmatrix}) : Z \mapsto (AZ + B)(CZ + D)^{-1}.$$

$\mathbb{H}_g$ is naturally a semialgebraic subset of its compact dual $\check{\mathbb{H}}_g$, which is the projective variety parametrizing Lagrangian planes in $\mathbb{C}^{2g}$.

The classification of bialgebraic subvarieties in Shimura varieties is known by [49]. These are the so-called weakly special subvarieties. The Ax–Lindemann–Weierstrass conjecture was proven by Pila for powers of the modular curve [35], by Pila–Tsimerman for A_g [38], and then by Klingler–Ulmo–Yafaev for general Shimura varieties [27]. The Ax–Schanuel conjecture was proven by Pila–Tsimerman [39] for powers of the modular curve and by Mok–Pila–Tsimerman for general Shimura varieties [33].

Importantly, Shimura varieties are moduli spaces of polarized pure integral Hodge structures which admit an algebraic structure, see Lecture 5.

1.2.19 Mixed Shimura Varieties

We will give fewer details in this case, but mixed Shimura varieties arise by allowing $\mathbf{G}$ to have a nontrivial unipotent radical. Mixed Shimura varieties are moduli spaces of graded polarized mixed integral Hodge structures which admit an algebraic structure.

Example 1.2.20. The (coarse) universal family of principally polarized abelian varieties X_g over A_g is a mixed Shimura variety. In this case X_g admits a uniformization $\pi : \mathbb{H}_g \times \mathbb{C}^g \to X_g$ realizing X_g as the quotient by a group Γ which is an extension of $\mathrm{Sp}_{2g}(\mathbb{Z})$ by $\mathbb{Z}^{2g}$.

The classification of bialgebraic subvarieties in mixed Shimura varieties is known by [18], and both the Ax–Lindemann–Weierstrass conjecture for mixed Shimura varieties and the Ax–Schanuel conjecture for the universal abelian variety have been proven by Gao [18, 19].

1.2.21 Period Spaces

Generalizing the case of Shimura varieties in a different direction, period spaces $\Gamma\backslash D$ parametrize pure polarized integral Hodge structures. Importantly, in this case $\Gamma\backslash D$ does not in general admit an algebraic structure, so the setup must be slightly modified (see Lecture 6). The proof of the Ax–Schanuel theorem (see Theorem 6.1.1 below) will be the main focus of these notes.

1.3   Arithmetic Applications

1.3.1   Special Point Problems

Suppose given a uniformization $\pi : X \to Y$ as in the previous section. Often there is a "special" set of points $Y_{\mathrm{sp}} \subset Y$ which have an interesting arithmetic interpretation in Y and whose *preimages* in X also have a simple arithmetic description.

Example 1.3.2. As in Example 1.2.16, take $Y = V/\Lambda$ a semiabelian variety, $X = V$, and $\pi : X \to Y$ the quotient map. Then we take Y_{sp} to be the set of torsion points, and $\pi^{-1}(Y_{\mathrm{sp}}) = \Lambda_{\mathbb{Q}}$.

Example 1.3.3. As in Example 1.2.17, take $Y = A_g$ the coarse moduli space of principally polarized abelian varieties, $X = \mathbb{H}_g$ the Siegel upper halfplane, and $\pi : \Omega \to Y$ the quotient. We take Y_{sp} to be the set of points corresponding to abelian varieties with CM. In this case, $\pi^{-1}(Y_{\mathrm{sp}})$ are points of $\check{\mathbb{H}}_g$ valued in number fields of bounded degree, with certain Galois properties.

Question 1.3.4. For an algebraic subvariety $V \subset Y$, denote $V_{\mathrm{sp}} := V \cap Y_{\mathrm{sp}}$. For what V do we have

$$(V_{\mathrm{sp}})^{\mathrm{Zar}} = V?$$

In the above contexts we expect that answer to be: only when V is bialgebraic. The property in Question 1.3.4 is in fact usually more restrictive, only holding for what are called *special* subvarieties, while bialgebraic subvarieties often turn out to be *weakly special*. For example, for $Y = (\mathbb{C}^*)^n$ and Y_{sp} the torsion points, the irreducible weakly special subvarieties are cosets of subtori, whereas the irreducible special subvarieties are *torsion* cosets of subtori.

Example 1.3.5. In the case of the exponential $\pi : \mathbb{C}^n \to (\mathbb{C}^*)^n$ with torsion points as the special points the above expectation is known as Lang's conjecture. Precisely: if $V \subset (\mathbb{C}^*)^n$ is an algebraic variety and V_{tor} is the set of torsion points on V, then Lang conjectures $(V_{\mathrm{tor}})^{\mathrm{Zar}}$ is a finite union of torsion cosets of subtori. For $n = 2$ this was proven by Lang [28].

Example 1.3.6. For $\pi : \mathbb{C}^n \to Y$ the uniformization of an abelian variety with torsion points as the special points, this is known as the Manin–Mumford conjecture. Precisely: if $V \subset Y$ is an algebraic variety and V_{tor} is the set of torsion points on V, they conjectured that $(V_{\mathrm{tor}})^{\mathrm{Zar}}$ is a finite union of torsion cosets of abelian subvarieties. Both the general form of Lang's conjecture and the Manin–Mumford conjecture were proven by Raynaud [42, 43].

Example 1.3.7. For $\pi : \Omega \to Y$ the uniformization of a Shimura variety, this is known as the André–Oort conjecture. Precisely, if $V \subset Y$ is an algebraic variety and V_{sp} is the set of special points on V, then they conjectured that $(V_{sp})^{Zar}$ is a finite union of special subvarieties. The conjecture was conditionally[2] proven in [26] and unconditionally for $Y = A_g$ by [48].

The proof of Raynaud proceeds by singling out a prime p and using different ingredients to deal with the "prime-to-p-parts" and "p-parts." For the former, Raynaud crucially uses the Frobenius at p in the Galois group. He observes that the Frobenius operator on prime-to-p roots of unity is closely related to the multiplication by p map (they are identical in the multiplicative case). This allows him to reduce from a variety X to $X \cap (p \cdot X)$, and conclude by induction. This argument is heavily relied upon in the conditional proof of André–Oort assuming the generalized Riemann hypothesis in [26]. For the "p-part" Raynaud proceeds using a p-adic deformation theory argument, which is generalized to the Shimura case by Moonen [32], allowing him to establish certain cases of André–Oort unconditionally.

The general hyperbolic case requires new ideas, and the proof of Tsimerman [48] builds on a strategy developed by Pila–Zannier which critically uses the Ax–Lindemann–Weierstrass theorem [35].

The Zilber–Pink Conjecture There is a wider set of conjectures, due to Bombier–Masser–Zannier in the multiplicative setting and Zilber–Pink more generally. Instead of only considering special points, one considers points of various "degrees" of specialness, and studies algebraic relations between such points. It is easiest to present in the multiplicative setting: for a point $x = (x_1, \ldots, x_n) \in (\mathbb{C}^*)^n$ define its rank $\mathrm{rk}(x)$ to be the rank as an abelian group of the span $\langle x_1, \ldots, x_n \rangle$ in $\mathbb{C}^*$. Observe that the rank is 0 precisely for torsion points. One consequence of the conjecture is the following:

Conjecture 1.3.8 (Consequence of Zilber–Pink [41, 53]). Let $V \subset (\mathbb{C}^*)^n$ be an irreducible algebraic subvariety of codimension d. Let V_m be the points of $V(\mathbb{C})$ of rank at most m and assume V_{d-1} is Zariski-dense in V. Then V is contained in a proper special subvariety. In other words, there is a nonconstant monomial which is identically 1 on V.

There is some progress on the conjecture above in the multiplicative case due to Habegger [22], Maurin [30], Bombieri–Masser–Zannier [10, 11], and others. We refer the interested reader to [37] for a more complete survey.

The Shafarevich Conjecture After Lawrence–Venkatesh Lawrence and Venkatesh [29] have outlined a strategy for proving instances of the Shafarevich conjecture which uses the functional transcendence of period maps. Briefly, let $\mathcal{O} = \mathcal{O}_{K,S}$ be the ring of integers $\mathcal{O}_K$ in a number field K away from a finite set S of primes and $\pi : Y \to X$ a smooth projective family defined over $\mathcal{O}$. Then assuming certain geometric properties of π one

---

[2]Conditional on the generalized Riemann hypothesis.

expects the number of integral points $X(\mathcal{O})$ to be finite, for example, when the family π has an immersive period map. The Shafarevich conjecture for moduli spaces of polarized abelian varieties was proven by Faltings in the landmark paper [17].

The strategy of Lawrence–Venkatesh uses the p-adic period map in the context of p-adic Hodge theory. Their argument requires a p-adic transcendence result which formally follows from the corresponding transcendence result for the complex analytic period map. Using this technique, they are able to show that when X is taken to be certain moduli spaces of hypersurfaces in $\mathbb{P}^n$, the integral points $X(\mathcal{O})$ are not Zariski dense in X.

2   o-Minimal Geometry

For background on o-minimal structures and o-minimal geometry, we refer to [50].

2.1   o-Minimal Structures

An o-minimal structure specifies "tame" subsets of euclidean space which can be used as local models for "tame" geometry. On the one hand, the tameness will rule out pathologies such as Cantor sets and space-filling curves; on the other hand, as we will see, the tameness hypothesis locally imposes remarkably few conditions on analytic functions.

Definition 2.1.1. A structure S is a collection $(S_n)_{n \in \mathbb{N}}$ where each S_n is a set of subsets of $\mathbb{R}^n$ satisfying the following conditions:

(1) Each S_n is closed under finite intersections, unions, and complements;
(2) The collection (S_n) is closed under finite Cartesian products and coordinate projection;
(3) For every polynomial $P \in \mathbb{R}[x_1, \dots, x_n]$, the zero set

$$(P = 0) := \{x \in \mathbb{R}^n \mid P(x) = 0\} \subset \mathbb{R}^n$$

is an element[3] of S_n.

We refer to the elements $U \in S_n$ as S-definable subsets of $\mathbb{R}^n$. For $U \in S_n$, and $V \in S_m$, we say a map $f : U \to V$ of S-definable sets is S-definable if

---

[3]One can work in greater generality by allowing structures without this assumption, but we will only require ones satisfying it.

the graph is. When the structure S is clear from context, we will often just refer to "definable" sets and functions.

The definable sets should be thought of as the sets that are "constructible" within the theory. From the axioms, it is easy to prove the following:

Proposition 2.1.2. *Let S be a structure.*

(1) *The image and preimage of a definable set under a definable map are definable;*
(2) *The composition of two definable maps is definable.*

Thus, for example, whereas we only required coordinate projections to be definable in Definition 2.1.1, it follows that all linear projections are definable. By definition, any structure S contains all real algebraic sets, but this is not enough:

Example 2.1.3. The collection S of real algebraic sets—that is, $S_n =$ the Boolean algebra generated by sets of the form $(P = 0)$ for $P \in \mathbb{R}[x_1, \ldots, x_n]$— is *not* a structure. Indeed, for any $P \in \mathbb{R}[x_1, \ldots, x_n]$, the image of the projection of $(x_0^2 = P)$ forgetting x_0 is $(P \geq 0)$.

Example 2.1.4. Let $\mathbb{R}_{\mathrm{alg}}$ be the collection of real semialgebraic subsets of $\mathbb{R}^n$—that is, $(\mathbb{R}_{\mathrm{alg}})_n$ is the Boolean algebra generated by sets of the form $(P \geq 0)$ for $P \in \mathbb{R}[x_1, \ldots, x_n]$. Then $\mathbb{R}_{\mathrm{alg}}$ *is* a structure. By the Tarski–Seidenberg theorem (see for example [50, Chapter 2]), coordinate projections of real semialgebraic sets are real semialgebraic, and the other axioms are easy to verify. $\mathbb{R}_{\mathrm{alg}}$ is therefore a structure, in fact the structure generated by real algebraic sets given Example 2.1.3.

Remark 2.1.5. Tarski–Seidenberg is usually phrased as quantifier elimination for the real ordered field, and structures as defined above are important in model theory. Indeed, the axioms say definable sets are closed under first order formulas, as intersections, unions, and complements correspond to the logical operators "and," "or," and "not," while the projection axiom corresponds to universal and existential quantifiers. Moreover, we can make the same definition for any real closed field, and base-change to these fields plays a similar role to base-changing to generic points of schemes in algebraic geometry. We won't say much about it, but it is a useful perspective to keep in mind.

While infinite unions or intersections of definable subsets are not definable, it is nonetheless the case that many topological constructions with respect to the euclidean topology are definable:

Proposition 2.1.6. *Let S be a structure, and endow $\mathbb{R}^n$ with the euclidean topology. Closures, interiors, and boundaries of definable sets are definable.*

Proof. We just show that the closure of a definable set $U \subset \mathbb{R}^n$ is defined by a first order formula and leave the rest as an exercise:

$$\overline{U} = \left\{ x \in \mathbb{R}^n \ \middle| \ \forall \epsilon > 0, \exists y \in U \text{ s.t. } \sum_i (x_i - y_i)^2 < \epsilon \right\}$$

$\square$

Remark 2.1.7. We have the following formal operations on structures.

(1) Given two structures S and S', we say S is contained in S', denoted $S \subset S'$, if $S_n \subset S'_n$ for all n. Note that any structure S contains $\mathbb{R}_{\text{alg}}$.
(2) Given structures $\{S^{(i)}\}_{i \in I}$ indexed by a set I, the intersection $(\bigcap S^{(i)})_n := \bigcap (S^{(i)})_n$ is evidently a structure. Thus, given a collection $(T_n)_{n \in \mathbb{N}}$ of sets of subsets of $\mathbb{R}^n$, we may speak of the structure S generated by the $(T_n)_{n \in \mathbb{N}}$ as the smallest structure S with $S_n \supset T_n$.
(3) Given an increasing chain

$$S^{(0)} \subset S^{(1)} \subset \cdots \subset S^{(i)} \subset \cdots$$

the union $(\bigcup S^{(i)})_n := \bigcup (S^{(i)})_n$ is a structure.

Thus far we have only specified the rules by which we can construct definable subsets from other definable subsets; we have not yet controlled how complicated definable sets are allowed to be. The crucial "tameness" property is o-minimality:

Definition 2.1.8. A structure S is said to be o-minimal if $S_1 = (\mathbb{R}_{\text{alg}})_1$—that is, if the S-definable subsets of the real line are exactly finite unions of intervals.

The intervals in the definition are allowed to be closed or open on either end, may extend to infinity, and may be zero length (i.e. points).

Example 2.1.9. $\mathbb{R}_{\text{alg}}$ is o-minimal, clearly.

Example 2.1.10. Let $\mathbb{R}_{\sin}$ be the structure generated by the graph of $\sin : \mathbb{R} \to \mathbb{R}$. $\mathbb{R}_{\sin}$ is not o-minimal as $\pi \mathbb{Z} = \sin^{-1}(0)$ is definable and infinite.

Example 2.1.11. Let $\mathbb{R}_{\exp}$ be the structure generated by the graph of the real exponential $\exp : \mathbb{R} \to \mathbb{R}$. $\mathbb{R}_{\exp}$ is o-minimal by a result of Wilkie [52]. Quantifier elimination does not hold for $\mathbb{R}_{\exp}$.

Example 2.1.12. Let $\mathbb{R}_{\text{an}}$ be the structure generated by the graphs of all restrictions $f|_{B(R)}$ of real analytic functions $f : B(R') \to \mathbb{R}$ on a finite radius $R' < \infty$ open euclidean ball (centered at the origin) to a strictly smaller radius $R < R'$ ball. Via the embedding $\mathbb{R}^n \subset \mathbb{R}\mathrm{P}^n$, this is equivalent to the structure of subsets of $\mathbb{R}^n$ that are subanalytic in $\mathbb{R}\mathrm{P}^n$. $\mathbb{R}_{\text{an}}$ is o-minimal by van-den-Dries [50], using Gabrielov's theorem of the complement. Note that while $\sin(x)$ is not $\mathbb{R}_{an}$-definable, its restriction to any finite interval is.

Example 2.1.13. Let $\mathbb{R}_{an,exp}$ be the structure generated by $\mathbb{R}_{an}$ and $\mathbb{R}_{exp}$. $\mathbb{R}_{an,exp}$ is o-minimal by a result of van-den-Dries–Miller [51]. Most of the applications to algebraic geometry currently use the structure $\mathbb{R}_{an,exp}$.

Remark 2.1.14. By Remark 2.1.7, there are maximal o-minimal structures, but not a unique one, as the structure generated by two o-minimal structures can fail to be o-minimal [44].

For the rest of this lecture, we fix an o-minimal structure S, and by "definable" we mean S-definable, unless explicitly otherwise stated.

2.2   Cylindrical Cell Decomposition

Sets that are definable in an o-minimal structure can be decomposed into graphs of definable functions in a systematic way. It would take us too far afield to prove the main existence result (Theorem 2.2.5 below), but it is important to keep in mind as it gives a clear picture of some of the finiteness properties that such definable sets possess.

We follow the treatment in [50] closely.

Definition 2.2.1. A *definable cylindrical cell decomposition* of $\mathbb{R}^n$ is a partition $\mathbb{R}^n = \bigsqcup D_i$ into finitely many pairwise disjoint definable subsets D_i, called cells. The cells have the following inductive description.

$\underline{n = 0}.$ There is exactly one definable cylindrical cell decomposition of $\mathbb{R}^0$. Its unique cell is all of $\mathbb{R}^0$.

$\underline{n > 0}.$ Write $\mathbb{R}^n = \mathbb{R}^{n-1} \times \mathbb{R}$. There is a definable cylindrical cell decomposition $\{E\}$ of $\mathbb{R}^{n-1}$ and for each E we have: an integer $m_E \in \mathbb{N}$ and continuous definable functions $f_{E,k} : E \to \mathbb{R}$ for each $0 < k < m_E$ such that

$$f_{E,0} := -\infty < f_{E,1} < \cdots < f_{E,m_E-1} < f_{E,m_E} := +\infty$$

The cells are:

- graphs: $\{(x, f_{E,k}(x)) \mid x \in E\}$ for each E and $0 < k < m_E$;
- bands: $(f_{E,k}, f_{E,k+1}) := \{(x,y) \mid x \in E \text{ and } y \in (f_{E,k}(x), f_{E,k+1}(x))\}$ for each E and $0 \leq k < m_E$.

Note that because of the inductive nature of the definition, we have implicitly chosen an ordering of the coordinates.

Example 2.2.2. The cylindrical cell decompositions of $\mathbb{R}$ are easy to understand. In this case, there is $m \in \mathbb{N}$ and $a_k \in \mathbb{R}$ for each $0 < k < m$ such that

$$a_0 := -\infty < a_1 < \cdots a_{m-1} < a_m := +\infty$$

and the cells are:

- $\{a_k\}$ for $0 < k < m$;
- (a_k, a_{k+1}) for $0 \le k < m$.

Such a cell decomposition is shown in Figure 1.

Example 2.2.3. Figure 2 shows a cylindrical cell decomposition of $\mathbb{R}^2$ that projects to the cell decomposition of Figure 1.

Remark 2.2.4. Each cell D in a definable cylindrical cell decomposition has a well-defined dimension $\dim_\mathbb{R} D$, and it is definably homeomorphic to $\mathbb{R}^{\dim_\mathbb{R} D}$ as follows. For $n = 0$ it is trivial, as it is inductively for the graph cells for

Fig. 1 A cell decomposition of $\mathbb{R}$.

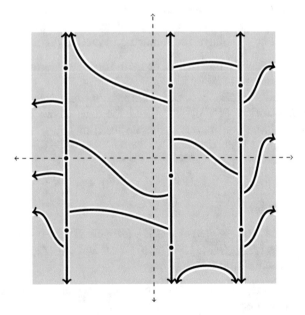

Fig. 2 A cell decomposition of $\mathbb{R}^2$ projecting to that of Figure 1.

$n > 0$. For band cells, given two definable $f, g : E \to \mathbb{R}$ with $f < g$, we have a definable homeomorphism $(f, g) \to E \times \mathbb{R}$ via

$$(x, y) \mapsto \left(x, \frac{1}{f(x) - y} + y + \frac{1}{g(x) - y} \right).$$

The main result is the following:

Theorem 2.2.5. *For any finite collection $U_j \subset \mathbb{R}^n$ of definable sets, there is a definable cylindrical cell decomposition of $\mathbb{R}^n$ such that each U_j is a union of cells.*

Every cell has a well-defined (real) dimension, so we have as a consequence:

Corollary/Definition 2.2.6. For any definable set $U \subset \mathbb{R}^n$ we define $\dim_{\mathbb{R}} U$ to be the largest dimension of its cells with respect to a definable cylindrical cell decomposition.

We won't give a proof of Theorem 2.2.5, but an essential ingredient is the following stronger version in a special case:

Lemma 2.2.7. *For every definable function $f : (a, b) \to \mathbb{R}$, there is a finite subdivision*

$$a_0 = a < a_1 < \cdots < a_m = b$$

such that each $f|_{(a_k, a_{k+1})}$ is either constant or strictly monotonic.

Proof. The proof is taken directly from [50]. We begin with the following:

Claim. There is a subinterval $J \subset (a, b)$ on which f is constant or f is strictly monotonic and continuous.

Proof. We may assume f is not constant on any subinterval of (a, b).

Step 1. f is injective on a subinterval J.

It follows from the above assumption that all fibers are finite. The function $g(y) = \min f^{-1}(y)$ is a definable section of f, for we may write its graph as

$$\{(f(x), x) \in \mathbb{R}^2 \mid x \in (a, b) \text{ s.t. } x \leq x' \text{ for all } x' \in (a, b) \text{ with } f(x') = f(x)\}.$$

The image of g is definable and not finite by assumption, so by the o-minimality property it contains an interval J, and on this interval $g \circ f = \mathrm{id}$, so f is injective.

Step 2. f is strictly monotonic on a subinterval J.

Assuming now that f is injective, for each $x \in (a, b)$ the sets

$$\{y \in (a, b) \mid f(y) < f(x)\}$$
$$\{y \in (a, b) \mid f(y) > f(x)\}$$

are a definable partition of $(a, b)\backslash\{x\}$. It follows that the sets

$$A =\{x \in (a, b) \mid \exists \epsilon > 0 \text{ s.t. } f|_{(x-\epsilon,x)} < f(x) < f|_{(x,x+\epsilon)}\}$$
$$B =\{x \in (a, b) \mid \exists \epsilon > 0 \text{ s.t. } f|_{(x-\epsilon,x)} > f(x) > f|_{(x,x+\epsilon)}\}$$
$$C =\{x \in (a, b) \mid \exists \epsilon > 0 \text{ s.t. } f|_{(x-\epsilon,x)} > f(x) < f|_{(x,x+\epsilon)}\}$$
$$D =\{x \in (a, b) \mid \exists \epsilon > 0 \text{ s.t. } f|_{(x-\epsilon,x)} < f(x) > f|_{(x,x+\epsilon)}\}$$

are a definable partition of (a, b).

We now claim that the last two sets are finite; it's enough to show D is, as the proof for C is similar. If the claim was false, then there would be a subinterval J for which every point is a local maximum. For $n \in \mathbb{N}$, consider the sets

$$J_n := \{x \in J \mid x \text{ is a maximum on } (x - 1/n, x + 1/n)\}$$

which are clearly definable and $J = \cup_n J_n$. The J_n can't all be finite, so one J_n contains an interval by o-minimality, and this is clearly nonsense.

Thus, one of A and B (say A) contains an interval $J = (c, d)$. But then for each $x \in J$,

$$\{y \in J \mid y > x \text{ and } f|_{(x,y)} > f(x)\}$$

must be all of (x, d).

Step 3. f is strictly monotonic and continuous on a subinterval J.

Restrict f to an interval whose image is an interval. Then it is strictly monotonic and bijective, hence continuous.   □

To finish, the set of points x for which either f is constant on a neighborhood of x or f is strictly monotonic and continuous in a neighborhood of x is definable, and hence is a finite set of points by the claim. This finishes the proof, since if for all x in some interval either f is constant on a neighborhood of x or f is strictly monotonic and continuous in a neighborhood of x, then the same is true on the entire interval.   □

By reasoning along the lines of Lemma 2.2.7 one can show that definable functions have limits away from definable sets of smaller dimension. This can be upgraded to the fact that definable functions are C^k off of a definable set of smaller dimension:

Corollary 2.2.8. *Let $U \subset \mathbb{R}^n$ be a definable set. Then for each k, U has a stratification by definable C^k-submanifolds.*

Corollary 2.2.9. *Let $f : U \to V$ be a definable map. Then for each $n \in \mathbb{N}$, the subset*

$$V_n := \{v \in V \mid \dim f^{-1}(v) = n\} \subset V$$

is definable.

Proof. Consider the graph, and order the coordinates *backwards*. As is clear from the inductive definition, each cell has constant dimension over its projection. $\qquad\qquad\square$

Corollary 2.2.10. *Let $f : U \to V$ be a definable map with finite fibers. Then for each $n \in \mathbb{N}$, the subset*

$$V_n := \{v \in V \mid \#f^{-1}(v) = n\} \subset V$$

is definable. Moreover, the size of the fibers is uniformly bounded.

Proof. As above, consider the graph and order the coordinates backwards. All of the cells are graphs over cells of V. $\qquad\qquad\square$

2.3 Definable Topological Spaces

Let M be a topological space. We can endow M with a geometry locally modeled on definable sets in the usual way using atlases.

Definition 2.3.1. A $(S\text{-})$definable topological space M is a topological space M, a finite open covering V_i of M, and homeomorphisms $\varphi_i : V_i \to U_i \subset \mathbb{R}^n$ such that

(1) The U_i and the pairwise intersections $U_{ij} := \varphi_i(V_i \cap V_j)$ are definable sets;
(2) The transition functions $\varphi_{ij} := \varphi_j \circ \varphi_i^{-1} : U_{ij} \to U_{ji}$ are definable.

We call the data (V_i, φ_i) a definable atlas. A morphism of definable spaces $f : M \to M'$ is a continuous map f such that for all i and i', the composition

$$(f \circ \varphi_i^{-1})^{-1}(V'_{i'}) \xrightarrow{\varphi_i^{-1}} f^{-1}(V'_{i'}) \xrightarrow{f} V'_{i'} \xrightarrow{\varphi'_{i'}} U'_{i'}$$

is S-definable. Note that this is a condition both on the map and the source. M is said to be a $(S\text{-})$definable manifold if the definable atlas additionally gives M the structure of a manifold.

We denote the category of S-definable topological spaces by $(S\text{-Top})$.

We will often use the term "$(S\text{-})$definable structure" as a shorthand for "structure as a $(S\text{-})$definable topological space" when no confusion is likely to arise, and likewise we will say a continuous map $f : M \to M'$ is "$(S\text{-})$definable" as shorthand for "a morphism of $(S\text{-})$definable topological spaces."

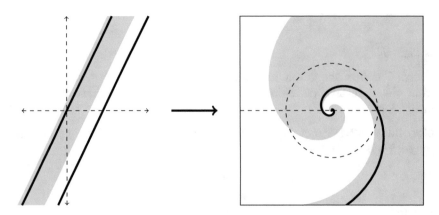

Fig. 3 The "slanted strip" definable structures considered in Examples 2.3.2 and 2.3.3.

We will ultimately be interested in definable structures on the topological spaces underlying complex analytic varieties, and all of the examples below are of this sort. Throughout we use the identification $\mathbb{C} \cong \mathbb{R}^2$ to speak about definable subsets of $\mathbb{C}^n$.

Example 2.3.2. (See Figure 3.) Let $\mathbb{C}^* \subset \mathbb{C}$ be the punctured plane and $e : \mathbb{C} \to \mathbb{C}^*$ the usual covering map $e(z) = e^{2\pi i z}$. We can endow $\mathbb{C}^*$ with a number of $\mathbb{R}_{\mathrm{alg}}$-definable structures:

(1) $\mathbb{C}^*$ is a (real) algebraic subset of $\mathbb{C}$, and we call this $\mathbb{R}_{\mathrm{alg}}$-definable topological space $\mathbb{G}_m^{\mathrm{def}}$.

(2) For $a \in \mathbb{R}$, define the following slope a "slanted strip" fundamental set for the covering action on $\mathbb{C}$:

$$F_a = \{z \in \mathbb{C} \mid a \cdot \operatorname{Im} z < \operatorname{Re} z < (1 + \epsilon) + a \cdot \operatorname{Im} z\}.$$

F_a is evidently semialgebraic, and thus has a natural $\mathbb{R}_{\mathrm{alg}}$-definable structure. A slightly thinner open strip will inject into $\mathbb{C}^*$, and taking translates of such a strip will then give a $\mathbb{R}_{\mathrm{alg}}$-definable atlas of $\mathbb{C}^*$. We call the resulting $\mathbb{R}_{\mathrm{alg}}$-definable topological space $\mathbb{C}_a^*$. By definition the map $e : F_a \to \mathbb{C}_a^*$ is a morphism of $\mathbb{R}_{\mathrm{alg}}$-definable topological spaces.

Evidently if S, S' are two structures with $S \subset S'$ and M is an S-definable topological space, then we have an induced structure as an S'-definable space. In particular, an $\mathbb{R}_{\mathrm{alg}}$-definable structure on M will induce an S-definable structure on M for any S.

Example 2.3.3. (See Figure 3.) Consider again the previous example.

(1) The spaces $\mathbb{C}_a^*$ are all isomorphic as $\mathbb{R}_{\mathrm{alg}}$-definable topological spaces, as, for instance, the map $x + iy \mapsto (x + ay) + iy$ yields an isomorphism $\mathbb{C}_0^* \xrightarrow{\cong} \mathbb{C}_a^*$.

(2) The identity map $\mathbb{C}_a^* \to \mathbb{G}_m^{\mathrm{def}}$ is *not* definable for any $a \neq 0$ in any o-minimal structure. Indeed, any ray is definable in $\mathbb{G}_m^{\mathrm{def}}$, but the preimage in F_a has infinitely many components for $a \neq 0$.

(3) The identity map $\mathbb{C}_0^* \to \mathbb{G}_m^{\mathrm{def}}$ is not $\mathbb{R}_{\mathrm{alg}}$-definable. This is equivalent to $e : F_0 \to \mathbb{G}_m^{\mathrm{def}}$ being definable, which would imply that the real and imaginary parts $e^{-2\pi y} \cos(2\pi x)$ and $e^{-2\pi y} \sin(2\pi x)$ are $\mathbb{R}_{\mathrm{alg}}$-definable as functions $[0,1] \times \mathbb{R}_{\geq 0} \to \mathbb{R}$, which is clearly false. In fact, they are not even $\mathbb{R}_{\mathrm{an}}$-definable, as otherwise $e^{2\pi y}$ would be $\mathbb{R}_{\mathrm{an}}$-definable, whereas one can show that any $\mathbb{R}_{\mathrm{an}}$-definable function has sub-exponential growth. It is however clearly $\mathbb{R}_{\mathrm{an,exp}}$-definable (and in fact an isomorphism of $\mathbb{R}_{\mathrm{an,exp}}$-definable spaces).

Thus, of the "slanted strip" fundamental domains considered in Examples 2.3.2 and 2.3.3, the vertical strip is the unique one for which the covering map $e : F_0 \to \mathbb{G}_m^{\mathrm{def}}$ is definable in an o-minimal structure.

Remark 2.3.4. While the $\mathbb{C}_a^*$ of Example 2.3.2 are all isomorphic as $\mathbb{R}_{\mathrm{alg}}$-definable spaces, $\mathbb{C}_a^*$ and $\mathbb{C}_b^*$ do not admit a *holomorphic* isomorphism as S-definable spaces for $a \neq b$ and any o-minimal structure S. Indeed, the only holomorphic automorphisms of $\mathbb{C}^*$ are q and q^{-1} up to scaling, and one can manually check that these do not give definable isomorphisms $\mathbb{C}_a^* \to \mathbb{C}_b^*$ for $a \neq b$. However, the identity $\mathbb{C}_0^* \to \mathbb{G}_m^{\mathrm{def}}$ does give a holomorphic $\mathbb{R}_{\mathrm{an,exp}}$-definable isomorphism.

Example 2.3.5. Let X be a real algebraic variety. Then the set of real points $X(\mathbb{R})$ equipped with the euclidean topology carries a canonical isomorphism class of $\mathbb{R}_{\mathrm{alg}}$-definable topological space structures, by covering by (finitely many) affine varieties. It is an easy exercise to see that any two (finite) affine coverings specify isomorphic $\mathbb{R}_{\mathrm{alg}}$-definable structures.

Likewise, as the complex points of an affine complex algebraic variety are naturally the real points of an affine real algebraic variety (by Weil restriction), for X a complex algebraic variety the same construction yields a canonical (unique up to isomorphism) $\mathbb{R}_{\mathrm{alg}}$-definable topological space structure on the set of complex points $X(\mathbb{C})$ with the euclidean topology.

Given a complex algebraic variety X, we define X^{eucl} to be $X(\mathbb{C})$ endowed with its euclidean topology.

Definition 2.3.6. Let X a complex algebraic variety. We define X^{def} to be the $(S\text{-})$definable topological space with underlying topological space X^{eucl} and the definable structure induced from the $\mathbb{R}_{\mathrm{alg}}$-definable structure constructed in Example 2.3.5. We refer to X^{def} as the $(S\text{-})$definabilization of X.

Note that the notation does not reflect the dependence of X^{def} on the structure S.

Let $(\mathrm{AlgVar}/\mathbb{C})$ be the category of complex algebraic varieties. It is not hard to see that we in fact have a "definabilization" functor

$$(\mathrm{AlgVar}/\mathbb{C}) \to (S\text{-}\mathrm{Top}) : X \mapsto X^{\mathrm{def}}.$$

Likewise for real algebraic varieties.

Let (Top) be the category of topological spaces. Every definable space has an underlying topological space, and we denote the resulting forgetful functor

$$(S\text{-}\mathrm{Top}) \to (\mathrm{Top}) : X \mapsto X^{\mathrm{top}}.$$

We then clearly have a diagram:

$$
\begin{array}{ccc}
(\mathrm{AlgVar}/\mathbb{C}) & \xrightarrow{\ (-)^{\mathrm{def}}\ } & (S\text{-}\mathrm{Top}) \\
& \searrow{\scriptstyle (-)^{\mathrm{eucl}}} \quad {\scriptstyle (-)^{\mathrm{top}}}\swarrow & \\
& (\mathrm{Top}) &
\end{array}
$$

There is likewise a similar picture over $\mathbb{R}$, but for us complex algebraic varieties will play a particularly important role.

Example 2.3.7. We have the following hyperbolic analog of Examples 2.3.2 and 2.3.3. Let $Y(2)$ be the full-level two modular curve, with analytic uniformization $Y(2)^{\mathrm{an}} := \Gamma(2)\backslash\mathbb{H}$ where

$$\Gamma(2) = \left\{ A \in \mathrm{PSL}_2(\mathbb{Z}) \;\middle|\; A \equiv \begin{pmatrix} 1 & 0 \\ 0 & 1 \end{pmatrix} \bmod 2 \right\}.$$

A fundamental domain F for the action of $\Gamma(2)$ on $\mathbb{H}$ is shown in Figure 4, corresponding to a choice of section of the quotient $\mathrm{PSL}_2(\mathbb{Z}) \to \mathrm{PSL}_2(\mathbb{F}_2)$. Let

$$F := \left\{ z \in \mathbb{C} \;\middle|\; |\operatorname{Re} z| < \frac{1}{2} + \epsilon \text{ and } |z|^2 > 1 - \epsilon \right\}$$

be a slight enlargement of the usual fundamental domain for the action of $\mathrm{PSL}_2(\mathbb{Z})$ on $\mathbb{H}$. Clearly F is real semialgebraic and injects into $Y(2)^{\mathrm{an}}$. The translates of F under the choosen lifts provide a cover of $Y(2)^{\mathrm{an}}$, and as the action of $\mathrm{PSL}_2(\mathbb{R})$ on $\mathbb{H}$ is algebraic, this is a (finite) cover by real semialgebraic sets with real semialgebraic transition functions. Thus, we have a $\mathbb{R}_{\mathrm{alg}}$-definable structure on $Y(2)^{\mathrm{an}}$ which we call $\mathcal{Y}(2)$.

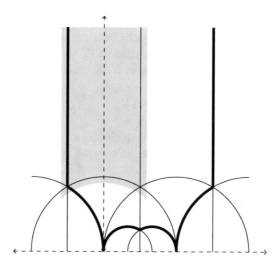

Fig. 4 The definable fundamental set for $Y(2)$ considered in Example 2.3.7.

The $\mathbb{R}_{\text{alg}}$-definable spaces $\mathcal{Y}(2)$ and $Y(2)^{\text{def}}$ are not isomorphic via a holomorphic map, and in fact, the induced $\mathbb{R}_{\text{an}}$- and $\mathbb{R}_{\text{exp}}$-definable structures are not even the same, just as in Remark 2.3.4. Indeed, the image of the horoball

$$\{z \in \mathbb{H} \mid \operatorname{Im} z > 1\}$$

gives a neighborhood of the cusp at ∞ holomorphically isomorphic to Δ^*. On the one hand, in $Y(2)^{\text{def}}$ there's an algebraic coordinate at the cusp which is $\mathbb{R}_{\text{alg}}$-definable, and which moreover extends holomorphically to the cusp. Thus, after shrinking Δ^*, the $\mathbb{R}_{\text{an}}$-definable structure induced by $Y(2)^{\text{def}}$ is that of $\Delta^* \subset \mathbb{G}_m^{\text{def}}$. On the other hand, the $\mathbb{R}_{\text{an}}$-definable structure induced by $\mathcal{Y}(2)$ is clearly $\Delta^* \subset \mathbb{C}_0^*$.

The two structures on $Y(2)^{\text{an}}$ *are* isomorphic over $\mathbb{R}_{\text{an,exp}}$. Indeed, by the previous example they are isomorphic in the cuspidal neighborhoods, whereas on the complement of the union of (slightly shrunken) cuspidal neighborhoods the two structures are clearly isomorphic over $\mathbb{R}_{\text{an}}$.

Remark 2.3.8. We can alternatively think of Example 2.3.7 (or indeed any of the above examples) in the following way. Let F' be an open semialgebraic fundamental set for the action of $\Gamma(2)$. The action of $\Gamma(2)$ on $\mathbb{H}$ induces a closed étale equivalence relation $R \subset \mathbb{H} \times \mathbb{H}$. Each component of this equivalence relation is evidently algebraic, and only finitely many components intersect $F' \times F'$. Thus, the restriction of the equivalence relation to F' is $\mathbb{R}_{\text{alg}}$-definable. One can show that quotients by closed étale definable equivalence relations exist in the category of definable topological spaces [6].

Example 2.3.9. Let X be a smooth proper complex algebraic variety. Then we may cover X^{eucl} by finitely many polydisks Δ^n. Endow each Δ^n with the $\mathbb{R}_{\mathrm{an}}$-definable structure coming from that of Δ in $(\mathbb{A}^1)^{\mathrm{def}}$. After shrinking the disks slightly the transition functions are evidently restricted analytic and therefore $\mathbb{R}_{\mathrm{an}}$-definable. This atlas gives a $\mathbb{R}_{\mathrm{an}}$-definable structure to X^{eucl} which is evidently X^{def} (over $\mathbb{R}_{\mathrm{an}}$).

Likewise, if X is a smooth complex algebraic variety (not necessarily proper), then let $\overline{X}$ be a log smooth algebraic compactification. $\overline{X}^{\mathrm{eucl}}$ can be covered by finitely many polydisks Δ^n whose intersection with X^{eucl} is of the form $(\Delta^*)^r \times \Delta^s$. This atlas then gives a $\mathbb{R}_{\mathrm{an}}$-definable structure to X^{eucl}, which is once again isomorphic to X^{def}.

Remark 2.3.10. The cylindrical cells of Section 2.2 depend on an embedding into $\mathbb{R}^n$, but there is a notion of cell decomposition for definable topological spaces for which the analogs of Corollaries 2.2.9 and 2.2.10 hold. See [6] for details.

3   Algebraization Theorems in o-Minimal Geometry

O-minimal geometry has found a number of applications to the functional transcendence theory of uniformizations of algebraic varieties because it allows one to ascend and descend algebraic structures along the uniformizing map by way of two important algebraization theorems.

3.1   The Counting Theorem of Pila–Wilkie

Definition 3.1.1. The (archimedean) *height* $H(r)$ of a rational number $r \in \mathbb{Q}$ is defined to be $\max(|a|, |b|)$, where $r = a/b$ for coprime integers a, b. Likewise, for $\alpha \in \mathbb{Q}^n$ we define the height to be $H(\alpha) = \max H(\alpha_i)$.

Note that there are finitely many points of $\mathbb{Q}^n$ of bounded height. Let $U \subset \mathbb{R}^n$ be a subset. We define the counting function as

$$N(U, t) := \#\{\alpha \in U \cap \mathbb{Q}^n \mid H(\alpha) \leq t\}.$$

Furthermore, we define the algebraic and transcendental parts

$$U^{\mathrm{alg}} := \bigcup_{\substack{Z \text{ connected semi-algebraic} \\ \dim Z > 0 \\ Z \subset U}} Z$$

$$U^{\mathrm{tr}} := U \smallsetminus U^{\mathrm{alg}}.$$

Note that U^{alg} may well *not* be definable in any o-minimal structure even if U is.

The counting theorem says that rational points can only accumulate along the algebraic part in a precise sense:

Theorem 3.1.2 (Counting theorem, Theorem 1.8 of [40]). *Let $U \subset \mathbb{R}^n$ be definable in an o-minimal structure. Then for any $\epsilon > 0$,*

$$N(U^{\mathrm{tr}}, t) = O(t^\epsilon).$$

Remark 3.1.3.

(1) The o-minimal hypothesis is essential: the graph $U \subset \mathbb{R}^2$ of $\sin(\pi x)$ contains polynomially many integer points.
(2) The general form of 3.1.2 builds on an earlier result of Bombieri–Pila [9], which asserts the conclusion of the theorem for $U = C$ a compact real analytic curve $C \subset \mathbb{R}^2$ containing no semialgebraic curves, which is obviously $\mathbb{R}_{\mathrm{an}}$-definable.
(3) There is a stronger form of 3.1.2 which is useful for applications. Informally, it states that for any $\epsilon > 0$ you can cover all the points of height at most t by at most $O(t^\epsilon)$ semialgebraic sets. In fact, it is this version which most naturally comes up in the proof of the Ax–Lindemann–Weierstrass and Ax–Schanuel theorems, as it is more naturally fits into inductive arguments.

Formally speaking, it says that for any $\epsilon > 0$ there is a finite number $J = J(U, \epsilon)$ of definable sets $W^{(i)} \subset \mathbb{R}^n \times \mathbb{R}^{m_i}$ such that each fiber $W_y^{(i)} \subset \mathbb{R}^n$ is semialgebraic and contained inside U, and a constant $c(U, \epsilon)$, such that all the rational points in U of height at most t are contained inside ct^ϵ many sets of the form $W_y^{(i)}$. See [40] for more details, refinements, and generalizations.

The counting theorem is often used to deduce from the presence of many rational points on U the existence of a *semialgebraic* subset $Z \subset U$ with many rational points, and this is why Theorem 3.1.2 is so powerful a tool in proving transcendence results. We will specifically need the following corollary of the strong form of Theorem 3.1.2 alluded to in the above remark:

Corollary 3.1.4. *If $N(U, t) \neq O(t^\epsilon)$ for some $\epsilon > 0$, then for any $N \in \mathbb{N}$ there is a semialgebraic subset $Z_N \subset U$ containing N rational points.*

We refer to [37] for a nice survey of the counting theorem and it's applications, but we say a few words about its role in the Pila–Zannier strategy to prove André–Oort type problems. Theorem 3.1.2 is used in two fundamentally different ways:

Let $\pi : \mathbb{H}_g \to A_g$ be the uniformizing map, $\pi_F : F \to A_g$ its restriction to a definable fundamental set, $V \subset A_g$ an algebraic subvariety, and $V_{\mathrm{sp}} \subset V$ the set of special points on V. As any subvariety with a Zariski dense set of special points is defined over a number field K, we may assume this is true for V, and thus V_{sp} is closed under the action of the Galois group $\mathrm{Gal}(\bar{K}/K)$. One has to show using arithmetic arguments that special points have Galois orbits which are "large," so that $\pi_F^{-1}(V_{\mathrm{sp}}) \subset \pi_F^{-1}(V)$ has many rational points in the sense of Theorem 3.1.2.[4] Applying Theorem 3.1.2 shows that $\pi^{-1}(V) \supset V'$ for some semialgebraic subvariety $V' \subset \mathbb{H}_g$. Next, by Corollary 1.2.13— whose proof also uses Theorem 3.1.2 as we'll see—it then follows there is a bialgebraic $L \subset \mathbb{H}_g$ such that $\pi^{-1}(V) \supset L \supset V'$. In particular, there are special subvarieties of V containing "most" points of V_{sp}. To finish, one has to apply an induction argument wherein special varieties are parametrized by special points on a lower-dimensional Shimura variety.

3.2   The Definable Chow Theorem of Peterzil–Starchenko

For X a complex algebraic variety, denote by X^{an} the complex points $X(\mathbb{C})$ with its natural structure of a complex analytic variety. Recall that Chow's theorem states that if X is a proper complex variety and $Y \subset X^{\mathrm{an}}$ is a closed complex analytic subvariety, then Y is algebraic. If the properness hypothesis on X is dropped, then the theorem is false: consider, for example, the graph of the complex exponential in $\mathbb{C} \times \mathbb{C}^*$.

The "definable Chow" theorem of Peterzil–Starchenko essentially states that the conclusion of Chow's theorem in the non-proper case holds if Y is additionally required to be definable with respect to an o-minimal structure.

Theorem 3.2.1 (Definable Chow, Theorem 5.1 of [34]). *Fix an o-minimal structure and let X be a complex algebraic variety. Then any closed complex analytic subvariety $Y \subset X^{\mathrm{an}}$ whose underlying set is definable in X^{def} is algebraic.*

Note that it is enough to assume Y is (analytically) irreducible of dimension d. Furthermore, we may replace X with a (nonempty) affine Zariski open subset U and algebraize $U^{\mathrm{an}} \cap Y$, for then Y is the closure of $U^{\mathrm{an}} \cap Y$. We can thus assume $X = \mathbb{A}^n$, and in the sequel we'll simply write $\mathbb{C}^n = (\mathbb{A}^n)^{\mathrm{an}}$.

We'll give two proofs, the first of which minimizes the explicit use of o-minimality, and the second that of complex analysis. The first proof relies on an important analyticity criterion of Bishop:

Theorem 3.2.2 (Theorem 3 of [8]). *Let $U \subset \mathbb{C}^n$ be an open subset and $Z \subset U$ a closed analytic subset. If $Y \subset U \backslash Z$ is a pure dimension d closed*

---

[4]This is classical in the case of the modular curve, much harder for A_g, and still open in general. See [48].

analytic subset of finite 2d-dimensional volume, then the closure $\overline{Y}$ of Y in U is an analytic subset.

Proof (First Proof of Theorem 3.2.1). Consider $\mathbb{C}^n \subset \mathbb{P}^n$ with complement $\mathbb{P}^{n-1}$ the plane at infinity. By the lemma below, Y has finite volume locally around $\mathbb{P}^{n-1}$, so by Bishop's theorem the closure $\overline{Y}$ of Y in $\mathbb{P}^n$ is an analytic subvariety, hence algebraic by the usual Chow theorem.   □

Lemma 3.2.3. *Any bounded k-dimensional definable $V \subset \mathbb{R}^m$ has finite k-volume.*

Proof. A bounded k-dimensional definable subset of $\mathbb{R}^k$ certainly has finite volume. The volume of $V \subset \mathbb{R}^m$ is bounded up to a constant by the maximum volume of its coordinate projections to $\mathbb{R}^k$—which is finite—times the maximum degree of these projections, which is also finite.   □

The second proof relies on the following fact using only elementary complex analysis.

Lemma 3.2.4. *Any definable holomorphic function $f : \mathbb{C}^n \to \mathbb{C}$ is algebraic.*

Proof.

Step 1.   An entire definable function $f : \mathbb{C} \to \mathbb{C}$ is algebraic.
    f cannot have an essential singularity at infinity or else it would have
    infinite fibers, by Casorati–Weierstrass.
Step 2.   Any definable holomorphic function $f : \mathbb{C}^n \to \mathbb{C}$ is algebraic.

Write $\mathbb{C}^n = \mathbb{C} \times \mathbb{C}^{n-1}$. For any $w \in \mathbb{C}^{n-1}$, $f(z, w)$ is a polynomial in z by Step 1. By Corollary 2.2.10, the degree of $f(z, w)$ in z is uniformly bounded,[5] so for some N,

$$f(z, w) = \sum_{k=0}^{N} \frac{\partial^k f}{\partial z^k}(0, w) \frac{z^k}{k!}.$$

By induction (using the previous step as the base case) the definable holomorphic functions $\frac{\partial^k f}{\partial z^k}(0, w) : \mathbb{C}^{n-1} \to \mathbb{C}$ are algebraic.

□

Second Proof of Theorem 3.2.1. We prove the claim by induction on the dimension d of Y, the base case being obvious.

Step 1.   The boundary $\partial Y := \overline{Y} \backslash Y \subset \mathbb{P}^{n-1}$ of Y in $\mathbb{P}^n$ is a definable subset
    of (real) dimension at most $2d - 1$.
    From cell decomposition, the boundary of a definable set always has
    smaller dimension.

---

[5] We might have to consider $f(z, w) - c$ to avoid multiplicity.

Step 2.   There is a linear projection $\pi : \mathbb{C}^n \to \mathbb{C}^d$ for which the restriction $\pi_Y : Y \to \mathbb{C}^d$ is proper.

Linear projections $\mathbb{C}^n \to \mathbb{C}^{n-1}$ are obtained by projecting from a point $p \in \mathbb{P}^{n-1}$ at infinity; the fibers of this projection are the lines through p (minus the point p itself). As $d < n$, by the previous step $\dim_{\mathbb{R}} \partial Y < \dim_{\mathbb{R}} \mathbb{P}^{n-1} = 2n-2$, so there is a projection $\mathbb{C}^n \to \mathbb{C}^{n-1}$ for which each fiber has bounded intersection with Y. The projection $Y \to \mathbb{C}^{n-1}$ is therefore proper, and the image is clearly definable and closed analytic by Remmert's proper mapping theorem. Now iterate.

Step 3.   The locus $Y_0 \subset Y$ where $\pi_Y : Y \to \mathbb{C}^d$ is not étale is a closed algebraic subvariety Y_0 of $\mathbb{C}^n$.

Y_0 is analytic of strictly smaller dimension than Y and evidently definable (as, for instance, it is the locus where the fiber size is nongeneric). By the inductive hypothesis we therefore have that Y_0 is algebraic.

Step 4. Y is algebraic.

Write $\mathbb{C}^n = \mathbb{C}^{n-d} \times \mathbb{C}^d$, so π is projection to the second factor. Let $Z^{\mathrm{an}} = \pi(Y_0)$, which is a closed algebraic subvariety of $\mathbb{C}^d$. Let N be the degree of the map $\pi_Y : Y \to \mathbb{C}^d$ and consider the function

$$F : \mathbb{C}^d \backslash Z^{\mathrm{an}} \to \mathrm{Sym}^N \mathbb{C}^{n-d} : z \mapsto \pi^{-1}(z).$$

Note that $\mathrm{Sym}^N \mathbb{C}^{n-d}$ is an affine algebraic variety. F is evidently definable and holomorphic, as well as locally bounded around Z^{an} (as π_Y is proper). Thus, the pullbacks of the coordinate functions of $\mathrm{Sym}^N \mathbb{C}^{n-d}$ extend to definable holomorphic functions $f : \mathbb{C}^d \to \mathbb{C}$, which are therefore algebraic by Lemma 3.2.4. It follows that $Y \backslash Z^{\mathrm{an}}$ is algebraic, and therefore that Y is.

$\square$

Remark 3.2.5. Neither of these proofs is the one given by Peterzil–Starchenko—as they prove it for arbitrary real closed fields—but the second proof is close to that of [34]: we've only really cheated by using Casorati–Weierstrass. Step 1 of the proof of Lemma 3.2.4 can be proven in general using a version of Liouville's theorem proven by Peterzil–Starchenko.

4   The Ax–Lindemann–Weierstrass Theorem

In this section, as a warm up for the proof of Theorem 6.1.1, we show how to use the Pila–Wilkie theorem to prove the Ax–Lindemann–Weierstrass theorem for the exponential map. Many of the same arguments will be used in the proof of Theorem 6.1.1. The notable exception is that the definable Chow theorem does not play a role in the proof of the Ax–Lindemann–Weierstrass theorem but is essential to the proof of the Ax–Schanuel theorem.

4.1   The Exponential Function

Let

$$\pi : \mathbb{C}^n \to (\mathbb{C}^*)^n : (z_1, \dots, z_n) \mapsto (e(z_1), \dots, e(z_n))$$

where $e(z) = e^{2\pi i z}$. Let's first give a proof of the classification of the bialgebraic subvarieties of $\mathbb{C}^n$ which only mildly uses some of the o-minimal machinery—and in particular will not use either of the algebraization theorems discussed in the previous lecture.

Consider an algebraic subvariety $M \subset (\mathbb{C}^*)^n$ and the induced map on fundamental groups

$$\pi_1(M) \to \pi_1((\mathbb{C}^*)^n) \cong \mathbb{Z}^n.$$

The important observation is that we can directly relate the size of the monodromy (that is, the image of $\pi_1(M)$) to the invariance of M.

Proposition 4.1.1. *If the image of $\pi_1(M)$ is not finite index in $\pi_1((\mathbb{C}^*)^n)$, then M is contained in a coset of a proper algebraic subtorus.*

Proof. Without loss of generality we may assume

$$\pi_1(M) \to 0 \oplus \mathbb{Z}^{n-1} \subset \mathbb{Z}^n.$$

Let

$$F = \{(z_1, \dots, z_n) \in \mathbb{C}^n \mid -\epsilon < \text{Re}(z_i) < 1 + \epsilon\} \tag{6}$$

which is a fundamental set for $\pi : \mathbb{C}^n \to (\mathbb{C}^*)^n$. Now, on the one hand, the function z_1 descends to a holomorphic function $f : M \to \mathbb{C}$ by the assumption on the monodromy. On the other hand, we may take a definable cell decomposition[6] of M. Each cell D is simply connected and therefore lifts to F, so z_1 has bounded real part on D. It then follows that z_1 has bounded real part on all of M, so f must be constant.   □

Corollary 4.1.2. *The closed irreducible bialgebraic subvarieties of $(\mathbb{C}^*)^n$ are precisely cosets of algebraic subtori.*

Proof. Equivalently, we must show that the closed irreducible bialgebraic subvarieties of $\mathbb{C}^n$ are translates of $\mathbb{C}$-subspaces defined over $\mathbb{Q}$. Suppose $L \subset \mathbb{C}^n$ is a closed irreducible bialgebraic subvariety, which we may assume is not contained in any translate of a $\mathbb{C}$-subspace defined over $\mathbb{Q}$. By the proposition

---

[6]We're not really using o-minimality here—just a statement about the topology of algebraic varieties.

(applied to $M = \pi(L)$), L is invariant under a finite index subgroup of $\mathbb{Z}^n$. As L is algebraic, its stabilizer under translation by vectors in $\mathbb{C}^n$ is an algebraic subgroup, which is therefore all of $\mathbb{C}^n$. $\qquad\qquad\qquad\qquad\qquad\qquad\Box$

We are now ready to prove the Ax–Lindemann–Weierstrass theorem, whose statement we recall.

Theorem 4.1.3 (Ax–Lindemann–Weierstrass). *Suppose there are algebraic subvarieties $V_1 \subset \mathbb{C}^n$ and $V_2 \subset (\mathbb{C}^*)^n$.*

(1) If $\pi(V_1) \subset V_2$, then there is a bialgebraic $M \subset (\mathbb{C}^*)^n$ with

$$\pi(V_1) \subset M \subset V_2;$$

(2) If $\pi(V_1) \supset V_2$, then there is a bialgebraic $M \subset (\mathbb{C}^*)^n$ with

$$\pi(V_1) \supset M \supset V_2.$$

Before the proof we make a crucial observation: both the fundamental set $F \subset \mathbb{C}^n$ and the restriction $\pi_F : F \to (\mathbb{C}^*)^n$ of the covering map are definable in the o-minimal structure $\mathbb{R}_{\mathrm{an,exp}}$ (*c.f.* Example 2.3.3).

Proof of Theorem 4.1.3. We start with the proof of (1). We can assume by taking closures and components that V_1 (resp. V_2) is a closed irreducible algebraic subvariety of $\mathbb{C}^n$ (resp. $(\mathbb{C}^*)^n$). We can further assume that V_2 is not contained in any proper subtorus, and that V_1 is a maximal closed irreducible algebraic subvariety of $\pi^{-1}(V_2)$. It remains to show that V_1 is bialgebraic.

Consider the set

$$I := \{v \in \mathbb{R}^n \mid \dim\left((V_1 + v) \cap \pi_F^{-1}(V_2)\right) = \dim V_1\}.$$

As V_1 is irreducible, we see that $v \in I$ if and only if the translate $V_1 + v$ meets F and $V_1 + v \subset \pi^{-1}(V_2)$.

Step 1. I is $\mathbb{R}_{\mathrm{an,exp}}$-definable.

Indeed, the universal translate

$$\mathcal{V}_1 := \{(v, z) \mid z \in V_1 + v\} \subset \mathbb{R}^n \times \mathbb{C}^n$$

is (real) algebraic so definable, as therefore is the universal intersection

$$\mathcal{U} := \mathcal{V}_1 \cap \left(\mathbb{R}^n \times \pi_F^{-1}(V_2)\right).$$

Applying Corollary 2.2.10 to the projection $\mathcal{U} \to \mathbb{R}^n$ yields the claim.

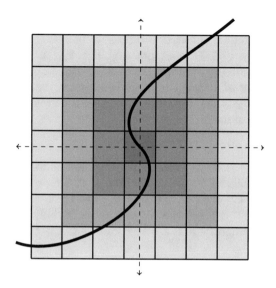

Fig. 5 V_1 must pass through at least one fundamental domain $F - v$ of each height.

Step 2. $\mathrm{Stab}_{\mathbb{Z}^n}(V_1)$ is infinite.

We may assume V_1 meets F, as $\pi^{-1}(V_2)$ is covered by integral translates[7] of F. Note that for any $v \in \mathbb{Z}^n$, V_1 meets $F - v$ if and only if $v \in I$, so the integral points of I correspond to fundamental domains that V_1 passes through. Observe that V_1 cannot be contained in any "height ball"

$$\bigcup_{\substack{v \in \mathbb{Z}^n \\ H(v) \leq r}} (F - v)$$

as then each coordinate z_i would have bounded real part and therefore be constant. For each $t \in \mathbb{Z}_{>0}$, the complement of the "height sphere"

$$\bigcup_{\substack{v \in \mathbb{Z}^n \\ H(v) = t}} (F - v)$$

has two connected components, so V_1 must pass through it (see Figure 5). Thus, we have

$$N(I, t) \geq t + 1.$$

---

[7]Strictly speaking we should take $\epsilon = 0$ and make F a fundamental *domain* for this argument.

By the strong form of the counting theorem, we have a (real) semialgebraic curve $C \subset I$ that contains at least two integral points.

If translation by $c \in C$ does not stabilize V_1, then $\bigcup_c (V_1 + c)$ is a real semialgebraic subset of $\pi^{-1}(V_2)$, and its $\mathbb{C}$-Zariski closure is a complex algebraic subvariety of $\pi^{-1}(V_2)$ of larger dimension than V_1, contradicting the maximality of V_1. Thus, $V_1 = V_1 + c$ for all $c \in C$, and V_1 is stabilized by a nonzero integer point.

Step 3.   Induction step.

Since $\mathrm{Stab}_{\mathbb{C}^n}(V_1)$ is an algebraic subgroup, it follows from the previous step that V_1 is stabilized *by a complex line* $\mathbb{C} \subset \mathbb{C}^n$ *defined over* $\mathbb{Q}$. Thus, there is a splitting $\mathbb{C}^n = \mathbb{C}^{n-1} \oplus \mathbb{C}$ defined over $\mathbb{Q}$ such that $V_1 = V_1' \times \mathbb{C}$. Let $V_2' = V_2 \cap (\mathbb{C}^*)^{n-1}$. Since the proposition is trivially true for $n = 1$, we may inductively assume there is a bialgebraic $L' \subset \mathbb{C}^{n-1}$ with

$$V_1' \subset L' \subset \pi^{-1}(V_2').$$

By the assumption on V_2, we must have $L' \neq \mathbb{C}^{n-1}$ (or else $V_2 = (\mathbb{C}^*)^n$), so we can apply the induction hypothesis again to $V_2'' = \pi(L' \oplus \mathbb{C}) \cap V_2$ and $V_1'' = V_1$. We conclude there is a bialgebraic $L'' \subset L' \oplus \mathbb{C}$ with

$$V_1 \subset L'' \subset \pi^{-1}(V_2'') \subset \pi^{-1}(V_2)$$

and so $V_1 = L''$ is bialgebraic, by the maximality of V_1.

The proof of part (2) is very similar, so we just sketch the argument. We may now assume V_1 is the $\mathbb{C}$-Zariski closure of a component of $\pi^{-1}(V_2)$ and apply the Pila–Wilkie theorem to

$$I := \{v \in \mathbb{R}^n \mid \dim \left((V_1 + v) \cap \pi_F^{-1}(V_2)\right) = \dim V_2\}.$$

We can then conclude that there is a real semialgebraic $C \subset I$, and if translation by $c \in C$ doesn't stabilize V_1, then $\bigcap_c (V_1 + c)$ would contain $\pi^{-1}(V_2)$, implying that the $\mathbb{C}$-Zariski closure of $\pi^{-1}(V_2)$ is smaller than V_1, a contradiction. We conclude that V_1 is invariant under a $\mathbb{C}$-line defined over $\mathbb{Q}$, and a similar induction yields the claim. $\square$

4.2   *Hyperbolic Uniformizations*

We give a sketch of how the above proof is adapted to the setting of Shimura varieties, but we first recall the basic structures associated with Shimura varieties (see [31] for details). These are:

- A connected semisimple algebraic $\mathbb{Q}$-group $\mathbf{G}$.

- A bounded symmetric domain

$$\Omega = \mathbf{G}(\mathbb{R})/K$$

where K is a maximal compact subgroup of $\mathbf{G}(\mathbb{R})$. Ω is a complex manifold and its biholomorphism group is $\mathbf{G}(\mathbb{R})$. It also carries a natural left-invariant Hermitian metric h which has negative sectional curvature. Note that the requirement that Ω have a holomorphic structure is a strong requirement on the group $\mathbf{G}$.
- The compact dual $\check{\Omega}$, which is

$$\check{\Omega} = \mathbf{G}(\mathbb{C})/B$$

where B is a maximal Borel subgroup. It is a homogeneous projective variety. The Harish–Chandra embedding theorem shows that for any choice of B containing K, Ω is realized as a semialgebraic subset of $\check{\Omega}$. Moreover, this embedding is unique up to the action of $\mathbf{G}(\mathbb{C})$.
- An arithmetic lattice $\Gamma \subset \mathbf{G}(\mathbb{Q})$, that is, a subgroup which is commensurable to the subgroup preserving an integral structure $H_{\mathbb{Z}}$ in a faithful representation $\mathbf{G}(\mathbb{Q}) \to \mathrm{GL}(H_{\mathbb{Q}})$. Γ is discrete and finite co-volume in $\mathbf{G}(\mathbb{R})$ (with respect to a left-invariant metric).
- The analytic quotient

$$Y = \Gamma\backslash\Omega = \Gamma\backslash\mathbf{G}(\mathbb{R})/K.$$

Y uniquely has the structure of an algebraic variety [4], and it is called a *Shimura variety.*

Example 4.2.1. For $\mathbf{G} = \mathrm{Sp}_{2g}$ and $\Gamma = \mathrm{Sp}_{2g}(\mathbb{Z})$ we have

$$H_{\mathbb{Z}} = \text{the unimodular symplectic lattice of rank } 2g$$

$$\Omega = \text{Siegel upper half-space } \mathbb{H}_g$$

$$\check{\Omega} = \text{the Lagrangian Grassmannian of } H_{\mathbb{C}}$$

$$Y = \Gamma\backslash\mathbb{H}_g = \text{the (coarse) moduli space of principally polarized}$$

$$g - \text{dimensional abelian varieties } A_g$$

We can now consider the uniformization $\pi : \Omega \to Y$. Recall that we say a complex analytic subvariety $V \subset \Omega$ is algebraic if there is an algebraic subvariety $\check{V} \subset \check{\Omega}$ with $V = \check{V} \cap \Omega$. We say an algebraic subvariety $V \subset \Omega$ is bialgebraic if $\dim V = \dim \pi(V)^{\mathrm{Zar}}$, as in Definition 1.2.8. The bialgebraic subvarieties are the so-called weakly special subvarieties:

Definition 4.2.2. A *weakly special* subvariety of Y is a Shimura variety Y' given as

$$Y' = \Gamma' \backslash \mathbf{G}'(\mathbb{R}) / K'$$

where $\mathbf{G}'$ is an algebraic $\mathbb{Q}$-subgroup of $\mathbf{G}$, $\Gamma' = \Gamma \cap \mathbf{G}'(\mathbb{Q})$ is an arithmetic lattice, and $K' = K \cap \mathbf{G}'(\mathbb{R})$. Evidently Y' is then an analytic subvariety of Y, and in fact it is (uniquely) an algebraic subvariety.

Proposition 4.2.3 (Theorem 1.2 of [49]). *Let Y be a Shimura variety. The closed irreducible bialgebraic subvarieties of Y are precisely the weakly special subvarieties.*

As in Proposition 4.1.1, the proof of Proposition 4.2.3 uses monodromy arguments and relies heavily on the work of André–Deligne [1, 16]. The Ax–Lindemann–Weierstrass theorem in this context was proven by Pila for powers of the modular curve [35], by Pila–Tsimerman for A_g [38], and then by Klingler–Ulmo–Yafaev for general Shimura varieties [27]:

Theorem 4.2.4 (Ax–Lindemann–Weierstrass, Theorem 1.6 of [27]). *Let Y be a Shimura variety uniformized by Ω. Suppose there are algebraic subvarieties $V_1 \subset \Omega$ and $V_2 \subset Y$.*

(1) If $\pi(V_1) \subset V_2$, then there is a bialgebraic $M \subset \Omega$ with

$$\pi(V_1) \subset M \subset V_2;$$

(2) If $\pi(V_1) \supset V_2$, then there is a bialgebraic $M \subset \Omega$ with

$$\pi(V_1) \supset M \supset V_2.$$

Sketch of Proof. We will only sketch the proof of (1), as (2) is similar. We can make the same assumptions on V_1 and V_2 as in the proof of Theorem 4.1.3— that is, that both are closed irreducible subvarieties. We can further assume V_2 is not contained in any bialgebraic subvariety, and that V_1 is a maximal algebraic subvariety of $\pi^{-1}(V_2)$.

We follow the same three steps as the proof of Theorem 4.1.3

Step 1.

We first need a definable fundamental set $F \subset \Omega$ for which the restriction $\pi_F : F \to Y$ is a definable quotient map. In [27], this is done using finitely many Siegel sets, which yield a semialgebraic fundamental set $\tilde{F} \subset \mathbf{G}(\mathbb{R})$ for the action of any arithmetic lattice $\Gamma \subset \mathbf{G}(\mathbb{Q})$. We can then take F as the image of $\tilde{F}$ in $\Omega = \mathbf{G}(\mathbb{R})/K$. It is then shown that $\pi_F : F \to Y$ is $\mathbb{R}_{\mathrm{an,exp}}$-definable using the theory of toroidal compactifications. In Lecture 5 we will instead use the local theory of degenerations of Hodge structures to produce a definable fundamental set.

It then follows in the same way that

$$I := \{g \in \mathbf{G}(\mathbb{R}) \mid \dim \left(gV_1 \cap \pi_F^{-1}(V_2)\right) = \dim V_1\}.$$

is $\mathbb{R}_{\mathrm{an,exp}}$-definable.

Step 2. $\mathrm{Stab}_{\mathbf{G}(\mathbb{Z})}(V_1)$ is infinite.

We would like to apply the Pila–Wilkie theorem to I as in Step 2 of the proof of Theorem 4.1.3, so we need

Claim. $N(I, t) \gg t^\epsilon$ for some $\epsilon > 0$.

We postpone until Lecture 7 the precise definition of the height of an element of $\mathbf{G}(\mathbb{Q})$ and the counting function. The above argument using "height balls" to produce polynomially many $\mathbb{Z}$-points of I (in the height) breaks down, essentially because the uniformizing group Γ and its action on Ω are now very complicated.

The problem is remedied in [27, 38] by instead using *metric balls*. Let Γ_V be the image of the monodromy representation $\pi_1(V_2) \to \mathbf{G}(\mathbb{Q})$. Recall that since $\pi^{-1}(V_2)$ is stable under Γ_V, it will be sufficient to show that V_1 passes through polynomially many (in the height of γ) integral translates $\gamma^{-1}F$ for $\gamma \in \Gamma_V$. We may assume V_1 meets F and take a basepoint $x_0 \in F \cap V_1$. Consider the metric balls $B_{x_0}(R)$ centered at x_0. By a result of Hwang–To, the volume achieved by V_1 in $B_{x_0}(R)$ is large:

Theorem 4.2.5 (Corollary 3 of [23]). *There is a constant $\beta > 0$ only depending on Ω such that for any closed positive-dimensional analytic subvariety $Z \subset B_{x_0}(R)$ we have*

$$\mathrm{vol}(Z) \gg \sinh(\beta R)^{\dim Z} \mathrm{mult}_{x_0} Z.$$

We will need a version of Theorem 4.2.5 for period domains, whose proof we sketch in Lecture 8.

To establish the claim, it now remains to show that:

(a) The only integral translates $\gamma^{-1}F$ meeting $B_{x_0}(R)$ have $H(\gamma) \ll e^{O(R)}$;
(b) V_1 has bounded volume intersection with all of the translates $\gamma^{-1}F$.

Indeed, the volume of $V_1 \cap B_{x_0}(R)$ is exponential in the radius by Theorem 4.2.5, so by (b) and the fact that the $\gamma^{-1}F$ cover $\pi^{-1}(V_2)$ with bounded overlaps we conclude that V_1 passes through exponentially many (in the radius) integral translates $\gamma^{-1}F$ in $B_{x_0}(R)$. It then follows from (a) that the number of these integral translates is polynomial in the height.

For (a), we need to compare the metric dilation of γ to its height, which is standard (see, for example, Lecture 7). For (b), it suffices to show that all translates gV_1 for $g \in \mathbf{G}(\mathbb{C})$ meet F with bounded volume, and since these translates form an algebraic family, we can use definability to get a uniform bound (see, for example, Proposition 5.5.1).

To finish, just as in the proof of Theorem 4.1.3, we obtain an algebraic family $\{g_c\}_{c \in C} \subset \mathbf{G}(\mathbb{C})$ with $g_c V_1 \subset \pi^{-1}(V_2)$ by applying the Pila–Wilkie theorem. If V_1 is a *maximal* irreducible algebraic subvariety of $\pi^{-1}(V_2)$, then we have $V_1 = \bigcup_{c \in C} g_c V_1$ and V_1 then is therefore invariant under $\{g_c\}_{c \in C}$ (which in particular contains a nontrivial integral point).

Step 3.   Induction step.

As the stabilizer of V_1 is an algebraic subgroup of $\mathbf{G}$ and we know from the previous step that $\mathrm{Stab}_{\mathbf{G}(\mathbb{Z})}(V_1)$ is infinite, it follows that V_1 is stabilized by a positive-dimensional connected $\mathbb{Q}$-subgroup $\mathbf{H}$ of $\mathbf{G}$, namely the identity component of the $\mathbb{Q}$-Zariski closure of $\mathrm{Stab}_{\mathbf{G}(\mathbb{Z})}(V_1)$. However, to make the induction work, one needs V_1 to be stabilized by a *normal* $\mathbb{Q}$-subgroup of $\mathbf{G}$, as this will imply $\mathbf{G}$ is isogeneous to a product. This problem is solved using Hecke correspondences in [27, 38]. In [33], the same problem is solved in a different way to prove the Ax–Schanuel theorem, essentially by using the definable Chow theorem to algebraize the family of algebraic deformations V_1' of V_1 that are contained in $\pi^{-1}(V_2)$, and then using the fact that algebraic families of varieties have large monodromy. We will use the same strategy in Lecture 6.

$\square$

5   Recollections from Hodge Theory

Shimura varieties are moduli spaces of very special polarized Hodge structures, and it is very natural to formulate the Ax–Schanuel conjecture (as well as the other transcendence statements) for general moduli spaces of polarized Hodge structures. We spend this lecture recalling the relevant notions from Hodge theory. We will be necessarily brief, and refer the interested reader to [13] and [20] for details.

5.1   *Preliminaries*

Definition 5.1.1. Fix an integer n. Let $H_\mathbb{Z}$ be a finite rank free $\mathbb{Z}$-module. A pure Hodge structure on $H_\mathbb{Z}$ of weight n is a decomposition into complex vector spaces

$$H_\mathbb{C} := H_\mathbb{Z} \otimes \mathbb{C} = \bigoplus_{p+q=n} H^{p,q} \tag{7}$$

satisfying $\overline{H^{p,q}} = H^{q,p}$. The dimensions $h^{p,q} = \dim_{\mathbb{C}} H^{p,q}$ are called the Hodge numbers. We say the Hodge structure is effective if $H^{p,q} = 0$ for $p > n$.

Note that the Hodge structure is determined by the *Hodge filtration*

$$F^p := \bigoplus_{r \geq p} H^{r,s}$$

as $H^{p,q} = F^p \cap \overline{F^q}$. Conversely, a descending filtration $F^\bullet$ determines a Hodge structure of weight n if it satisfies

$$F^p \cap \overline{F^{n-p+1}} = 0 \tag{8}$$

for all p.

Example 5.1.2. A pure weight 1 (or -1) Hodge structure is equivalent to a complex torus T. We canonically have an embedding

$$H_1(T, \mathbb{Z}) \to H^0(T, \Omega_T^1)^\vee \oplus H^0(T, \overline{\Omega}_T^1)^\vee : \gamma \mapsto \int_\gamma$$

which yields a decomposition

$$H_1(T, \mathbb{C}) = H^{-1,0} \oplus H^{0,-1}$$

with $H^{-1,0} = H^0(T, \Omega_T^1)^\vee$ and $H^{0,-1} = \overline{H^{-1,0}}$. Projecting $H_1(T, \mathbb{Z})$ to $H^{-1,0}$ we can recover T canonically by the albanese

$$T \xrightarrow{\cong} H^0(T, \Omega_T^1)^\vee / H_1(T, \mathbb{Z}) : p \mapsto \int_0^p .$$

The weight -1 Hodge structure on $H_1(T, \mathbb{Z})$ naturally induces a weight 1 Hodge structure on $H^1(T, \mathbb{Z})$.

Definition 5.1.3. Suppose $H_{\mathbb{Z}}$ carries a weight n Hodge structure, and let $q_{\mathbb{Z}}$ be a $(-1)^n$-symmetric bilinear form—that is, $q_{\mathbb{Z}}$ is symmetric if n is even and skew-symmetric if n is odd.

(1) The Weil operator $C \in \mathrm{End}(H_{\mathbb{R}})$ is the real endomorphism satisfying

$$C_{\mathbb{C}} = \bigoplus_{p,q} i^{p-q} \cdot \mathrm{id}_{H^{p,q}} .$$

(2) The *Hodge form* is the Hermitian form h on $H_{\mathbb{C}}$ defined by

$$h(u, v) = q_{\mathbb{C}}(Cu, \overline{v}).$$

(3) We say the Hodge structure is *polarized* by $q_\mathbb{Z}$ if the Hodge form is positive-definite and the decomposition (7) is h-orthogonal.

If the Hodge structure is polarized by $q_\mathbb{Z}$, then the Hodge filtration $F^\bullet$ is $q_\mathbb{C}$-isotropic: we have $(F^\bullet)^\perp = F^{n+1-\bullet}$. Conversely, a $q_\mathbb{C}$-isotropic Hodge filtration satisfying (8) determines a $q_\mathbb{Z}$-polarized Hodge structure if the Hodge form is positive-definite.

Example 5.1.4. A polarized weight 1 (or -1) Hodge structure is equivalent to a polarized abelian variety A. A skew-symmetric integral form $q_\mathbb{Z}$ on $H_1(A, \mathbb{Z})$ can be thought of as an element $h \in H^2(A, \mathbb{Z})$. By the Lefschetz $(1,1)$ theorem, the $q_\mathbb{C}$-isotropicity condition on the Hodge decomposition implies $h = c_1(L)$ for a line bundle L on A, and the positivity condition implies L is ample.

Example 5.1.5. We have the following broad generalization of the previous example, which was the original motivation for their introduction. Let Y be a proper Kähler manifold (for example, a smooth complex projective variety). After choosing a Kähler form ω, we obtain a weight n Hodge structure on degree n singular cohomology

$$H^n(Y, \mathbb{C}) = \bigoplus_{p+q=n} H^{p,q}(Y) \tag{9}$$

by decomposing harmonic representatives of de Rham cohomology classes into (p, q) parts. Furthermore, suppose Y is a smooth complex projective variety with ample bundle L and set $h = c_1(L)$. The singular cohomology $H^*(Y, \mathbb{Q})$ decomposes into polarized Hodge structures as follows. For $n \leq d = \dim X$, let

$$H^{d-n}_{\mathrm{prim}}(Y, \mathbb{Z}) := \ker\left(h^{n+1}\cup : H^{d-n}(Y, \mathbb{Z})_{\mathrm{tf}} \to H^{d+n+2}(Y, \mathbb{Z})_{\mathrm{tf}}\right).$$

Where $(-)_{\mathrm{tf}}$ denotes the torsion-free quotient. We have

$$H^n(Y, \mathbb{Q}) = \bigoplus_{0 \leq k \leq n/2} h^k \cup H^{n-2k}_{\mathrm{prim}}(Y, \mathbb{Q}).$$

$H^n_{\mathrm{prim}}(Y, \mathbb{Z})$ carries a natural integral form

$$q_n(a, b) := \int_Y h^{\dim Y - 2n} \cup a \cup b.$$

The decomposition (9) (associated to the Kähler class h) then induces a weight n Hodge structure on $H_{\mathrm{prim}}(Y, \mathbb{Z})$ polarized by q_n.

Remark 5.1.6. Note that if $H_{\mathbb{Z}}$ carries a pure Hodge structure, then so too will any tensor power, symmetric power, wedge power, etc. of $H_{\mathbb{Z}}$. The same is true of pure polarized Hodge structures.

5.2 Period Domains and Period Maps

Define the algebraic $\mathbb{Q}$-group $\mathbf{G}(\mathbb{Q}) = \mathrm{Aut}(H_{\mathbb{Q}}, q_{\mathbb{Q}})$; we will often denote $\mathbf{G}(\mathbb{Z}) = \mathrm{Aut}(H_{\mathbb{Z}}, q_{\mathbb{Z}})$. It is then not hard to see that the space D of $q_{\mathbb{Z}}$-polarized pure weight n Hodge structures on $H_{\mathbb{Z}}$ with specified Hodge numbers $h^{p,q}$ is a homogeneous space for $\mathbf{G}(\mathbb{R})$. Indeed, choosing a reference Hodge structure, we have

$$D = \mathbf{G}(\mathbb{R})/V$$

where V is a subgroup of the compact unitary subgroup $K = \mathbf{G}(\mathbb{R}) \cap U(h)$ of $\mathbf{G}(\mathbb{R})$ with respect to the hodge form of the reference Hodge structure. Moreover, D is canonically an open subset (in the euclidean topology) of $\check{D} = \mathbf{G}(\mathbb{C})/P$, the flag variety parametrizing $q_{\mathbb{C}}$-isotropic Hodge filtrations $F^{\bullet}$ on $H_{\mathbb{C}}$ with $h^{p,n-p} = \dim F^p/F^{p+1}$.

Definition 5.2.1. Such a D is called a *polarized period domain.*

Example 5.2.2. Given a smooth projective morphism $f : Y \to X$, consider the local system $R^k f_* \mathbb{Q}$ for some k. In the notation of Example 5.1.5, $R^n f_* \mathbb{Z}$ can be decomposed into primitive pieces, and each fiber of $R^n_{\mathrm{prim}} f_* \mathbb{Z}$ carries a pure weight n Hodge structure. By a theorem of Griffiths, the resulting map

$$\varphi : X^{\mathrm{an}} \to \mathbf{G}(\mathbb{Z}) \backslash D : y \mapsto [H^n_{\mathrm{prim}}(X_y, \mathbb{Z})]$$

is holomorphic and locally liftable to D.

The fundamental observation of Griffiths is that we cannot get arbitrary maps to $\mathbf{G}(\mathbb{Z})\backslash D$ from geometry as in Example 5.2.2. Indeed, only certain tangent directions of D are accessible to algebraic families. To make this precise, fix a point $x \in D$ and note that a deformation of the Hodge filtration at x in particular yields a deformation of each F_x^p, so we have a natural map

$$T_x D \to \bigoplus_p \mathrm{Hom}(F_x^p, H_{\mathbb{C}}/F_x^p). \tag{10}$$

Definition 5.2.3. The Griffiths transverse subspace $T_x^{GT} D \subset T_x D$ is the inverse image of $\bigoplus_p \mathrm{Hom}(F_x^p, F^{p-1}/F_x^p)$ under the map in (10).

In other words, to first order each F^p is only deformed inside F^{p-1}. The Griffiths transverse subspaces assemble into a holomorphic subbundle $T^{GT}D \subset TD$.

Remark 5.2.4. Each pure polarized Hodge structure $x \in D$ on $H_{\mathbb{Z}}$ naturally induces a pure polarized Hodge structure on the Lie algebra $\mathfrak{g}_{\mathbb{R}} \subset \mathrm{End}(H_{\mathbb{R}})$ of weight 0, which we call $\mathfrak{g}_x$. Denote its Hodge filtration by $F_x^{\bullet}\mathfrak{g}_{\mathbb{C}}$. The Lie algebra of the stabilizer $P_x \subset \mathbf{G}(\mathbb{C})$ of $x \in \check{D}$ is then naturally $F_x^0\mathfrak{g}_{\mathbb{C}}$. Thus, the tangent space T_xD is naturally (and holomorphically) identified with $\mathfrak{g}_{\mathbb{C}}/F_x^0\mathfrak{g}_{\mathbb{C}}$. The Griffiths transverse subspace is $F_x^{-1}\mathfrak{g}_{\mathbb{C}}/F_x^0\mathfrak{g}_{\mathbb{C}}$.

Definition 5.2.5. By a period map we mean a holomorphic locally liftable Griffiths transverse map

$$\varphi : X^{\mathrm{an}} \to \Gamma\backslash D$$

for a smooth complex algebraic variety X and a finite index $\Gamma \subset \mathbf{G}(\mathbb{Z})$.

Remark 5.2.6. A period map $\varphi : X^{\mathrm{an}} \to \mathbf{G}(\mathbb{Z})\backslash D$ is equivalent to the data of a pure polarized integral variation of Hodge structures on X. This consists of:

- A local system $\mathscr{H}_{\mathbb{Z}}$ with a flat quadratic form $Q_{\mathbb{Z}}$.
- A holomorphic locally split filtration $F^{\bullet}$ of $\mathscr{H}_{\mathbb{Z}} \otimes_{\mathbb{Z}} \mathcal{O}_{X^{\mathrm{an}}}$ such that the flat connection ∇ satisfies Griffiths transversality:

$$\nabla(F^p) \subset F^{p-1} \text{ for all } p.$$

- We moreover require that $(\mathscr{H}_{\mathbb{Z}}, Q_{\mathbb{Z}}, F^{\bullet})$ is fiberwise a pure polarized integral Hodge structure.

The period map lifts to $\Gamma\backslash D$ if Γ contains the image of the monodromy representation of $\mathscr{H}_{\mathbb{Z}}$.

Definition 5.2.7. Let $\overline{X}$ be a log smooth compactification of X. For any irreducible boundary divisor $E \subset \overline{X}$, the local monodromy operator $\gamma \in \mathbf{G}(\mathbb{Z})$ of E is the monodromy of the local system $\mathscr{H}_{\mathbb{Z}}$ along a small loop around E, which is defined up to conjugation (in $\mathbf{G}(\mathbb{Z})$).

The following result on the monodromy of variations of Hodge structures is of pervasive importance:

Theorem 5.2.8. *Any period map $\varphi : X^{\mathrm{an}} \to \Gamma\backslash D$ has quasiunipotent local monodromy.*

Corollary 5.2.9. *For any period map $\varphi : X^{\mathrm{an}} \to \Gamma\backslash D$, there is a finite étale cover $f : X' \to X$ such that the period map $\varphi' = \varphi \circ f : X' \to \Gamma\backslash D$ has unipotent local monodromy.*

Proof. Note that any quasiunipotent $\gamma \in \mathbf{G}(\mathbb{Z})$ has eigenvalues which are roots of unity of bounded order. Let $\Gamma(n) \subset \mathbf{G}(\mathbb{Z})$ be the full-level n subgroup

$$\Gamma(n) := \left\{ \gamma \in \mathbf{G}(\mathbb{Z}) \;\middle|\; \gamma \equiv \begin{pmatrix} 1 & 0 \\ 0 & 1 \end{pmatrix} \bmod n \right\}.$$

Since the roots of unity of bounded order inject $\bmod\, p$ for sufficiently large p, it follows that every quasiunipotent element of $\Gamma(p)$ is in fact unipotent for sufficiently large p. Now take X' to be the pullback of the finite étale cover $\Gamma(p)\backslash D \to \mathbf{G}(\mathbb{Z})\backslash D$ (technically as stacks).   □

5.3 The Mumford–Tate Group and Weakly Special Subvarieties

Definition 5.3.1. Suppose $H_{\mathbb{Z}}$ carries a pure weight $2k$ Hodge structure. An integral (resp. rational) class $v \in H_{\mathbb{Z}}$ (resp. $v \in H_{\mathbb{Q}}$) is *Hodge* if $v \in H^{k,k}$.

Note that an integral class $v \in H_{\mathbb{Z}}$ has pure Hodge type if and only if it is a Hodge class. Moreover, v is Hodge if and only if $v \in F^k$.

Example 5.3.2. The motivation for considering Hodge classes again comes from geometry. Given a smooth projective complex algebraic variety Y and a closed algebraic subvariety $Z \subset Y$, the fundamental class $[Z] \in H^{2\,\mathrm{codim}\,Z}(Y, \mathbb{Z})$ is a Hodge class. The Hodge conjecture says that moreover all rational Hodge classes arise from cycles (up to rational scaling).

The Hodge classes of a particular Hodge structure are described by the Mumford–Tate group:

Definition 5.3.3. Suppose $H_{\mathbb{Q}}$ carries a pure Hodge structure H. The (special) Mumford–Tate group $\mathbf{MT}_H$ of H is the algebraic $\mathbb{Q}$-subgroup of $\mathbf{End}(H_{\mathbb{Q}})$ with the following property: for any tensor power $H' = H^{\otimes k} \otimes (H^{\vee})^{\otimes \ell}$, the rational Hodge classes of H' are precisely the rational vectors fixed by $\mathbf{MT}_H$.

For simplicity we suppress the proof that such a group exists, as well as the relation to the Deligne torus, and we instead refer to [13] for details. Note that if the Hodge structure H is polarized by $q_{\mathbb{Q}}$, then $\mathbf{MT}_H \subset \mathbf{Aut}(H_{\mathbb{Q}}, q_{\mathbb{Q}})$.

Definition 5.3.4. Let D be a polarized period domain.

(1) A weak Mumford–Tate subdomain D' of D is an orbit $\mathbf{M}(\mathbb{R})x$ where $x \in D$ and $\mathbf{M}$ is a normal algebraic $\mathbb{Q}$-subgroup of $\mathbf{MT}_x$. In fact, D' is a smooth complex submanifold of D, and it is an irreducible component of the locus of Hodge structures H such that $\mathbf{MT}_H \supset \mathbf{M}$.

(2) If moreover $\mathbf{M} = \mathbf{MT}_x$, then $D' = \mathbf{M}(\mathbb{R})x$ is called a Mumford–Tate subdomain.
(3) Let $\pi : D \to \Gamma\backslash D$ be the quotient map. For $D' \subset D$ a (weak) Mumford–Tate subdomain, $\pi(D') \subset \Gamma\backslash D$ is a complex analytic subvariety which we call a (weak) Mumford–Tate subvariety. Likewise, given a period map $\varphi : X^{\mathrm{an}} \to \Gamma\backslash D$, we call $\varphi^{-1}\pi(D')$ a (weak) Mumford–Tate subvariety of X.

Given Definition 5.3.3, we see that we can also think of a Mumford–Tate subdomain as a component of the locus of Hodge structures for which some number of rational tensors are Hodge.

Theorem 5.3.5 (Theorem 1.6 of [15]). *Let $\varphi : X^{\mathrm{an}} \to \Gamma\backslash D$ be a period map. Then any weak Mumford–Tate subvariety of X is algebraic.*

Remark 5.3.6. In the special case of $f : Y \to X$ a smooth projective family, and the period map corresponding to the variation of Hodge structures on $R^{2k}_{\mathrm{prim}} f_* \mathbb{Z}$, the Hodge conjecture implies Theorem 5.3.5. Indeed, the locus $\mathrm{Hdg}_k(X) \subset X$ where $H^{2k}_{\mathrm{prim}}(Y_x, \mathbb{Q})$ acquires Hodge classes is the image of the codimension k relative Hilbert scheme $\mathrm{Hilb}(Y/X)$, hence a countable union of algebraic subvarieties.

Definition 5.3.7. Suppose $\varphi : X^{\mathrm{an}} \to \Gamma\backslash D$ is a period map. The $\mathbb{Q}$-Zariski closure of the image of the monodromy representation $\varphi_* : \pi_1(X, x) \to \mathbf{G}(\mathbb{Q})$ is called the *algebraic monodromy group*.

The following theorem is a consequence of the theorem of the fixed part [13, Theorem 13.1.10], which asserts that the trivial sub-local system of a variation of Hodge structures naturally supports a Hodge sub-variation.

Theorem 5.3.8. *The identity component of the algebraic monodromy group of a period map is a $\mathbb{Q}$-factor of the very general Mumford–Tate group.*[8]

5.4 Definable Fundamental Sets of Period Maps

We will need a slightly different definition of what a definable fundamental set of a period map is.

Definition 5.4.1. Let $\varphi : X^{\mathrm{an}} \to \Gamma\backslash D$ a period map. A definable fundamental set for φ is a definable space F whose underlying space is a complex analytic variety together with a commutative diagram of holomorphic maps

$$F \xrightarrow{\tilde{\varphi}} D$$

$$p \downarrow \qquad \downarrow \pi$$

$$X^{\mathrm{an}} \xrightarrow{\varphi} \Gamma \backslash D$$

such that p realizes X^{def} as a quotient by a closed étale definable equivalence relation and $\tilde{\varphi}$ is definable.

A crucial observation for the proof of the Ax–Schanuel conjecture is the following:

Proposition 5.4.2. *Any period map with unipotent monodromy admits a* $\mathbb{R}_{\mathrm{an,exp}}$-*definable fundamental set.*

The proof of Proposition 5.4.2 is not hard—it follows easily from the local description of degenerations of Hodge structures, as we will see below. For Proposition 5.4.2 the assumption on the monodromy is not necessary, but given Corollary 5.2.9 it is sufficient for our purposes to restrict to this case.

By a *local* period map we mean a holomorphic locally liftable Griffiths transverse map

$$\varphi : (\Delta^*)^r \times \Delta^s \to \Gamma \backslash D.$$

Given such a map, let $\mu : \mathbb{H}^r \times \Delta^s \to (\Delta^*)^r \times \Delta^s$ be the standard covering map, and consider a lift of the period map The covering group of μ is $\mathbb{Z}^r$,

$$\mathbb{H}^r \times \Delta^s \xrightarrow{\tilde{\varphi}} D$$

$$\mu \downarrow \qquad \downarrow \pi$$

$$(\Delta^*)^r \times \Delta^s \xrightarrow{\varphi} \Gamma \backslash D$$

generated by the real translations

$$t_i : \mathbb{H}^r \to \mathbb{H}^r : (z_1, \ldots, z_i, \ldots, z_r) \mapsto (z_1, \ldots, z_i + 1, \ldots, z_r)$$

on the ith $\mathbb{H}$ factor, for $1 \le i \le r$. Let $\gamma_i \in \mathbf{G}(\mathbb{Z})$ be the corresponding unipotent monodromy operator, so that

$$\tilde{\varphi} \circ (t_i \times \mathrm{id}_{\Delta^s}) = \gamma_i \tilde{\varphi}$$

---

[8]That is, the Mumford–Tate group at a very general point.

Let

$$N_i := \log \gamma_i = -\sum_k \frac{(1-\gamma_i)^k}{k} \in \mathfrak{g}_{\mathbb{R}}$$

be the nilpotent logarithms of T_i, which makes sense since each T_i is unipotent. It follows that the map $\tilde{\psi} : \mathbb{H}^r \times \Delta^s \to D$ defined by "untwisting" the monodromy

$$\tilde{\psi} := \exp\left(-\sum_i z_i N_i\right) \tilde{\varphi}$$

descends to a map $\psi : (\Delta^*)^r \times \Delta^s \to D$.

Theorem 5.4.3 (Corollary 8.35 of [45]). *For any local period map, ψ as defined above extends to a holomorphic map $\overline{\psi} : \Delta^n \to \check{D}$.*

Remark 5.4.4. Given a variation $(\mathcal{H}_{\mathbb{Z}}, Q_{\mathbb{Z}}, F^\bullet)$ of pure polarized integral Hodge structures over a smooth algebraic base X with unipotent local monodromy, the Deligne extension is a canonical extension of the associated flat bundle $\mathcal{O}_{X^{\mathrm{an}}} \otimes_{\mathbb{Z}} \mathcal{H}_{\mathbb{Z}}$ to a log smooth compactification $\overline{X}$ as a holomorphic vector bundle. For v_i a (multivalued) flat frame for $\mathcal{H}_{\mathbb{Z}}$ in a polydisk $(\Delta^*)^r \times \Delta^s$, the extension is defined using the frame

$$\tilde{v}_i := \exp\left(-\sum_i z_i N_i\right) v_i.$$

One then shows that these extensions patch to form a global extension of $\mathcal{O}_{X^{\mathrm{an}}} \otimes \mathcal{H}_{\mathbb{Z}}$ to $\overline{X}$ (see [12]). Theorem 5.4.3 then implies that the Hodge filtration $F^\bullet$ extends holomorphically to the Deligne extension.

Let $\Sigma \subset \mathbb{H}$ be the bounded vertical strip

$$\Sigma := \{z \in \mathbb{H} \mid -\epsilon < \operatorname{Re} z < 1 + \epsilon\}$$

with its $\mathbb{R}_{\mathrm{alg}}$-definable structure as a semialgebraic subset of $\mathbb{C}$. For $\delta > 0$ define

$$\Delta_\delta := \{q \in \Delta \mid |q| < 1 - \delta\}$$
$$\mathbb{H}_\delta := \{z \in \mathbb{H} \mid \operatorname{Im} z > \delta\}$$
$$\Sigma_\delta := \Sigma \cap \mathbb{H}_\delta.$$

Corollary 5.4.5. *For all sufficiently small $\delta > 0$,*

$$\tilde{\varphi} : \Sigma_\delta^r \times \Delta_\delta^s \to D$$

is $\mathbb{R}_{\mathrm{an,exp}}$-definable.

Proof. We have $\varphi = \exp(z \cdot N)\tilde{\psi}$. By Theorem 5.4.3, $\psi : \Delta_\delta^n \to D$ is restricted analytic, hence $\mathbb{R}_{\mathrm{an}}$-definable. It follows that $\tilde{\psi} : \Sigma_\delta^r \times \Delta_\delta^s \to D$ is $\mathbb{R}_{\mathrm{an,exp}}$-definable since μ is $\mathbb{R}_{\mathrm{an,exp}}$-definable. Now, $\mathbf{G}(\mathbb{C})$ (with its canonical definable structure) acts algebraically on $\check{D}$, and $\exp(z \cdot N)$ is in fact an algebraic map $\Sigma^r \to \mathbf{G}(\mathbb{C})$, hence $\mathbb{R}_{\mathrm{alg}}$-definable. Thus, $\tilde{\varphi}$ is $\mathbb{R}_{\mathrm{an,exp}}$-definable. $\square$

Proof of Proposition 5.4.2. Take an algebraic log smooth compactification $\bar{X}$ of X, and a finite cover of X by polydisks of the form $f_i : (\Delta^*)^{r_i} \times \Delta^{s_i} \to X$. For each such polydisk, take $F_i = \Sigma_\delta^{r_i} \times \Delta_\delta^{s_i}$, and let $p_i = f_i \circ \mu$. Finally, take $F = \bigsqcup_i F_i$, with $p = \bigsqcup_i p_i : F \to X$. For sufficiently small $\delta > 0$ the map

$$p : F \to X^{\mathrm{an}}$$

realizes X^{def} as a $\mathbb{R}_{\mathrm{an,exp}}$-definable quotient of F. By Corollary 5.4.5, the lifted period map $\tilde{\varphi} : F \to D$ is $\mathbb{R}_{\mathrm{an,exp}}$-definable. $\square$

5.5 Intersections with Definable Fundamental Sets

Given a definable fundamental set for a period map as in the last subsection, we evidently have a natural diagram where $X_D := X \times_{\Gamma \backslash D} D$. Fix a left-

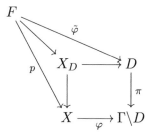

invariant metric h_D and let $\Phi = \tilde{\varphi}(F)$. For the proof in the next section, it will be important that a given algebraic subvariety $Z \subset \check{D}$ has bounded volume intersection with all translates of Φ under the action by $\mathbf{G}(\mathbb{Z})$. We in fact have the stronger statement:

Proposition 5.5.1 (Proposition 3.2 of [5]). *Let $Z \subset \check{D}$ be a closed algebraic subvariety. For all $\gamma \in \mathbf{G}(\mathbb{C})$, $\mathrm{vol}(Z \cap \gamma\Phi) = O(1)$.*

Proof. Evidently it is enough to show $\mathrm{vol}(Z' \cap \Phi) = O(1)$ for all Z' in the same connected component of the Hilbert scheme of $\check{D}$ as Z. Further, it suffices to show $\mathrm{vol}(\tilde{\varphi}^{-1}(Z')) = O(1)$ for each local period map $\tilde{\varphi} : \mathbb{H}_\delta^r \times \Delta_\delta^s \to D$ considered above, where the volume is computed with respect to $\tilde{\varphi}^* h_D$.

For any holomorphic Griffiths transverse map $f : M \to \Gamma\backslash D$ we have $f^* h_D \ll \kappa_M$ where κ_M is the Kobayashi metric of M. In particular, for $M = \mathbb{H}^r \times \Delta^s$ the metric κ_M is the maximum over the coordinate-wise Poincaré metrics. The factors in $\Sigma_\delta^r \times \Delta_\delta$ have finite volume with respect to the Kobayashi metric of $\mathbb{H}^r \times \Delta^s$, and thus it is enough to uniformly bound the degree of the projection of $\tilde{\varphi}^{-1}(Z')$ to any subset of coordinates. This in turn follows by applying Corollary 2.2.10 to the pullback of the universal family. $\square$

6   The Ax–Schanuel Theorem for Period Maps

In this section we give the proof of the Ax–Schanuel conjecture for period maps from [5]. The proof follows the same strategy as the proof of Mok–Pila–Tsimerman [33] for Shimura varieties.

6.1   *Statement of the Main Theorem*

Let X be a smooth complex algebraic variety over $\mathbb{C}$ supporting a pure polarized integral variation of Hodge structures $\mathcal{H}_\mathbb{Z}$. Let $\mathbf{MT}_{\mathcal{H}_\mathbb{Z}}$ be the generic Mumford–Tate group—that is, the Mumford–Tate group at a very general point—and let $\Gamma \subset \mathbf{MT}_{\mathcal{H}_\mathbb{Z}}(\mathbb{Q})$ be the image of the monodromy representation $\pi_1(X) \to \mathbf{MT}_{\mathcal{H}_\mathbb{Z}}(\mathbb{Q})$ after possibly passing to a finite cover. Let $\mathbf{G}$ be the identity component of the $\mathbb{Q}$-Zariski closure of Γ. Let $D = D(\mathbf{G})$ be the associated weak Mumford–Tate domain and $\varphi : X \to \Gamma\backslash D$ the period map of $\mathcal{H}_\mathbb{Z}$. The compact dual $\check{D}$ of D is a projective variety containing D as an open set in the archimedean topology.

Consider the fiber product

$$X \times D \supset X_D := \!\!=\!\! X \times_{\Gamma\backslash D} D \xrightarrow{\ \tilde{\varphi}\ } D$$

$$\begin{array}{ccc} & & \downarrow{\scriptstyle \pi} \\ X & \xrightarrow{\ \varphi\ } & \Gamma\backslash D. \end{array}$$

Theorem 6.1.1 (Ax–Schanuel, Theorem 1.1 of [5]). *In the above setup, let $V \subset X \times \check{D}$ be an algebraic subvariety, and let U be an irreducible analytic component of $V \cap X_D$ such that*

$$\mathrm{codim}_{X \times D}(U) < \mathrm{codim}_{X \times \check{D}}(V) + \mathrm{codim}_{X \times D}(X_D).$$

Then the projection of U to X is contained in a proper weak Mumford–Tate subvariety.

The theorem, for example, implies that the (analytic) locus in X where the periods satisfy a given set of algebraic relations must be of the expected codimension unless there is a reduction in the generic Mumford–Tate group. See [25] for some related discussions.

Corollary 6.1.2 (Ax–Lindemann–Weierstrass). *Assume the above setup.*

(1) *For any algebraic $V \subset D$, the Zariski closure of $\varphi^{-1}\pi(V)$ is a weak Mumford–Tate subvariety.*
(2) *For any algebraic $V \subset X$, the Zariski closure of any component V_0 of $\pi^{-1}\varphi(V)$ is a weak Mumford–Tate subdomain.*

6.2   Setup for the Proof

Given a period map $\varphi : X^{\mathrm{an}} \to \Gamma \backslash D$ and a subvariety $V \subset X \times D$, we define its type as the tuple

$$(\dim X, \dim V - \dim(V \cap X_D), -\dim(V \cap X_D))$$

ordered lexicographically. We say a closed algebraic $V \subset X \times D$ is *bad* at $p \in V \cap X_D$ if

$$\mathrm{codim}_p(V \cap X_D) < \mathrm{codim}(V) + \mathrm{codim}(X_D)$$

in which case we also say that both p and V are *bad*.

We proceed by induction and assume the theorem for all smaller types. Suppose V_0 is bad with $N_0 = \dim(V_0 \cap X_D)$. Let $M \subset \mathrm{Hilb}(X \times \check{D})$ be the connected component of the Hilbert scheme containing V_0, let $\mathcal{V} \subset (X \times \check{D}) \times M$ be the universal subscheme, and let $\mathcal{V}_{X \times D} \subset (X \times D) \times M$ be the restriction of the universal family to $X \times D \subset X \times \check{D}$. We will refer to points of $\mathcal{V}_{X \times D}$ as pairs (p, V), with $V \in M$ and $p \in V \cap (X \times D)$.

Let $\mathcal{V}_{X_D}$ be the universal intersection of $\mathcal{V}_{X \times D}$ with X_D. The set of "equally" bad points

$$\mathcal{B} := \{(p, V) \in \mathcal{V}_{X_D} \mid \dim_p(V \cap X_D) = N_0\} \subset \mathcal{V}_{X \times D}$$

is naturally a complex analytic subvariety which is moreover closed because of the inductive hypothesis (as $\dim_p(V \cap X_D)$ is semicontinuous). If $\mathcal{B} \to M$

is the projection $(p, V) \mapsto V$, the base-change $\mathcal{V}_\mathcal{B} \to \mathcal{B}$ of the universal family $\mathcal{V}_{X \times D}$ along $\mathcal{B} \to M$ is naturally the family of "equally" bad varieties V.

6.3 Ingredients for the Proof

Recall that we have a definable fundamental set in the sense of Definition 5.4.1:

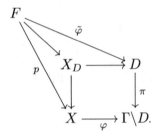

Given a bad V, we would like to apply the Pila–Wilkie theorem to the set

$$I := \{g \in \mathbf{G}(\mathbb{R}) \mid \dim(gV \cap F) = N_0\} \tag{11}$$

of translates of V that meet F badly, just as in the proof of Theorem 4.1.3. Let Γ_X be the image of the monodromy representation $\pi_1(X) \to \mathbf{G}(\mathbb{Z})$. Once again, X_D is covered by fundamental sets $\gamma^{-1}F$ with $\gamma \in \mathbf{G}(\mathbb{Z})$, and if U is a N_0-dimensional component of $V \cap X_D$ then for each $\gamma^{-1}F$ that U meets we certainly have $\gamma \in I$ (see Figure 6). We would like to argue that U passes through many fundamental sets, and therefore I has many integral points.

Like in the Shimura variety case, however, the monodromy is now very complicated and we cannot make the "height balls" argument work, so we instead use metric balls. We may assume U meets F and take a basepoint $x_0 \in F \cap U$. Let y_0 be the image in $\check{D}$, and consider the radius r ball $B_{p_0}(r)$ centered at y_0 with respect to the natural left-invariant metric on D. In the following we always measure volumes of subsets of $X \times D$ with respect to a left-invariant volume form on the second factor.

For $\gamma \in \Gamma_X$ we have $V \cap \gamma^{-1}F = U \cap \gamma^{-1}F$, as the component of X_D containing U is fixed by Γ_X. Now, by Proposition 5.5.1, U meets each $\gamma^{-1}F$ with bounded volume, while the $\gamma^{-1}F$ meet each other with bounded multiplicity, and it follows that the number of $\gamma^{-1}F$ that U passes through in $X \times B_{p_0}(r)$ is at least as much (up to a constant) as its volume in $X \times B_{y_0}(r)$. Given the following theorem, this volume grows exponentially in r:

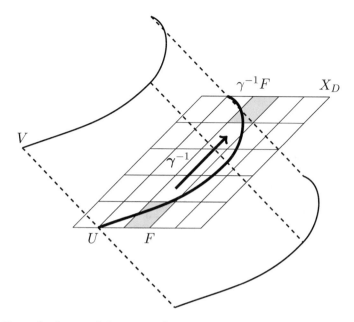

Fig. 6 Every fundamental domain $\gamma^{-1}F$ that U passes through yields an integral translate γV that meets F badly

Theorem 6.3.1 (Theorem 1.2 of [5]). *There are constants $\beta, R > 0$ such that for any closed positive-dimensional Griffiths transverse analytic subvariety $Z \subset B_{y_0}(r)$ for $r > R$ we have*

$$\mathrm{vol}(Z) \gg e^{\beta r} \, \mathrm{mult}_{y_0} Z.$$

On the other hand, the fundamental sets $\gamma^{-1}F$ which intersect $X \times B_{y_0}(r)$ have height which is at most exponential in the radius:

Theorem 6.3.2 (Theorem 4.2 of [5]). *For any $\gamma \in \mathbf{G}(\mathbb{Z})$ with*

$$\gamma^{-1}F \cap (X \times B_{y_0}(r)) \neq \varnothing$$

we have $H(\gamma) = e^{O(r)}$.

Putting Theorems 6.3.1 and 6.3.2 together we therefore obtain:

Proposition 6.3.3. *For some $\epsilon > 0$,*

$$N(I,t) \gg t^\epsilon.$$

We postpone a precise definition of the height function on $\mathbf{G}(\mathbb{Q})$ and $N(I,t)$ until the next lecture. In the remainder of this section, we prove

Theorem 6.1.1 assuming Proposition 6.3.3, and discuss the proofs of Theorems 6.3.2 and 6.3.1 in Lectures 7 and 8, respectively.

6.4   The Counting Step

We can now adapt the argument of Lecture 4 to first show:

Proposition 6.4.1. $\mathrm{Stab}_{\mathbf{G}(\mathbb{Z})}(V)$ *is infinite for any fiber* V *of* $\mathcal{V}_{\mathcal{B}}$.

Proof. $\mathbf{G}$ is an algebraic group, so $\mathbf{G}(\mathbb{R})$ has a natural definable structure. Exactly as in the proof of Theorem 4.1.3, the set I from (11) is $\mathbb{R}_{\mathrm{an,exp}}$-definable, and therefore by Proposition 6.3.3 and the Pila–Wilkie theorem we conclude that I contains a semialgebraic curve $C \subset I$ containing arbitrarily many integer points, in particular at least 2 integer points.

If cV is constant in $c \in C$, then it follows that V is stabilized by a non-identity integer point and we are done (since Γ is torsion free). So we assume that cV varies with $c \in C$. Note that since C contains an integer point that $\tilde{\varphi}(cV \cap X_D)$ is not contained in a weak Mumford–Tate subdomain for at least one $c \in C$, and thus for all but a countable subset of C (since there are only countably many families of weak Mumford–Tate subdomains).

We now have two cases to consider (see Figure 7). On the one hand, assume there is no fixed N_0-dimensional component U of $cV \cap X_D$ as $c \in C$ varies. Then we may replace V by $\bigcup_{c \in C} cV$ and increase both $\dim V$ and $\dim(V \cap X_D)$ by one, thus lowering the type and contradicting the inductive hypothesis. On the other hand, if there is such a component, then replacing V with $\bigcap_{c \in C} cV$ we lower $\dim V$ without changing $\dim(V \cap X_D)$, again contradicting the inductive assumption. This completes the proof. $\qquad\square$

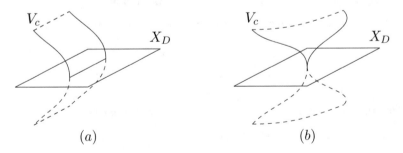

(a)   (b)

Fig. 7 If V is not stabilized by $c \in C$, then we get a counterexample with smaller type by replacing V with either (a) $\bigcup_c cV$ or (b) $\bigcap_c cV$.

6.5 The Definable Chow Step

Now we would like to control how many bad points of X there are. Obviously such points are Zariski dense, for replacing X by the Zariski closure we contradict the inductive assumption on $\dim X$. However, the bad points may *a priori* be quite sparse.

Proposition 6.5.1. *The projection $\mathcal{B} \to X$ is surjective.*

Proof. The universal intersection $\mathcal{V}_{X_D}$ is proper over X_D, and the restriction $\mathcal{V}_F \subset F \times M$ has a canonical definable structure as a restriction of an algebraic subvariety of $(X \times \check{D}) \times M$ to $F \times M$. The quotient $\mathcal{V}_X := \Gamma_X \backslash \mathcal{V}_{X_D}$ is a complex analytic space, proper over X, which thereby inherits a unique definable structure for which the quotient map $\mathcal{V}_F \to \mathcal{V}_X$ is definable.

Likewise, the subset

$$\mathcal{B}_F := \{(p, V) \in \mathcal{V}_F \mid \dim_p(V \cap F) = N_0\}$$

is a definable closed complex analytic subset of $\mathcal{V}_F$, and the quotient is a definable closed complex analytic subset $\mathcal{B}_X \subset \mathcal{V}_X$ which is proper over X.

To finish, the projection $\mathcal{B}_X \to X$ is a proper definable complex analytic map, and by Remmert–Stein and Proposition 2.1.2 the image $Z \subset X$ is a definable closed complex analytic subvariety of X, and therefore algebraic by Theorem 3.2.1. We must then have $Z = X$, by the induction hypothesis. ☐

Corollary 6.5.2. *The image of $\pi_1(\mathcal{B}_X) \to \pi_1(X)$ is finite index.*

Proof. $\mathcal{B}_X \to X$ is a proper surjective map of complex analytic varieties. ☐

6.6 The Induction Step

In the final step we produce a contradiction to Proposition 6.4.1.

Proposition 6.6.1. $\mathrm{Stab}_{\mathbf{G}(\mathbb{Z})}(V)$ *is finite for a very general fiber V of $\mathcal{V}_{\mathcal{B}}$.*

The crucial point is that Hodge theory relates the monodromy of a variation of Hodge structures to the Mumford–Tate group of a very general fiber. Theorem 5.3.8 will therefore imply a reduction in the Mumford–Tate group which cannot occur by the inductive hypothesis.

Proof. By the construction in the previous step we have $\mathcal{B}_X = \Gamma_X \backslash \mathcal{B}$, and the fundamental group $\pi_1(\mathcal{B}_X)$ naturally acts on $\mathcal{B}$. Explicitly, if ρ is the composition $\pi_1(\mathcal{B}_X) \to \pi_1(X) \to \mathbf{G}(\mathbb{Z})$, then for $\gamma \in \pi_1(\mathcal{B}_X)$ this action is $(p, V) \mapsto (\rho(\gamma)p, \rho(\gamma)V)$. Let $\Gamma_{\mathcal{B}}$ be the image of ρ, and note that by Corollary 6.5.2 that $\Gamma_{\mathcal{B}}$ is $\mathbb{Q}$-Zariski dense in $\mathbf{G}$.

As $\mathbf{G}(\mathbb{R})$ acts on $X \times \check{D}$ by algebraic automorphisms, given $g \in \mathbf{G}(\mathbb{R})$ the locus in M of varieties V stabilized by g is an algebraic subvariety of M. It follows that for the fibers of the family $\mathcal{V}_\mathcal{B} \to \mathcal{B}$ outside of a countable collection of proper subvarieties of $\mathcal{B}$—that is, for the very general fiber V— the stabilizer under $\mathbf{G}(\mathbb{Z})$ is a fixed group $\Gamma_\mathcal{V}$. Furthermore, for a very general fiber V, γV is also very general for any $\gamma \in \Gamma_\mathcal{A}$, and it follows that $\Gamma_\mathcal{V}$ is normalized by $\Gamma_\mathcal{B}$. Letting $\boldsymbol{\Theta}$ be the identity component of the $\mathbb{Q}$-Zariski closure of $\Gamma_\mathcal{V}$, we conclude that $\boldsymbol{\Theta}$ is a normal $\mathbb{Q}$-subgroup of $\mathbf{G}$.

It suffices to show that the $\mathbb{Q}$-Zariski closure of $\Gamma_\mathcal{V}$ is finite, or that: □

Claim. $\boldsymbol{\Theta}$ is the identity subgroup.

Proof. Since $\boldsymbol{\Theta}$ is a normal $\mathbb{Q}$-subgroup by construction, $\mathbf{G}$ is isogenous to $\boldsymbol{\Theta}_1 \times \boldsymbol{\Theta}_2$ with $\boldsymbol{\Theta}_2 = \boldsymbol{\Theta}$. We have a splitting of weak Mumford–Tate domains $D = D_1 \times D_2$ with $D_i = D(\boldsymbol{\Theta}_i)$. Replacing X by a finite cover we also have a splitting of the period map [20, Theorem III.A.1]

$$\varphi = \varphi_1 \times \varphi_2 : X \to \Gamma_1 \backslash D_1 \times \Gamma_2 \backslash D_2.$$

Moreover, φ_1, φ_2 satisfy Griffiths transversality (see the proof of [20, Theorem III.A.1]). Note that $V \subset X \times D$ by assumption, and as V is invariant under $\boldsymbol{\Theta}_2$ it is of the form $V_1 \times D_2$ where $V_1 \subset X \times D_1$.

Consider the period map $X \to \Gamma_1 \backslash D_1$, the resulting $X_{D_1} \subset X \times D_1$, and the subvariety $V_1 \subset X \times D_1$. Let U be a N_0-dimensional component of $V \cap X_D$ and let U_1 be the component of $V_1 \cap X_{D_1}$ onto which U projects. By assumption the theorem applies in this situation, and as U_1 cannot be contained in a proper weak Mumford–Tate subdomain (for then U would as well), we must have

$$\mathrm{codim}_{X \times D_1}(U_1) = \mathrm{codim}_{X \times \check{D}_1}(V_1) + \mathrm{codim}_{X \times D_1}(X_{D_1}).$$

Note that the projection $X_D \to X_{D_1}$ has discrete fibers, so $\dim X = \dim X_{D_1}$ and $\dim U = \dim U_1$, whereas $\mathrm{codim}\, V_1 = \mathrm{codim}\, V$, which is a contradiction if φ_2 is nonconstant.   □

7 Heights and Distances

In this section we establish the comparison between heights and metric dilation needed in Theorem 6.3.2. Recall that we have a period map $\varphi :$

$X^{\mathrm{an}} \to \Gamma \backslash D$ and definable fundamental set $p : F \to X$ in the sense of Definition 5.4.1 consisting of a union of unwrapped polydisks[9] $\Sigma^r \times \Delta^s$:

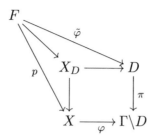

Letting $\Phi = \tilde{\varphi}(F)$ and fixing a basepoint $x_0 \in \Phi \subset D$, we identify $D \cong \mathbf{G}(\mathbb{R})/V$ for a compact subgroup $V \subset \mathbf{G}(\mathbb{R})$. Thinking of D as a space of Hodge structures on the fixed integral lattice $(H_{\mathbb{Z}}, q_{\mathbb{Z}})$, as before we denote by h_x the induced Hodge metric on $H_{\mathbb{C}}$ corresponding to $x \in D$.

Definition 7.0.1. For $\gamma \in \mathbf{G}(\mathbb{Z})$ let $H(\gamma)$ be the height of γ with respect to the representation $\rho_{\mathbb{Z}} : \mathbf{G}(\mathbb{Z}) \to \mathrm{GL}(H_{\mathbb{Z}})$. For $g \in \mathbf{G}(\mathbb{R})$, we denote by $\|\rho_{\mathbb{R}}(g)\|$ the maximum archimedean size of the entries of $\rho_{\mathbb{R}}(g)$, so that if $\gamma \in \mathbf{G}(\mathbb{Z})$ we have $H(\gamma) = \|\rho_{\mathbb{R}}(\gamma)\|$.

Remark 7.0.2. We can now precisely define the counting function used in the previous lecture. For $U \subset \mathbf{G}(\mathbb{R})$ a definable subset (where $\mathbf{G}(\mathbb{R})$ is given the canonical definable structure coming from the algebraic group structure), we define

$$N(U, t) := \#\{\gamma \in U \cap \mathbf{G}(\mathbb{Q}) \mid H(\gamma) \le t\}.$$

By fixing a V-invariant Hermitian metric at x_0, we obtain a left-invariant Hermitian metric on $\mathbf{G}(\mathbb{R})/V$. This metric is explicitly described as follows. For any point $x \in D$ we have seen in Remark 5.2.4 that the Lie algebra $\mathfrak{g}_{\mathbb{R}}$ inherits a polarized Hodge structure $\mathfrak{g}_x$, and that the tangent space $T_x D$ is identified with $T_x D = \mathfrak{g}_{\mathbb{C}}/F^0 \mathfrak{g}_{\mathbb{C}}$. The space $\mathfrak{g}^{<0} := \bigoplus_{p<0} \mathfrak{g}^{p,-p} \subset \mathfrak{g}_{\mathbb{C}}$ provides a real analytic lift of $T_x D$, and we endow $T_x D$ with the restriction of the hodge metric h_x on $\mathfrak{g}_x$. One can easily check that this metric is left-invariant.

For any $R > 0$ let $B_{x_0}(R) \subset D$ be the ball of radius R centered at x_0. The main goal of this section is to establish the following:

---

[9]Recall from the discussion following Theorem 5.4.3 that $\Sigma \subset \mathbb{H}$ is a bounded vertical strip

$$\Sigma := \{z \in \mathbb{H} \mid -\epsilon < \mathrm{Re}\, z < 1 + \epsilon\}.$$

Theorem 7.0.3 (Theorem 4.2 of [5]). *Any $\gamma \in \mathbf{G}(\mathbb{Z})$ with*

$$\gamma^{-1}\Phi \cap B_{x_0}(r) \neq \varnothing$$

has $H(\gamma) = e^{O(r)}$.

Define $d_0(x) = d(x, x_0)$. We write $f \preceq g$ if $|f| \ll |g|^{O(1)} + O(1)$, and $f \asymp g$ if $f \preceq g$ and $g \preceq f$.

Lemma 7.0.4. *Let $\lambda(x, x')$ be the maximal eigenvalue of h_x with respect to $h_{x'}$. Then*

(1) For all $g \in \mathbf{G}(\mathbb{R})$ we have $||\rho_{\mathbb{R}}(g)|| \asymp e^{d_0(gx_0)}$;
(2) $\lambda(x, x') \asymp e^{d(x, x')}$.

Proof. Let $K = U(h_{x_0}) \cap \mathbf{G}(\mathbb{R})$ be the subgroup of $\mathbf{G}(\mathbb{R})$ acting unitarily with respect to h_{x_0}. Then K is a maximal compact subgroup of $\mathbf{G}(\mathbb{R})$ containing V, and the above left-invariant metric on $G(\mathbb{R})/V$ descends to the symmetric space $\mathbf{G}(\mathbb{R})/K$. Note that the diameters of the fibers of $\mathbf{G}(\mathbb{R})/V \to \mathbf{G}(\mathbb{R})/K$ are bounded. Choosing a K-orthogonal split maximal torus $A \subset \mathbf{G}(\mathbb{R})$ and a basis A_i of the Lie algebra $\mathfrak{a}$ of A, the induced metric on A is up to scaling the unique left-invariant metric, which is identified with the euclidean metric on the Lie algebra $\mathfrak{a}$. We therefore have for any $g \in \mathbf{G}(\mathbb{R})$ with KAK decomposition $g = k_1 a k_2$

$$\sqrt{\sum_i t_i^2} \ll d_0(gx_0) = d_0(ax_0) + O(1) \ll \sqrt{\sum_i t_i^2} + O(1)$$

where $a = \exp(\sum_i t_i A_i)$. As

$$\max_i \exp(|t_i|) \preceq \rho_{\mathbb{R}}(g) \preceq \max_i \exp(|t_i|)$$

part (1) follows.

For part (2), note that by $\mathbf{G}(\mathbb{R})$-invariance we may restrict to the case $x' = x_0$. Setting $\rho = \rho_{\mathbb{R}}$ for convenience, note that $\operatorname{tr}(\rho(g)^* \rho(g))$ is a sum of the eigenvalues of h_{gx_0} with respect to h_{x_0}, where $\rho(g)^*$ is the adjoint of $\rho(g)$ with respect to h_{x_0}. Thus $\operatorname{tr}(\rho(g)^* \rho(g)) \asymp \lambda(gx_0, x_0)$. As $\operatorname{tr}(\rho(g)^* \rho(g))$ is the sum of the squares of the entries of $\rho(g)$, part (2) follows from part (1). $\qquad\square$

We define a function $\mu : D \to \mathbb{R}$ measuring proximity to the boundary by the minimal period length:

$$\mu(x) = \min_{v \in H_{\mathbb{Z}} \setminus \{0\}} h_x(v).$$

For any $v \in H_{\mathbb{C}}$ we have $\log \frac{h_{x_0}(v)}{h_x(v)} \ll d_0(x) + O(1)$ by part (2) of Lemma 7.0.4, and so we deduce the following:

Corollary 7.0.5. $-\log \mu \ll d_0 + O(1)$.

Proof. There is some $v \in H_{\mathbb{Z}} \backslash \{0\}$ with $\log \mu(x) = \log h_x(v)$ and thus

$$-\log \mu = -\log h_x(v) \ll \log \frac{h_{x_0}(v)}{h_x(v)} + O(1) \ll d_0(x) + O(1)$$

where we have used that h_{x_0} is comparable to a standard Hermitian metric on $H_{\mathbb{C}}$, so that $h_{x_0}(v) \gg 1$ for any $v \in H_{\mathbb{Z}} \backslash \{0\}$. $\qquad \square$

We in fact have a comparison in the other direction once we restrict to Φ:

Lemma 7.0.6. *For $x \in \Phi$ we have $d_0(x) \ll -\log \mu(x) + O(1)$.*

The proof uses the asymptotics of hodge norms, which we now recall. Given a local period map $\varphi : (\Delta^*)^r \times \Delta^s \to \Gamma \backslash D$ with unipotent local monodromy let $\tilde{\varphi} : \Sigma^r \times \Delta^s \to D$ be a lift and $N_1, \ldots, N_r$ the nilpotent monodromy logarithms. The monodromy logarithms (with the implicit chosen ordering of the coordinates of $(\Delta^*)^r$) define r weight filtrations $W^{(j)} = W(N_1, \ldots, N_j)$. For a given $v \in H_{\mathbb{Z}}$, let $w^{(j)}$ be its weight with respect to $W^{(j)}$—that is, for each j, we take $w^{(j)}$ to be the unique w such that $v \in W_w^{(j)}$ and $\operatorname{gr}_w^{W^{(j)}}(v) \neq 0$. By [14], on the region

$$\operatorname{Im} z_1 \gg \cdots \operatorname{Im} z_r \gg 1$$

the hodge norm $h_{\varphi(z)}(v)$ of v at $\varphi(z)$ is then given asymptotically by

$$h_{\tilde{\varphi}(z)}(v) \sim \left(\frac{\operatorname{Im} z_1}{\operatorname{Im} z_2} \right)^{w^{(1)}} \cdots \left(\frac{\operatorname{Im} z_{r-1}}{\operatorname{Im} z_r} \right)^{w^{(r-1)}} \cdot (\operatorname{Im} z_r)^{w^{(r)}}$$

where "$\sim$" means "within a bounded function of."

Proof of Lemma 7.0.6. It is enough to prove the statement for the image of a single $\Sigma^r \times \Delta^s$. Moreover, we may cover F with finitely many regions S_π of the form $\operatorname{Im} z_{\pi(1)} \gg \cdots \operatorname{Im} z_{\pi(\ell)} \gg 1$ where π ranges over all permutations of $\{1, \ldots, r\}$. Thus, we may assume Φ is the image of S_{id}.

Take v_i to be a basis of $H_{\mathbb{Z}}$ descending to a basis of the multi-graded module associated to the r weight filtrations $W^{(j)}$ as above, where we take each grading centered at 0. Let $w_i^{(j)}$ for $j = 1, \ldots, r$ be the weights of v_i with respect to $W(\pi)^{(j)}$. As above, on S_{id} we therefore have

$$h_{\tilde{\varphi}(z)}(v_i) \sim \left(\frac{\operatorname{Im} z_1}{\operatorname{Im} z_2} \right)^{w_i^{(1)}} \cdots \left(\frac{\operatorname{Im} z_{r-1}}{\operatorname{Im} z_r} \right)^{w_i^{r-1}} \cdot (\operatorname{Im} z_r)^{w_i^{(r)}}.$$

As the set of weights is preserved under negation, it follows that $\max_i h_{\tilde{\varphi}(z)}(v_i) \sim (\min_i h_{\tilde{\varphi}(z)}(v_i))^{-1}$, and so by Lemma 7.0.4,

$$d_0(\tilde{\varphi}(z)) \ll \max_i \log h_{\tilde{\varphi}(z)}(v_i) \ll -\log \mu(\tilde{\varphi}(z)) + O(1)$$

uniformly on every such region.   □

Proof of Theorem 6.3.2. Suppose $x \in B_0(R) \cap \gamma^{-1}\Phi$ for $\gamma \in \mathbf{G}(\mathbb{Z})$. Putting together Lemma 7.0.6 and Corollary 7.0.5 we have

$$d_0(\gamma x) \ll -\log \mu(\gamma x) + O(1) = -\log \mu(x) + O(1) \ll d_0(x) + O(1)$$

and since

$$d_0(\gamma x_0) \leq d(\gamma x, \gamma x_0) + d(\gamma x, x_0) \leq d_0(x) + d_0(\gamma x)$$

we are finished by part (1) of Lemma 7.0.4.   □

8   Volume Bounds

In this lecture we outline the proof of the volume bound in Theorem 6.3.1. To warm up for the proof, we first give a simple proof in the euclidean case.

8.1   *Euclidean Space*

Endow $\mathbb{C}^n$ with the standard Hermitian metric

$$h_{\mathrm{eucl}} = \sum_i dz_i \otimes d\bar{z}_i.$$

The real part

$$\mathrm{Re}\, h_{\mathrm{eucl}} = \sum_i dx_i^2 + dy_i^2$$

is the usual euclidean metric on $\mathbb{C}^n = \mathbb{R}^{2n}$, and the associated Kähler form is

$$\omega_{\mathrm{eucl}} := -\mathrm{Im}\, h_{\mathrm{eucl}} = \frac{1}{2} \sum_i i dz_i \wedge d\bar{z}_i = \sum_i dx_i \wedge dy_i.$$

Given a locally closed analytic subvariety $Z \subset \mathbb{C}^n$, its euclidean volume can be computed as

$$\mathrm{vol}_{\mathrm{eucl}}(Z) := \frac{1}{(\dim Z)!} \int_Z (\omega_{\mathrm{eucl}})^{\dim Z}.$$

Finally, for $z_0 \in \mathbb{C}^n$ denote by

$$B_{z_0}^{\mathrm{eucl}}(R) := \{z \in \mathbb{C}^n \mid |z - z_0|^2 < R\}$$

the radius R ball around z_0 with respect to h_{eucl}.

Theorem 8.1.1. *For any $z_0 \in \mathbb{C}^n$ and any closed analytic subvariety $Z \subset B_{z_0}^{\mathrm{eucl}}(R) \subset \mathbb{C}^n$, we have*

$$\mathrm{vol}(Z) \geq (\pi R^2)^{\dim Z} \cdot \mathrm{mult}_{z_0} Z.$$

The theorem is originally due to Federer (see, for example, [47]). Note moreover that the bound is sharp, as a union of N affine linear spaces through z_0 will realize the bound. Hwang–To [23] have generalized the theorem to bounded symmetric domains, and it is their approach that we follow—and in fact that will generalize to the period domain setting.

The proof hinges on two observations: on the one hand, the "distance to z_0" function $\nu_{z_0}(z) := |z - z_0|^2$ provides a potential for ω_{eucl},

$$\omega_{\mathrm{eucl}} = \frac{i}{2} \partial \bar\partial \nu_{z_0}$$

while on the other hand, the log-distance $\log \nu_{z_0}$ is the potential for a form (strictly speaking, a current) that computes the multiplicity, by the Poincaré–Lelong formula.

Proof of Theorem 8.1.1. We may as well assume $z_0 = 0$ and set $\nu := \nu_{z_0}$. Set $Z(r) := Z \cap B_0^{\mathrm{eucl}}(r)$. By Stokes' theorem we have

$$\begin{aligned}
\mathrm{vol}^{\mathrm{eucl}}(Z(r)) &= \int_{Z(r)} (\tfrac{i}{2}\partial\bar\partial\nu)^{\dim Z} \\
&= \int_{\partial Z(r)} \tfrac{1}{2} d^c \nu \wedge (\tfrac{i}{2}\partial\bar\partial\nu)^{\dim Z - 1} \\
&= r^2 \cdot \int_{\partial Z(r)} \tfrac{1}{2} d^c \log \nu \wedge (\tfrac{i}{2}\partial\bar\partial\nu)^{\dim Z - 1} \\
&= r^2 \cdot \int_{Z(r)} \tfrac{i}{2}\partial\bar\partial \log \nu \wedge (\tfrac{i}{2}\partial\bar\partial\nu)^{\dim Z - 1}.
\end{aligned}$$

Note that in going from the second to the third line, we used that ν is constant on $\partial Z(r)$, so for any function $f : \mathbb{R} \to \mathbb{R}$ we have

$$d^c f(\nu)|_{\partial Z(r)} = (f'(\nu) d^c \nu)|_{\partial Z(r)} = f'(r) \cdot d^c \nu|_{\partial Z(r)}.$$

Carrying out the same manipulation for each $\frac{i}{2} \partial \bar{\partial} \nu$ term we arrive at

$$\mathrm{vol}^{\mathrm{eucl}}(Z(r)) = r^{2\dim Z} \cdot \int_{Z(r)} (\tfrac{i}{2} \partial \bar{\partial} \log \nu)^{\dim Z}. \tag{12}$$

Without getting into the details (see, for example, [23]), we briefly remark that some care must be taken in the above wedge product as $\partial \bar{\partial} \log \nu$ must be interpreted as a current in order for Stoke's theorem to apply.

For the remaining part of the argument, let's for simplicity assume Z is a curve, so that we have a normalization of the form $g : \Delta \to Z(\epsilon)$ (with $g(0) = 0$), for some sufficiently small $\epsilon > 0$. Now, as $\log \nu$ is plurisubharmonic we have

$$\int_{Z(r)} \tfrac{i}{2} \partial \bar{\partial} \log \nu \geq \int_{Z(\epsilon)} \tfrac{i}{2} \partial \bar{\partial} \log \nu$$

$$= \int_{\Delta} \tfrac{i}{2} \partial \bar{\partial} \log g^* \nu$$

$$= \int_{\Delta} \tfrac{i}{2} \partial \bar{\partial} \log |t|^{2 \, \mathrm{mult}_0 \, Z}$$

$$= \frac{i}{2} \int_{S^1} \frac{d\bar{t}}{\bar{t}} \cdot \mathrm{mult}_0 \, Z$$

$$= \pi \cdot \mathrm{mult}_0 \, Z.$$

$\square$

From the proof, we can conclude the following statement about the growth of the volume:

Proposition 8.1.2. *In the situation of Theorem 8.1.1,*

$$\frac{\mathrm{vol}^{\mathrm{eucl}}(Z \cap B_{z_0}^{\mathrm{eucl}}(r))}{r^{2\dim Z}}$$

is a nondecreasing function of r for $0 < r < R$.

Proof. Immediate from (12), as $\log \nu$ is plurisubharmonic and thus

$$\int_{Z(r)} (\tfrac{i}{2} \partial \bar{\partial} \log \nu)^{\dim Z}$$

is a nondecreasing function of r.

$\square$

Remark 8.1.3. Let's say a few words about the last step of the above proof for those who are unfamiliar with multiplicity in the analytic category. Suppose z_i are the standard coordinates of $\mathbb{C}^n$, and suppose $g : \Delta \to Z(\epsilon)$ is the normalization considered in the proof. Let $\mathcal{O}_{\mathbb{C}^n,0}$ be the local ring of germs of analytic functions at 0, $m_{\mathbb{C}^n,0} \subset \mathcal{O}_{\mathbb{C}^n,0}$ the ideal of the origin, and $I_{Z,0} \subset \mathcal{O}_{\mathbb{C}^n,0}$ the ideal of Z. We have

$$\mathrm{mult}_0\, Z := \max\{k \in \mathbb{N} \mid m_{\mathbb{C}^n,0}^k \supset I_{Z,0}\}$$

$$= \min_i \mathrm{ord}_0\, g^* z_i.$$

8.2 Period Domains

Let D be a polarized period domain equipped with its natural left-invariant Hermitian metric and associated positive $(1,1)$ form ω. We would now like to adapt the ideas from the previous subsection to prove:

Theorem 8.2.1. *There are constants $\beta, \rho > 0$ (only depending on D) such that for any $R > \rho$, any $x_0 \in D$, and any positive-dimensional Griffiths transverse closed analytic subvariety $Z \subset B_{x_0}(R) \subset D$, we have*

$$\mathrm{vol}(Z) \geq e^{\beta R}\, \mathrm{mult}_{x_0}\, Z$$

where $B_{x_0}(R)$ is the radius R ball centered at x_0 and $\mathrm{vol}(Z)$ the volume with respect to the natural left-invariant metric on D.

The crux of the proof is to find an exhaustion function $\varphi_0 : D \to \mathbb{R}$ which on the one hand defines balls

$$B^{\varphi_0}(R) : \{X \in D \mid \varphi_0(x) < R\}$$

that are comparable to the metric balls $B_{x_0}(R)$ and on the other hand is a potential for a $(1,1)$ form that is comparable to ω in the Griffiths transverse directions. The difficulty is that unlike in the euclidean case (or indeed even the bounded symmetric domain case) Theorem 8.2.1 fails without the Griffiths transverse assumption, as D contains compact subvarieties in the vertical directions. Thus, the function φ_0 must necessarily treat the Griffiths transverse directions in a special way.

We state the precise properties of the function φ_0 in the following proposition, but first introduce some notation.

Definition 8.2.2.

(1) Given a real $(1,1)$ form α on D, we say $\alpha \geq_{\mathrm{trans}} 0$ if at point $x \in D$ and any Griffiths transverse $X \in T_x^{1,0} D$, we have

$$-i\alpha_x(X, \bar{X}) \geq 0.$$

(2) Given two real $(1, 1)$ forms α, β on D, we say that $\alpha = O_{\text{trans}}(\beta)$ if for some positive constant $C > 0$, we have

$$C\beta - \alpha \geq_{\text{trans}} 0.$$

Now fix $x_0 \in D$ and denote by $d_0 : D \to \mathbb{R}$ the distance function to x_0.

Proposition 8.2.3. *There is a smooth function $\varphi_0 : D \to \mathbb{R}$ with the following properties:*

(1) $d_0(x) \ll \varphi_0(x) + O(1)$ and $\varphi_0(x) \ll d_0(x) + O(1)$;
(2) $i\partial\bar{\partial}\varphi_0 \geq_{\text{trans}} 0$ and $i\partial\bar{\partial}\varphi_0 >_{\text{trans}} 0$ at x_0;
(3) $i\partial\bar{\partial}\varphi_0 = O_{\text{trans}}(\omega)$ and $|\partial\varphi_0|^2 = O_{\text{trans}}(i\partial\bar{\partial}\varphi_0)$.

Proof. See [5].   □

Assuming Proposition 8.2.3, we can now complete the proof of Theorem 6.3.1. For any closed Griffiths transverse analytic subvariety $Z \subset B(R) \subset D$ of dimension d, define

$$\text{vol}^{\varphi_0}(Z) := \frac{1}{d!} \int_Z (i\partial\bar{\partial}\varphi_0)^d.$$

We begin with the following:

Proposition 8.2.4. *There is a constant $\beta > 0$ such that for any $R > 0$ and any positive-dimensional Griffiths transverse closed analytic subvariety $Z \subset B^{\varphi_0}(R)$,*

$$e^{-\beta r} \text{vol}^{\varphi_0}(Z \cap B^{\varphi_0}(r))$$

is a nondecreasing function in $r \in [0, R]$.

Proof. Let $d = \dim Z$. Let $\psi_0 = -e^{-\beta\varphi_0}$ for $\beta > 0$ the constant such that

$$i\partial\bar{\partial}\varphi_0 - \beta|\partial\varphi_0|^2 \geq_{\text{trans}} 0$$

which is guaranteed by Proposition 8.2.3(3). We then have

$$i\partial\bar{\partial}\psi_0 = \beta e^{-\beta\varphi_0}\left(i\partial\bar{\partial}\varphi_0 - \beta|\partial\varphi_0|^2\right) \geq_{\text{trans}} 0.$$

By Stokes' theorem we have

$$\text{vol}^{\varphi}(Z \cap B^{\varphi_0}(r)) = \int_{Z \cap B^{\varphi_0}(r)} (i\partial\bar{\partial}\varphi_0)^d$$

$$= \int_{Z \cap \partial B^{\varphi_0}(r)} d^c\varphi_0 \wedge (i\partial\bar{\partial}\varphi_0)^{d-1}$$

$$= \beta^{-1} e^{\beta r} \int_{Z \cap \partial B^{\varphi_0}(r)} d^c \psi_0 \wedge (i \partial \overline{\partial} \varphi_0)^{d-1}$$

$$= \beta^{-1} e^{\beta r} \int_{Z \cap B^{\varphi_0}(r)} i \partial \overline{\partial} \psi_0 \wedge (i \partial \overline{\partial} \varphi_0)^{d-1}$$

$$= \beta^{-d} e^{\beta d r} \int_{Z \cap B^{\varphi_0}(r)} (i \partial \overline{\partial} \psi_0)^d$$

which implies the claim, as $\psi_0|_Z$ is plurisubharmonic. □

Proof (Proof of Theorem 8.2.1). Choose a fixed euclidean ball B centered around x_0 with respect to some coordinate system. By Theorem 8.1.1 we have an inequality of the form

$$\mathrm{vol}^{\mathrm{eucl}}(Z \cap B) \gg \mathrm{mult}_{x_0} Z$$

Choose a fixed radius ρ such that $B \subset B^{\varphi_0}(\rho)$. After possibly shrinking B, $i \partial \overline{\partial} \varphi_0$ is comparable to the euclidean Kähler form on B in Griffiths transverse directions by Proposition 8.2.3(2), and combining this with the previous proposition we have

$$\mathrm{vol}^{\varphi_0}(Z \cap B^{\varphi_0}(r)) \gg e^{\beta r} \, \mathrm{vol}^{\varphi_0}(Z \cap B^{\varphi_0}(\rho)) \gg e^{\beta r} \, \mathrm{mult}_{x_0} Z$$

for all $r > \rho$.

Now, by Proposition 8.2.3(1), after possibly increasing ρ, there is a constant $C > 0$ such that

$$B_{x_0}(r) \supset B^{\varphi_0}(Cr)$$

for all $r > \rho$, so

$$\mathrm{vol}^{\varphi_0}(Z \cap B_{x_0}(r)) \gg e^{\beta r} \, \mathrm{mult}_{x_0} Z$$

for all $r > \rho$. Finally, by Proposition 8.2.3(3) we have

$$\mathrm{vol}(Z \cap B_{x_0}(r)) \gg \mathrm{vol}^{\varphi_0}(Z \cap B_{x_0}(r))$$

and the claim follows. □

Remark 8.2.5. Theorem 8.2.1 has a number of interesting applications in its own right. They lie outside the scope of these notes, but we briefly describe one to give a flavor. We say a point $x \in \Gamma \backslash D$ has injectivity radius R if the ball $B_x(R) \subset D$ injects into $\Gamma \backslash D$. For a period map $\varphi : X \to \Gamma \backslash D$, Theorem 8.2.1 then says that the Seshadri constant of the Hodge bundle at a point $x \in X$ can be bounded by the injectivity radius of $\varphi(x)$. In particular, these Seshadri constants can be made to grow in the level covers of X. See [24] for some related applications in the context of Shimura varieties using the volume bounds of Hwang–To.

9   Further Directions

9.1   Derivatives

One can generalize the transcendence statements by considering not only automorphic functions, but also their derivatives. For example, in the case of the modular curve one has the parametrization $j : \mathbb{H} \to Y(1)$, and j satisfies a 3rd degree differential equation. In this context, building on work of Pila [36], the paper [33] proves the following generalization of the modular Ax–Schanuel statement:

Theorem 9.1.1. *Let* $z_1, \ldots, z_n$ *be meromorphic germs in auxiliary variables* t_i *at some point of* $\mathbb{H}^n$, *and assume that none of the* z_i *is constant, nor are* $\mathrm{SL}_2(\mathbb{Q})$ *translates of each other. Then*

$$\mathrm{trdeg}_{\mathbb{C}} \, \mathbb{C} \left(z_1, j(z_1), j'(z_1), j''(z_1), \ldots, z_n, j(z_n), j'(z_n), j''(z_n) \right) \geq 3n + \mathrm{rk} \left(\frac{\partial z_j}{\partial t_i} \right).$$

Note that the above is much stronger than the usual Ax–Schanuel as it includes the algebraic independence of the derivatives of j as well. One may also generalize (as [33] does) to arbitrary Shimura varieties, but in that generality one cannot easily pick out distinguished variables. Therefore the paper adopts the language of jet spaces to formulate the above statement. The proofs are much the same, except one has to keep track of jet spaces in all the geometric constructions.

9.2   Definability of Period Maps

In Lecture 4 we showed that weakly special subvarieties of Shimura varieties were algebraic in two steps: first by using the existence of a definable fundamental set to argue that weakly special subvarieties are definable complex analytic subvarieties and second by appealing to the definable Chow theorem.

To use the same argument to reprove Theorem 5.3.5, we must have two ingredients:

(1) $\mathbf{G}(\mathbb{Z})\backslash D$ must be given a S-definable structure for some o-minimal S, and (weak) Mumford–Tate subvarieties must be shown to be definable with respect to this structure.
(2) Period maps $\varphi : X^{\mathrm{an}} \to \mathbf{G}(\mathbb{Z})\backslash D$ from a complex algebraic variety X must be shown to be definable with respect to this definable structure.

Accomplishing (1) and (2) is the content of [7]. For (1), we define an arithmetic quotient (of a homogeneous space) to be

$$\Gamma\backslash\mathbf{G}(\mathbb{R})/V$$

for $\mathbf{G}$ a connected semisimple algebraic $\mathbb{Q}$-group, $\Gamma \subset \mathbf{G}(\mathbb{Q})$ an arithmetic lattice, $V \subset \mathbf{G}(\mathbb{R})$ a connected compact subgroup. We moreover define a morphism

$$\Gamma\backslash\mathbf{G}(\mathbb{R})/V \to \Gamma'\backslash\mathbf{G}'(\mathbb{R})/V'$$

of arithmetic quotients to be a map arising from a morphism $f : \mathbf{G} \to \mathbf{G}'$ of algebraic $\mathbb{Q}$-groups sending Γ to Γ' and V to V'.

Theorem 9.2.1 (Theorem 1.1 of [7]). *Every arithmetic quotient has a natural $\mathbb{R}_{\mathrm{alg}}$-definable structure with respect to which every morphism of arithmetic quotients is $\mathbb{R}_{\mathrm{alg}}$-definable.*

Briefly, the definable structure is built by using a Siegel set to construct a definable fundamental set. Theorem 9.2.1 is easily seen to imply the required statement about weak Mumford–Tate subvarieties of arithmetic quotients of period domains.

Theorem 9.2.2 (Theorem 1.3 of [7]). *Let X be a smooth complex algebraic variety. Any period map*

$$\varphi : X^{\mathrm{an}} \to \mathbf{G}(\mathbb{Z})\backslash D$$

is $\mathbb{R}_{\mathrm{an,exp}}$-definable with respect to the $\mathbb{R}_{\mathrm{an,exp}}$-definable structure[10] on $\mathbf{G}(\mathbb{Z})\backslash D$ induced from Theorem 9.2.1.

The crux of the proof of Theorem 9.2.2 is to show that lifts of local period maps (as in 5.4) land in *finitely* many Siegel sets. In addition to the norm asymptotics discussed in Lecture 7, the primary ingredient is the SL_2-orbit theorem of Schmid [45].

Corollary 9.2.3 (Theorem 1.6 of [7]). *Every weak Mumford–Tate subvariety of X is algebraic.*

---

[10] And the canonical $\mathbb{R}_{\mathrm{an,exp}}$-definable structure on X.

9.3   Definable GAGA

Let S be an o-minimal structure. There is a natural notion of S-definable
complex analytic varieties—loosely speaking, they are complex analytic
varieties with a finite holomorphic atlas by S-definable complex analytic
subvarieties of $\mathbb{C}^n$ with S-definable holomorphic transition functions. As first
examples we have $\mathbb{G}_m^{\mathrm{def}}$ and $\mathbb{C}_a^*$ for each $a \in \mathbb{R}$ from Example 2.3.2. Some
care is needed to define the sheaf of S-definable holomorphic functions, as it
will only satisfy the sheaf axiom with respect to S-definable—in particular
finite—covers. Thus, it is naturally a sheaf on the *S-definable site* of the
underlying S-definable space. The category of definable complex analytic
varieties is introduced in [6].

Let $(\mathrm{AlgSp}/\mathbb{C})$ be the category of separated algebraic spaces[11] that
are finite type over $\mathbb{C}$, $(\mathrm{An}/\mathbb{C})$ the category of complex analytic spaces,
and $(S\text{-}\mathrm{An}/\mathbb{C})$ the category of S-definable complex analytic spaces. The
definabilization functor of Lecture 2 can be upgraded to a functor

$$(\mathrm{AlgSp}/\mathbb{C}) \to (S\text{-}\mathrm{An}/\mathbb{C}) : X \mapsto X^{\mathrm{def}}$$

which fits into a diagram

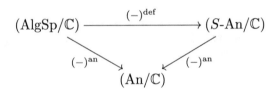

where $(\mathrm{AlgSp}/\mathbb{C}) \to (\mathrm{An}/\mathbb{C}) : X \mapsto X^{\mathrm{an}}$ is now the usual analytification
functor. Moreover, there is a natural definabilization functor on coherent
sheaves

$$(-)^{\mathrm{def}} : \mathbf{Coh}(X) \to \mathbf{Coh}(X^{\mathrm{def}}).$$

Recall that GAGA says that for X a proper separated algebraic space of
finite type over $\mathbb{C}$, the analytification functor on coherent sheaves

$$(-)^{\mathrm{an}} : \mathbf{Coh}(X) \to \mathbf{Coh}(X^{\mathrm{an}})$$

is an equivalence of categories. As a companion to the definable Chow theorem
in Lecture 3, we have the following definable GAGA:

---

[11]One could consider the category of schemes that are of finite type over $\mathbb{C}$ for
simplicity.

Theorem 9.3.1 (Theorem 1.3 of [6]). *Let S be an o-minimal structure. Let X be a separated algebraic space of finite type over $\mathbb{C}$ and X^{def} the associated definable analytic space. The definabilization functor $(-)^{\mathrm{def}}$: $\mathbf{Coh}(X) \to \mathbf{Coh}(X^{\mathrm{def}})$ is fully faithful, exact, and its essential image is closed under subobjects and quotients.*

Thus in particular definable coherent subsheaves of algebraic coherent sheaves are algebraic.

Note that $(-)^{\mathrm{def}}$ is *not* essentially surjective in general. The reason for this is as follows. By definable cell decomposition, it is not hard to see that there is a definable cover of X^{eucl} by simply connected (definable) subspaces. It follows that any $\mathbb{C}$-local system L is definable, and therefore that the coherent sheaf $F := L \otimes_{\mathbb{C}_X} \mathcal{O}_{X^{\mathrm{def}}}$ is definable, but analytic sections with the prescribed monodromy may easily fail to be definable. See [6, Example 3.2] for details.

9.4 Definable Images

By combining the definable GAGA theorem with algebraization theorems of Artin, it is proven in [6] that proper definable images of algebraic varieties are algebraic:

Theorem 9.4.1 (Theorem 1.4 of [6]). *Let S be an o-minimal structure. Let X be a separated algebraic space of finite type over $\mathbb{C}$, S a definable analytic space, and $\varphi : X^{\mathrm{def}} \to S$ a proper definable analytic map. Then $\varphi : X^{\mathrm{def}} \to \varphi(X^{\mathrm{def}})$ is (uniquely) the definabilization of a map of algebraic spaces.*

This can be used to resolve a conjecture of Griffiths [21, pg.259] on the quasiprojectivity of images of period maps. For a pure polarized integral variation of Hodge structures $(\mathcal{H}_{\mathbb{Z}}, F^{\bullet}, q_{\mathbb{Z}})$, we define the Griffiths bundle to be $L := \bigotimes^i \det F^i$.

Theorem 9.4.2 (Theorem 1.1 of [6]). *Let X be a reduced separated algebraic space of finite type over $\mathbb{C}$ and $\varphi : X^{\mathrm{an}} \to \Gamma \backslash \Omega$ a period map as above. Then*

(1) φ factors (uniquely) as $\varphi = \iota \circ f^{\mathrm{an}}$ where $f : X \to Y$ is a dominant map of (reduced) finite-type algebraic spaces and $\iota : Y^{\mathrm{an}} \to \Gamma \backslash \Omega$ is a closed immersion of analytic spaces;

(2) the Griffiths $\mathbb{Q}$-bundle L restricted to Y is the analytification of an ample algebraic $\mathbb{Q}$-bundle, and in particular Y is a quasi-projective variety.

Theorem 9.4.2 in turn has a number of applications; we refer to [6] for related discussions.

References

1. Yves André. Mumford-Tate groups of mixed Hodge structures and the theorem of the fixed part. *Compositio Math.*, 82(1):1–24, 1992.
2. James Ax. On Schanuel's conjectures. *Ann. of Math. (2)*, 93:252–268, 1971.
3. James Ax. Some topics in differential algebraic geometry. I. Analytic subgroups of algebraic groups. *Amer. J. Math.*, 94:1195–1204, 1972.
4. W. L. Baily, Jr. and A. Borel. Compactification of arithmetic quotients of bounded symmetric domains. *Ann. of Math. (2)*, 84:442–528, 1966.
5. B. Bakker and J. Tsimerman. The Ax-Schanuel conjecture for variations of Hodge structures. *Invent. Math.*, 217(1):77–94, 2019.
6. B. Bakker, Y. Brunebarbe, and J. Tsimerman. o-minimal GAGA and a conjecture of Griffiths. arXiv:1811.12230, 2018.
7. B. Bakker, B. Klingler, and J. Tsimerman. Tame topology of arithmetic quotients and algebraicity of Hodge loci. *J. Amer. Math. Soc.*, 33(4):917–939, 2020.
8. Errett Bishop. Conditions for the analyticity of certain sets. *Michigan Math. J.*, 11:289–304, 1964.
9. E. Bombieri and J. Pila. The number of integral points on arcs and ovals. *Duke Math. J.*, 59(2):337–357, 1989.
10. E. Bombieri, D. Masser, and U. Zannier. Intersecting a curve with algebraic subgroups of multiplicative groups. *Internat. Math. Res. Notices*, (20):1119–1140, 1999.
11. E. Bombieri, D. Masser, and U. Zannier. On unlikely intersections of complex varieties with tori. *Acta Arith.*, 133(4):309–323, 2008.
12. A. Borel, P.-P. Grivel, B. Kaup, A. Haefliger, B. Malgrange, and F. Ehlers. *Algebraic D-modules*, volume 2 of *Perspectives in Mathematics*. Academic Press, Inc., Boston, MA, 1987.
13. J. Carlson, S. Müller-Stach, and C. Peters. *Period mappings and period domains*, volume 85 of *Cambridge Studies in Advanced Mathematics*. Cambridge University Press, Cambridge, 2003.
14. E. Cattani, A. Kaplan, and W. Schmid. Degeneration of Hodge structures. *Ann. of Math. (2)*, 123(3):457–535, 1986.
15. E. Cattani, P. Deligne, and A. Kaplan. On the locus of Hodge classes. *J. Amer. Math. Soc.*, 8(2):483–506, 1995.
16. Pierre Deligne. Théorie de Hodge. II. *Inst. Hautes Études Sci. Publ. Math.*, (40):5–57, 1971.
17. G. Faltings. Die Vermutungen von Tate und Mordell. *Jahresber. Deutsch. Math.-Verein.*, 86(1):1–13, 1984.
18. Ziyang Gao. Towards the André–Oort conjecture for mixed Shimura varieties: The Ax–Lindemann theorem and lower bounds for Galois orbits of special points. *J. Reine Angew. Math.*, 732:85–146, 2017.
19. Z. Gao. Mixed Ax-Schanuel for the universal abelian varieties and some applications. arXiv:1806.01408, 2018.
20. M. Green, P. Griffiths, and M. Kerr. *Mumford-Tate groups and domains*, volume 183 of *Annals of Mathematics Studies*. Princeton University Press, Princeton, NJ, 2012.
21. Phillip A. Griffiths. Periods of integrals on algebraic manifolds: Summary of main results and discussion of open problems. *Bull. Amer. Math. Soc.*, 76:228–296, 1970.
22. P. Habegger. On the bounded height conjecture. *Int. Math. Res. Not. IMRN*, (5):860–886, 2009.
23. J. Hwang and W. To. Volumes of complex analytic subvarieties of Hermitian symmetric spaces. *American Journal of Mathematics*, 124(6):1221–1246, 2002.

24. J.-M. Hwang and W.-K. To. Uniform boundedness of level structures on abelian varieties over complex function fields. *Mathematische Annalen*, 335(2):363–377, 2006.

25. B. Klingler. Hodge loci and atypical intersections: conjectures. Motives and Complex Multiplication (Monte-Verita, 2016), to appear.

26. Bruno Klingler and Andrei Yafaev. The André-Oort conjecture. *Ann. of Math. (2)*, 180(3):867–925, 2014.

27. B. Klingler, E. Ullmo, and A. Yafaev. The hyperbolic Ax-Lindemann-Weierstrass conjecture. *Publ. Math. Inst. Hautes Études Sci.*, 123:333–360, 2016.

28. Serge Lang. *Algebra*, volume 211 of *Graduate Texts in Mathematics*. third edition, 2002.

29. B. Lawrence and A. Venkatesh. Diophantine problems and p-adic period mappings. *Invent. Math.*, 221(3):893–999, 2020.

30. Guillaume Maurin. Courbes algébriques et équations multiplicatives. *Math. Ann.*, 341(4):789–824, 2008.

31. J. S. Milne. Shimura varieties and moduli. In *Handbook of moduli. Vol. II*, volume 25 of *Adv. Lect. Math. (ALM)*, pages 467–548. Int. Press, Somerville, MA, 2013.

32. Ben Moonen. Linearity properties of Shimura varieties. II. *Compositio Math.*, 114(1):3–35, 1998.

33. N. Mok, J. Pila, and J. Tsimerman. Ax-Schanuel for Shimura varieties. *Ann. of Math. (2)*, 189(3):945–978, 2019.
    N. Mok, J. Pila, and J. Tsimerman. Ax-Schanuel for Shimura varieties. `arXiv:1711.02189`, 2017.

34. Y. Peterzil and S. Starchenko. Expansions of algebraically closed fields. II. Functions of several variables. *J. Math. Log.*, 3(1):1–35, 2003.

35. J. Pila. O-minimality and the André-Oort conjecture for $\mathbb{C}^n$. *Ann. of Math. (2)*, 173(3):1779–1840, 2011.

36. Jonathan Pila. Modular Ax-Lindemann-Weierstrass with derivatives. *Notre Dame J. Form. Log.*, 54(3-4):553–565, 2013.

37. Jonathan Pila. O-minimality and Diophantine geometry. In *Proceedings of the International Congress of Mathematicians—Seoul 2014. Vol. 1*, pages 547–572. Kyung Moon Sa, Seoul, 2014.

38. J. Pila and J. Tsimerman. Ax-Lindemann for $\mathscr{A}_g$. *Ann. of Math. (2)*, 179(2):659–681, 2014.

39. J. Pila and J. Tsimerman. Ax-Schanuel for the j-function. *Duke Math. J.*, 165(13):2587–2605, 2016.

40. J. Pila and A. J. Wilkie. The rational points of a definable set. *Duke Math. J.*, 133(3):591–616, 2006.

41. Richard Pink. A common generalization of the conjectures of André-Oort, Manin-Mumford, and Mordell-Lang. Preprint, available from the authors website, 2005.

42. M. Raynaud. Courbes sur une variété abélienne et points de torsion. *Invent. Math.*, 71(1):207–233, 1983.

43. M. Raynaud. Sous-variétés d'une variété abélienne et points de torsion. In *Arithmetic and geometry, Vol. I*, volume 35 of *Progr. Math.*, pages 327–352. Birkhäuser Boston, Boston, MA, 1983.

44. J.-P. Rolin, P. Speissegger, and A. J. Wilkie. Quasianalytic Denjoy-Carleman classes and o-minimality. *J. Amer. Math. Soc.*, 16(4):751–777, 2003.

45. W. Schmid. Variation of Hodge structure: the singularities of the period mapping. *Invent. Math.*, 22:211–319, 1973.

46. A. Seidenberg. Abstract differential algebra and the analytic case. *Proc. Amer. Math. Soc.*, 9:159–164, 1958.

47. G. Stolzenberg. *Volumes, limits, and extensions of analytic varieties*. Lecture Notes in Mathematics, No. 19. Springer-Verlag, Berlin-New York, 1966.

48. J. Tsimerman. The André-Oort conjecture for A_g. *Ann. of Math. (2)*, 187(2):379–390, 2018.
49. E. Ullmo and A. Yafaev. A characterization of special subvarieties. *Mathematika*, 57(2):263–273, 2011.
50. Lou van den Dries. *Tame topology and o-minimal structures*, volume 248 of *London Mathematical Society Lecture Note Series*. Cambridge University Press, Cambridge, 1998.
51. Lou van den Dries and Chris Miller. Geometric categories and o-minimal structures. *Duke Math. J.*, 84(2):497–540, 1996.
52. A. J. Wilkie. A theorem of the complement and some new o-minimal structures. *Selecta Math. (N.S.)*, 5(4):397–421, 1999.
53. Boris Zilber. Exponential sums equations and the Schanuel conjecture. *J. London Math. Soc. (2)*, 65(1):27–44, 2002.

Arithmetic Aspects of Orbifold Pairs

Frédéric Campana

MSC codes 11A451, 11J97, 11G32, 11G35, 14D10, 14E05, 14E22, 14E99, 14G05, 14H25, 14H30, 14J20, 14J99, 14M22, 32A22, 32J25, 32Q45

1 Introduction

Let X be a smooth connected projective manifold of dimension n defined over a number field k, let $k' \supset k$ be a larger number field. We denote by $X(k')$ the set of k'-rational points of X. Diophantine geometry aims at describing, in terms of the 'geometry' of $X(\mathbb{C})$, the qualitative structure of $X(k')$ when k' is sufficiently large, depending on X. When k is too small, the paucity of $X(k)$ may indeed be related not only to the geometry of $X(\mathbb{C})$, but also to the coefficients of the equations[1] defining X, as seen on the rational curve $x^2 + y^2 + 1 = 0$ for $k = \mathbb{Q}$, and $k' = \mathbb{Q}(\sqrt{-1})$.

Definition 1.1. *We say that X/k is 'potentially dense' if $X(k')$ is Zariski dense[2] in X for some $k' \supset k$, k' depending on X.*

---

[1] However, even when solving in $\mathbb{Q}$ the Fermat equations $x^n + y^n = z^n$, the arithmetic and analytic methods used during 3 centuries only gave partial answers. Its solution by Wiles rests on the parametrisation of elliptic curves over $\mathbb{Q}$ by modular curves, a geometric approach suggested only 23 years earlier in 1972 by Hellegouarch's curve $y^2 = x(x - a^p)(x - b^p)$, where $(\frac{a}{b})^p + (\frac{b}{c})^p = 1$ is a putative solution for $p > 3$ prime. The reason why this curve is usually called the 'Frey-curve' (appeared only 14 years later for the same purpose) is a mystery for me.

[2] One can also ask for density in the analytic topology, and expect that this will then hold after a further finite enlargement of k.

F. Campana (✉)
Université de Lorraine, Institut Elie Cartan, Nancy, France
e-mail: frederic.campana@univ-lorraine.fr

© Springer Nature Switzerland AG 2020 69
M.-H. Nicole (ed.), *Arithmetic Geometry of Logarithmic Pairs and Hyperbolicity of Moduli Spaces*, CRM Short Courses,
https://doi.org/10.1007/978-3-030-49864-1_2

The opposite property is X being 'Mordellic',[3] which means the existence of a nonempty Zariski open subset $U \subset X$ such that $(X(k') \cap U)$ is finite for any $k' \supset k$.

A curve is thus either Mordellic or potentially dense, according to whether $X(k')$ is finite for any k'/k, or infinite for some k'/k. A curve X/k of genus g is potentially dense if and only if $g = 0, 1$, curves of genus $g \geq 2$ being 'Mordellic', by Faltings' theorem (=Mordell's conjecture).

In higher dimension, X may be neither potentially dense nor 'Mordellic', as seen from the (exceedingly simple) product $X := F \times C$ of two curves, if $g(F) \leq 1, g(C) \geq 2$, equipped with the projection $c : X \to C$ onto C: $X(k')$ is concentrated on the finitely many fibres lying over $C(k')$, while the points in these fibres coincide with those of $F(k')$, which are thus Zariski dense there for k'/k large enough.

The aim of the present notes is to present, following [11], a conjectural description 'in geometric terms' (the meaning will be made precise below), for any X/k, of the qualitative structure of $X(k')$, similar to the previous product of curves, by means of its 'Core Map' $c : X \to C$, defined over k and conjectured to split X into its 'Potentially Dense' part (the fibres), and its 'Mordellic' part (the 'Orbifold' Base (C, Δ_c) of the Core Map c, which encodes its multiple fibres). The expectation is that $X(k')$ is concentrated on finitely many fibres of c outside of $c^{-1}(W)$ for some fixed Zariski closed $W \subsetneq C$, and that $X(k')$ is Zariski dense in the fibres contained in $c^{-1}(W)$ for $k' \supset k$ sufficiently large. In the previous example, the core map is simply the projection $c : F \times C \to C$.

The core map indeed splits any $X(\mathbb{C})$ geometrically, according to the positivity/negativity of its cotangent bundle Ω^1_X. The 'Mordellicity' of X is conjecturally equivalent to the maximal positivity, called 'Bigness', of its canonical bundle K_X. The 'Potential density' of X/k is conjectured to be equivalent to the 'Specialness' of X, a suitable notion of non-maximal positivity of its cotangent bundle Ω^1_X.

- Preservation by birational and étale equivalences.

Let us notice that the qualitative structure of $X(k')$ (and in particular being 'potentially dense' or 'Mordellic') is preserved under birational equivalence and unramified covers (due to the Chevalley–Weil theorem). The geometric properties conjectured to describe potential density and Mordellicity must be birational and preserved by unramified covers. This is indeed the case for their conjectural geometric counterparts: specialness, general type and the core map.

- Positivity/negativity of the canonical bundle (§4, §5).

---

[3]The term is due to S. Lang.

The fundamental principle of birational geometry, based on increasingly convincing evidence, is that the qualitative geometry of a projective[4] manifold X_n can be deduced from the positivity/negativity of its canonical bundle K_X. The birational and étale evaluation of this positivity is made by means of the 'Kodaira' dimension $\kappa(X_n) \in \{-\infty, 0, \ldots, n\}$ which measures the rate of growth of the number of sections of $K_X^{\otimes m}$ when $m \to +\infty$. For curves, we have $\kappa = -\infty$ (resp. $\kappa = 0$, resp. $\kappa = 1$) if $g = 0$ (resp. $g = 1$, resp. $g \geq 2$). In higher dimension n, curves of genus at least 2 generalise to manifolds with $\kappa = n$, said to be of 'general type'. The higher dimensional generalisations of curves of genus $0, 1$ are the 'special' manifolds, defined by a suitable notion of non-positivity of their cotangent bundles.

The 'core map' then decomposes (see §8) any X into these two fundamental 'building blocks': special vs general type.

- General type and Mordellicity (§8.6).

Mordell's conjecture claiming that curves of genus at least 2 are not potentially dense has been generalised in arbitrary dimension by S. Lang, who conjectured in [36] that X/k is 'Mordellic' if and only if it is of 'general type'. Lang's conjecture is still widely open, even for surfaces. It has been subsequently extended to the quasi-projective case by Vojta, replacing the canonical bundle by the Log-canonical bundle. Vojta also gave quantitative versions of this conjecture, relating it in a precise manner to its Nevanlinna analogues (see [47]). We propose in §8.6 an orbifold version of Lang's conjecture, Vojta's conjecture being the particular case when the boundary divisor is reduced.

- Specialness and Potential Density (§7).

We conjecture here (following [11]) that X/k is 'potentially dense' if and only if it is 'special'. This (new) 'specialness' property is defined by the absence of 'big' line subbundles of the exterior powers of the cotangent bundle of X. The two main classes of special manifolds are those which are either rationally connected or with $\kappa = 0$, generalising, respectively, rational and elliptic curves. Special manifolds are exactly the manifolds not dominating any 'orbifold' of general type. They may have, however, any κ strictly smaller than their dimension.

We conjecture that special manifolds have a virtually abelian[5] fundamental group, which leads to the following conjectural topological obstruction to potential density: 'the (topological) fundamental group of a potentially dense manifold X/k is virtually abelian'.

---

[4]Everything proved or conjectured here either extends, or should extend, to compact Kähler manifolds, except of course for the arithmetic versions.

[5]Recall that 'virtually abelian' means that some finite index subgroup is Abelian.

- The Core map (§8).

We show that any X admits a unique canonical and functorial fibration (its 'core map') with 'special' fibres, and 'general type' 'orbifold' base.

The 'orbifold base' (Z, Δ_f) of a fibration $f : X \to Z$ is simply its base Z equipped with a suitable 'orbifold divisor' Δ_f of Z (Δ_f effective with $\mathbb{Q}$-coefficients), encoding the multiple fibres of f. This orbifold base can be thought of as a 'virtual' ramified cover of Z eliminating the multiple fibres of f by the base-change $(Z, \Delta_f) \to Z$.

- 'Building Blocks' of projective manifolds (§8.4, §8.6).

It turns out that the 'building blocks' for constructing arbitrary X are **not only** manifolds but, more generally, 'orbifold pairs' with a negative, zero or positive canonical bundle $K_Z + \Delta_f$. In the birational category, this translates, respectively, to: $\kappa^+ = -\infty, \kappa = 0 \kappa(X) = dim(X)$. The study of geometric, arithmetic and hyperbolicity properties of any projective X thus essentially reduces, but also requires, to extend the definition and study of the corresponding invariants to orbifold pairs.

For this reason, we not only need to extend Lang's conjectures to orbifold pairs of general type but also to conjecture the potential density of orbifold pairs having either $\kappa^+ = -\infty$ or $\kappa = 0$. Since such orbifolds are the building blocks for all special manifolds, this justifies the expectation that all special manifolds should be potentially dense.

- Orbifold pairs: geometry and integral points (§2, §3).

A (smooth) orbifold pair (X, Δ) consists of a smooth projective X together with an effective $\mathbb{Q}$-divisor $\Delta := \sum_j (1 - \frac{1}{m_j}).D_j$ for distinct prime divisors D_j of X whose union D is of simple normal crossings, and 'multiplicities' $m_j \in (\mathbb{Z}^+ \cup \{+\infty\})$. They interpolate between $\Delta = 0$ and $\Delta = D$, corresponding, respectively, to the projective and quasi-projective cases. The usual invariants of quasi-projective manifolds can be attached to them, including the fundamental group and integral points if defined over $\overline{\mathbb{Q}}$. These integral points are modelled after the notion of 'orbifold morphisms' $h : C \to (X, \Delta)$ from a smooth connected curve C to (X, Δ), obtained by imposing conditions on the orders of contact between $h(C)$ and the $D'_j s$. These conditions appear in two different versions (gcd or inf), according to whether one compares positive integers according to divisibility or Archimedean order. The first notion is the one used classically in stack and moduli theories, but is not appropriate in birational geometry, and we thus consider the second one, here. This 'inf' version of integral points leads, even for orbifold pairs over $X = \mathbb{P}^1$ to an orbifold version of Mordell's conjecture which is presently open, implied by the abc-conjecture, but possibly much more accessible. This orbifold Mordell conjecture is in fact merely the one-dimensional case of the orbifold version of Lang's conjecture that we formulate in §8.6.

- Link with hyperbolicity and entire curves (§9, §10, §11).

The Lang and Vojta conjectures establish an equivalence between geometry, arithmetic and hyperbolicity of (quasi)-projective manifolds of general type. We formulate an analogous equivalence for special manifolds first, and then for all X's via the Core map, in the last two sections. Since entire curves are much easier to construct than infinite sets of k'-rational points, we can show more cases of these conjectures for entire curves, especially for rationally connected manifolds, for which analytic analogues of the Weak Approximation Property and of the Hilbert Property can be obtained.

- The material in these notes mainly comes from [11]. Unpublished observations are: Proposition 9.1 proving the conditional equivalence between entire curves and countable sequences of k'-rational points, and the last section (qualitative description of the Kobayashi pseudodistance on any X, using the 'core map').

These notes can be complemented by many texts, including: [1], the books [31] and [41] for arithmetic notions and proofs, [42], [46] on the geometric side and the references in [13] for more recent developments in birational complex geometry. The reference [9], which contains everything needed on the arithmetic side, including proofs and much more, deserves a special mention.

- These notes are an extended version of a mini-course given at UQÀM in December 2018, and part of the workshop 'Géométrie et arithmétique des orbifoldes' organised by M.H. Nicole, E. Rousseau and S. Lu. I thank them for the invitation, and also K. Ascher, H. Darmon, L. Darondeau, A. Turchet, J. Winkelmann for interesting discussions (and collaboration in the case of L.D, E.R and J.W) on this topic. Many thanks also to P. Corvaja for several exchanges and explanations he gave me on arithmetic aspects of birational geometry. In particular, §10 originates from his joint text with U. Zannier [23], the connection made there with the Weak Approximation Property is due to him. Many thanks also to Lionel Darondeau also for making my original drawings computer compatible. Thanks to the referee who read carefully the text, suggesting improvements and complementing references.

Conventions In the whole text, X will be a connected n-dimensional projective (smooth) manifold defined either over $\mathbb{C}$ or over a number field k, of which a finite extension will be denoted k'. A fibration $f : X \to Z$ is a regular surjective map with connected fibres over another projective manifold Z (of dimension usually denoted $p > 0$). A dominant rational map will be denoted $f : X \dashrightarrow Z$. We denote here always by K_X the canonical line bundle of X, which is the major invariant of the birational classification.

2   Orbifold Pairs and Their Integral Points

This section is aimed at the definition of integral points on orbifolds for potential readers with a complex geometric background. We thus try to avoid the conceptual notions of schemes, and models. The readers familiar with them can skip this section or alternatively consult either [1] or [2], where all definitions are given in this language.

2.1   Integral Points Viewed as Maps from a Curve

We shall describe a standard geometric way of seeing rational points on an n-dimensional manifold defined over a number field k as sections from an 'arithmetic curve' $Spec(\mathcal{O}_k)$ to the 'arithmetic $(n+1)$-dimensional manifold' $X(\mathcal{O}_{k,S})$ fibred over $Spec(\mathcal{O}_k)$. This description is modelled after the cases, which we describe first, of holomorphic maps from a curve, and then of function fields, in which rational points are seen as sections of a suitable fibration.

- **Morphisms from a curve.**

Let C be a smooth connected complex curve (the important cases here are when $C = \mathbb{C}, \mathbb{P}^1, \mathbb{D}$ (the complex unit disk), or a complex projective curve. Let M be a smooth connected complex manifold. Let $Hol(C, M)$ be the set of holomorphic maps from C to M. When $h \in Hol(C, M)$ is non-constant we say that h is a (parametrized) rational (resp. entire) curve on M if $C = \mathbb{P}^1$ (resp. $C = \mathbb{C}$).

We may identify any $h \in Hol(C, M)$ with its graph in $X := C \times M$, and thus with a section of the projection $f : X \to C$ onto the first factor. More generally, we can replace the product $C \times M$ with any proper holomorphic map with connected fibres $f : X \to C$ from a complex manifold X. Manifolds over a function field provide such examples.

- **Function field version of integral points.**

When X and C are projective, the preceding construction makes sense over any field, not only $\mathbb{C}$ and leads to the 'function field' version.

Let $f : X \to C$ be a holomorphic fibration (i.e.: surjective with connected fibres) from X onto C, where X is now a smooth complex projective manifold of dimension $(n+1)$. This is a 'model' of an n-dimensional manifold over the field $K := \mathbb{C}(C)$, the field of rational (or meromorphic) functions on C, with 'generic fibre' X_c, if c is a generic point of C.

More precisely, X can be embedded in $\pi_N : \mathbb{P}_N \times C = \mathbb{P}_N(K) \to C$, the first projection, for some $N \geq n$. The rational points of $\mathbb{P}_N(K)$ are thus the $N + 1$-tuples $[f_0, f_1, \ldots, f_n]$ of elements of K, up to K^*-homothety, or equivalently, sections of π_N. The elements of $X(K)$ are then those of $\mathbb{P}_N(K)$ which are contained in X, hence those which satisfy the equations defining X in $\mathbb{P}_N(K)$ over K. Said differently: $X(K)$ are the sections of f.

The set of points of C coincide with the set of inequivalent valuations (or 'places') of the field K with field of constants $\mathbb{C}$. If $S \subset C$ is any (nonempty) finite set, $C \setminus S$ also coincide with the set of maximal ideals of the ring $\mathcal{O}_{K,S}$ of rational functions on C regular outside S.

- **Integral points: the arithmetic version.**

If X is defined over the number field k, the role of the curve C will be played by $Spec(\mathcal{O}_k)$, the set of (non-archimedean) places of k.

Let k be a number field, $\mathcal{O}_k$ be its ring of integers and S a finite set of non-archimedean 'places' (i.e.: prime ideals $\mathfrak{p}$ of the ring of integers). Let $C := Spec(\mathcal{O}_{k,S}) = Spec(\mathcal{O}_k) \setminus S$ be the set of prime (=maximal) ideals $\mathfrak{p}$ of the ring $\mathcal{O}_k$ localised at S.

Let X be defined over k. Assume (in order to avoid the use of a 'model') that $X \subset \mathbb{P}_N$ is defined by homogeneous equations with coefficients in k.

An element x of $\mathbb{P}_N(k) = \mathbb{P}_N(\mathcal{O}_{k,S})$ is an $(N+1)$-tuple $[x_0, \ldots, x_N]$ of elements of either k, or equivalently $\mathcal{O}_{k,S}$, not all zero, up to $\mathcal{O}_{k,S}^*$-homothety equivalence. The elements of $X(k)$ are those satisfying the equations defining X.

The 'arithmetic projective N-space over $Spec(\mathcal{O}_{k,S})$' is the map $\pi_N : \mathbb{P}_N(\mathcal{O}_{k,S}) \to Spec(\mathcal{O}_{k,S})$, where for each prime ideal $\mathfrak{p}$ of $\mathcal{O}_{k,S}$, the fibre of π_N over $\mathfrak{p}$ is $\mathbb{P}_N(F_\mathfrak{p})$, where $F_\mathfrak{p} = \mathcal{O}_k/\mathfrak{p}$, the residue field of $\mathcal{O}_k$ by its prime (i.e.: maximal) ideal $\mathfrak{p}$.

The above point $x = [x_0 : \cdots : x_N]$ of $\mathbb{P}_N(k)$ is identified with the section of π_N which sends, for each $\mathfrak{p} \in Spec(\mathcal{O}_k)$, x to its reduction $x_\mathfrak{p}$ modulo $\mathfrak{p}$, which is the image of x by the map: $\mathbb{P}_N(\mathcal{O}_k) \to \mathbb{P}_N(F_\mathfrak{p})$. This map is well-defined, since $[x_0 : \cdots : x_N]$ may be chosen in such a way that no $\mathfrak{p}$ divides all x_j simultaneously.

Then $X(\mathcal{O}_{k,S})$ is the subset of $\mathbb{P}_N(\mathcal{O}_{k,S})$ consisting of the sections of π_N which satisfy the equations defining X, or equivalently, which take, for each $\mathfrak{p}$, their values in $X(F_\mathfrak{p})$, the reduction of X modulo $\mathfrak{p}$.

When $X = \overline{X} \setminus D$ is quasi-projective, complement of a Zariski closed subset D in the projective $\overline{X}$, everything being defined over k, the set of S-integral points of X is simply the subset of $X(\mathcal{O}_{k,S})$ which do not take their values in $D(F_\mathfrak{p})$, for each $\mathfrak{p} \in \mathcal{O}_{k,S}$ (Figure 1).

2.2 Orbifold Pairs

The birational classification requires the consideration of more general objects: 'orbifold pairs', which interpolate between the projective and quasi-projective cases.

Section $x = \frac{10}{21}$ of the arithmetic surface $\mathbb{P}^1_{\mathbb{Z}}$:

Intersections with: (0) ● (1) ● (∞) ●

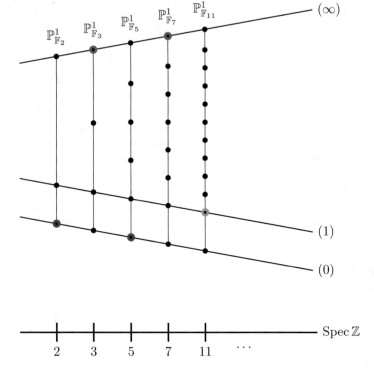

Fig. 1 The arithmetic section $\frac{10}{21}$

Definition 2.1. *An orbifold pair* (X, Δ) *consists of an irreducible normal projective variety together with an effective* $\mathbb{Q}$-*divisor* $\Delta := \sum_j c_j . D_j$ *in which the* $D'_j s$ *are irreducible pairwise distinct (Weil) divisors on* X, *and the* $c_j \in]0, 1]$ *are rational numbers of the form* $c_j = 1 - \frac{1}{m_j}$ *for integers* $m_j > 1$ *(or* $m_j = +\infty$ *if* $c_j = 1$).

The support of Δ *(denoted* $Supp(\Delta)$, *or* $\lceil \Delta \rceil$) *is* $\cup_j D_j$.

The orbifold pair (X, Δ) is smooth if X is smooth and if $Supp(\Delta)$ is SNC (i.e.: of simple normal crossings)

The canonical bundle of (X, Δ) is $K_X + \Delta$ (if $K_X + \Delta$ is $\mathbb{Q}$-Cartier, which is the case if (X, Δ) is smooth). The Kodaira dimension of (X, Δ) is defined as $\kappa(X, K_X + \Delta)^6$ if $K_X + \Delta$ is $\mathbb{Q}$-Cartier.

When $\Delta = 0$, the orbifold pair $(X, 0)$ is identified with X. When $\Delta = Supp(\Delta)$ (i.e.: $m_j = +\infty, \forall j$, or equivalently, $c_j = 1, \forall j$)), (X, Δ) is identified with the quasi-projective variety $(X \setminus \Delta)$.

The general case interpolates between the projective and quasi-projective cases, and plays the rôle of a virtual ramified cover of X ramifying at order m_j over each D_j. These orbifold pairs appear naturally in order to encode multiple fibres of fibrations (see Subsection 2.3).

The usual geometric invariants of manifolds (such as cotangent bundles, jet differentials, fundamental group in particular) can be defined for orbifold pairs as well. We shall define S-integral points on them when they are defined over a number field k (i.e. when X and Δ are both defined over k, and thus invariant under $Gal(\overline{\mathbb{Q}}/k)$).

Before defining S-integral points of an orbifold pair, we give our motivation[7] for the notion of orbifold pairs.

2.3   The Orbifold Base of a Fibration

Let $f : X \to Z$ be a fibration, with X, Z smooth projective. Let $E \subset Z$ be an irreducible divisor, and let $f^*(E) := \sum_h t_h.F_h + R$ be its scheme-theoretic inverse image in X, with $codim_Z(f(R)) \geq 2$. For each E, we define $m_f(E) := inf_h\{t_h\}$. This is the multiplicity of the generic fibre of f over E. We next define the 'orbifold base' of f as being (Z, Δ_f) with $\Delta_f := \sum_E (1 - \frac{1}{m_f(E)}).E$.

- Notice that the sum is finite, since $m_f(E) = 1$ if E is not contained in the discriminant locus of f.

The pair (Z, Δ_f) should be thought of as a virtual ramified cover $u : Z' \to Z$ ramifying at order $m_f(E)$ over each of the components of Δ_f, so as to eliminate in codimension 1 the multiple fibres of f by the base-change $u : Z' \to Z$.

We have, of course: $dim(Z) \geq \kappa(Z, K_Z + \Delta_f) \geq \kappa(Z)$

- **'Classical multiplicities'**: denoted by $m_f^*(E)$, they are defined by replacing inf by gcd in the definition of $m_f(E)$ above, which leads to the 'classical orbifold base' (Z, Δ_f^*) of f, $\Delta_f^* := \sum_E (1 - \frac{1}{m_f^*(E)}).E$.

The difference between the two notions is quite essential in the sequel.

---

[6]See Definition 4.1 below (or any text, such as [46]).

[7]The Log Minimal Model Program introduced these very same objects for apparently different reasons: adjunction formula and induction on the dimension.

Remark 2.2. *A birational base-change $Z' \to Z$ gives a new 'orbifold base' $(Z'\Delta_{f'})$, with $\kappa(Z', K_{Z'} + \Delta_{f'}) \leq \kappa(Z, K_Z + \Delta_f)$. The inequality is strict in general. By flattening[8] and desingularisation, one gets 'neat birational models' of f for which the orbifold base has minimal κ. See [11] for details.*

2.4   Orbifold Morphisms from Curves

We shall next define the two versions of orbifold morphisms from a smooth connected curve C to an orbifold pair (Z, Δ). The main examples over $\mathbb{C}$ are $C = \mathbb{C}, \mathbb{P}^1, \mathbb{D}$ (the unit disk in $\mathbb{C}$). The following example indicates a necessary condition for the functoriality of the definition.

Let (Z, Δ_f) (resp. (Z, Δ_f^*)) be the orbifold base of a fibration $f : X \to Z$ as above, with Z smooth. Let $h : C \to X$ be any holomorphic map. Consider the composite map: $f \circ h : X \to Z$. One immediately checks the following property:

Lemma 2.3. *Let $a \in C$ be such that $f \circ h(a) \in D_j$. Let $t > 0$ be the order of contact (or intersection multiplicity, see also [1], or [2]) of $f \circ h(C)$ with D_j (i.e.: $(f \circ h)^*(D_j) = t.\{a\} + R$, where R is a divisor on C supported away from a).*

Then $t \geq m_j$ (resp. m_j divides t).

The following simple example shows that any $m \geq m_j$ may occur:

Example 2.4. *Let $f : \mathbb{A}_2 \to \mathbb{A}_1$ be the fibration given by: $f(x, y) = x^2.y^3 = 0$. For any $m \geq 2$, the map $h : t \to (x, y) := (t^a, t^b)$ is such that $f \circ h(t) = t^m$, if $2a + 3b = m$, since $(f \circ h)^*(z) = t^{2a+3b}$. We may choose $a := \frac{m}{2}, b = 0$ if m is even, $a := [\frac{m}{2}] - 1, b := 1$ if m is odd.*

If the multiplicities $2 < 3$ are replaced by $p < q$, then any $t \geq t_0(p, q)$ may occur, but in general $t_0(p, q) > p$.

The preceding Lemma 2.3 shows that the functoriality of morphisms from curves to orbifold pairs requires to define them as follows:

Definition 2.5. *A non-constant regular map $h : C \to (X, \Delta)$ is an orbifold morphism (i.e.: a Δ-morphism) (resp. a 'classical orbifold morphism') if:*

1. *$h(C)$ is not contained in the support of Δ.*
2. *For any $a \in C$, and any j such that $h(a) \in D_j$, we have: $t_{a,j} \geq m_j$ (resp. $t_{a,j}$ is divisible by m_j). Here $t_{a,j}$ is the order of contact at $a \in C$ of $h(C)$ with D_j, as defined in Lemma 2.3, namely by the equality: $h^*(D_j) = t_{a,j}.\{a\} + \dots$.*

---

[8]This replaces f by a birational model with equidimensional fibres. We shall always implicitly consider these models in order to avoid birational technicalities.

We denote by $Hol(C, (X, \Delta))$ (resp. $Hol^*(C, (X, \Delta))$ the set of orbifold morphisms (resp. of classical orbifold morphisms) from C to (X, Δ).

When $C = \mathbb{C}$ (resp. $C = \mathbb{P}^1$), we say that h is a Δ-entire curve (resp. a Δ-rational curve). When $C = \mathbb{C}$, we allow h to be holomorphic transcendental in the definitions.

The Δ-morphisms are thus the usual ones when $\Delta = 0$, and are the morphisms from C to $(X \backslash D)$ when $\Delta = D := Supp(\Delta)$, with all multiplicities equal to $+\infty$.

In the general case, we have:

$$Hol(C, (X \setminus D)) \subset Hol^*(C, (X, \Delta)) \subset Hol(C, (X, \Delta)) \subset Hol(C, X).$$

We now describe this notion in the case of function fields, and next in the definition of Δ-integral points.

2.5 The Function Field Version

Let $f : X \to C$ be a regular map with connected fibres (a 'fibration') from the connected projective manifold X onto the projective curve C. We present here a geometric version of the notion of orbifold integral points. A more conceptual approach based on the notion of schemes and models can be found in [1] and [2], §2.3.

Let $\Delta = \sum_j (1 - \frac{1}{m_j}).\{D_j\}$ be an orbifold divisor on X, with $f(D_j) = C, \forall j$ (i.e.: with horizontal support). The orbifold pair (X, Δ) has as generic 'orbifold fibre' the smooth orbifold pair (X_s, Δ_s) over $s \in C$ generic,[9] if Δ_s is simply the restriction of Δ to X_s. Notice that (X_s, Δ_s) is indeed smooth for $s \in C$ generic.

Let $S \subset C$ be a finite subset containing the points of 'bad reduction' of (X, Δ) over C (i.e.: the finitely many points over which either (X_s, Δ_s) is not smooth). In this situation, the integral points of $X/(C \setminus S)$ are simply the sections $\sigma : C \setminus S \to X$ of f (i.e.: such that $f \circ \sigma = id_{(C \setminus S)}$).

We define the S-integral (resp. the 'classical' S-integral) points of $(X, \Delta)/C$ to be the sections of f which are orbifold (resp. 'classical' orbifold) morphisms from $(C \setminus S)$ to (X, Δ) over $(C \setminus S)$. We denote this set

---

[9]Let us stress that we do not use here the language of schemes, so our points are always 'closed' points, the generic point of a projective irreducible variety Z is any (closed) point outside some Zariski closed strict subset of Z. A 'general' point lies in a countable intersection of such open subsets if the base field is uncountable. We thus use 'general' in the sense we already introduced in 1980, instead of the terminology 'very general' introduced much later with the same meaning.

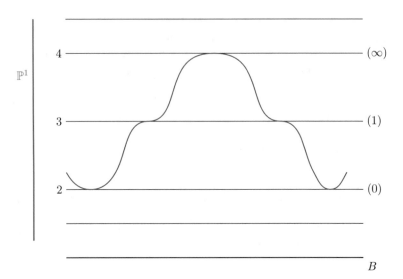

Fig. 2 A function field 'orbifold' section (see below)

with $(X, \Delta)(\mathcal{O}_{K,S})$ (resp. $(X, \Delta)^*(\mathcal{O}_{K,S})$), where K is the field of rational functions on C (Figure 2).

When $\Delta = 0$ and $S = \emptyset$, we thus recover the rational points of X over K, and when $\Delta = Supp(\Delta)$, we recover the sections of f avoiding $Supp(\Delta)$. In the general case, we have:

$$(X \setminus \Delta)(\mathcal{O}_{K,S}) \subset (X, \Delta)^*(\mathcal{O}_{K,S}) \subset (X, \Delta)(\mathcal{O}_{K,S}) \subset X(\mathcal{O}_{K,S}).$$

2.6 *Integral Points on Arithmetic Orbifolds*

We will now model the definition of the S-integral points of the orbifold (X, Δ) on their function field definition, replacing K by a number field k, and the curve C , which is the set of 'places' (i.e., non-equivalent valuations of K) by $Spec(\mathcal{O}_k)$, the ring of integers of k. The rôle of order of contact will be played by arithmetic intersection numbers.

Let k be a number field, $\mathcal{O}_k$ be its ring of integers and S a finite set of 'places' (i.e.: prime ideals $\mathfrak{p}$ of the ring of integers). Let $B := Spec(\mathcal{O}_{k,S}) = Spec(\mathcal{O}_k) \setminus S$ be the set of prime (=maximal) ideals of the ring $\mathcal{O}_k$ localised at S.

Let $f : \mathcal{X}_k \to Spec(\mathcal{O}_k)$ be the arithmetic manifold (of dimension $(n+1)$ if $dim(X) = n$) whose fibre over each prime ideal $\mathfrak{p}$ is the reduction in the quotient field $\mathcal{O}_k/\mathfrak{p}$ of X. The orbifold pair (X, Δ) being given, we define similarly the fibres of the arithmetic orbifold $(\mathcal{X}, \mathcal{D})$ over $Spec(\mathcal{O}_k))$ to be the reductions $(X_{\mathfrak{p}}, \Delta_{\mathfrak{p}})$ of (X, Δ) mod $\mathfrak{p}$. Then (X, Δ) has good reduction at $\mathfrak{p}$ if the fibre of $(\mathcal{X}, \mathcal{D})$ over $\mathfrak{p}$ is a smooth orbifold pair.

- Arithmetic intersection numbers: Let $f_S : \mathcal{X}_{k,S} \to Spec(\mathcal{O}_{k,S})$ be the 'arithmetic manifold' associated with X, as above, assuming $S \subset Spec(\mathcal{O}_k)$, finite and sufficiently large, so as to fulfil the conditions below. Any $x \in X(k)$ defines a section of f mapping any $\mathfrak{p} \notin S$ to the image of $x_{\mathfrak{p}}$ in $X_{\mathfrak{p}}$. Assume that $x \notin D_j, \forall j$. Let S be any finite set of 'places' of k containing those where (X, Δ) has 'bad reduction'. For each j, there thus exists on X a function g_j generically defining D_j reduced, g_j regular and non-vanishing at x. The reduction of g_j modulo $\mathfrak{p}$ thus does not vanish identically at $x_{\mathfrak{p}}$. The arithmetic intersection number $(x, D_j)_{\mathfrak{p}}$ is the largest integer t such that $\mathfrak{p}^t$ divides $g_j(x)$. This integer does not depend on the choice of g_j, which is well-defined up to a unit in the ring of rational functions on X regular at x.

Notice that $(x, D_j)_{\mathfrak{p}} \geq 1$ if and only if $x_{\mathfrak{p}} \in (D_j)_{\mathfrak{p}}$, this happening only for the finitely many $\mathfrak{p}'s$ which divide $g_j(x)$. See [2], §2.3 for a more conceptual definition.

Definition 2.6. *Let (X, Δ) be a smooth orbifold pair defined over k, with S a finite set of places of k containing those over which (X, Δ) has bad reduction.*

- *A point $x \in X(k)$ is (S, Δ)-integral if, for any j, $x \notin D_j$, and if $(x, D_j)_{\mathfrak{p}} \geq m_j$ for each $\mathfrak{p} \notin S$ such that $(x, D_j)_{\mathfrak{p}} \geq 1$.*
- *A point $x \in X(k)$ is a 'classical (S, Δ)-integral' if $x \notin D_j, \forall j$, and if m_j divides $(x, D_j)_{\mathfrak{p}}$ for each $\mathfrak{p} \notin S$ such that $(x, D_j)_{\mathfrak{p}} \geq 1$.*

We shall denote by $(X, \Delta)(k, S)$ (resp. $(X, \Delta)^*(k, S)$ the set of (S, Δ)-integral points (resp. of 'classical (S, Δ) integral' points) of X.

Let D be the support of Δ, we have obvious inclusions and equalities:

$$(X, D)(k, S) \subset (X, \Delta)^*(k, S) \subset (X, \Delta)(k, S) \subset X(k, S).$$

Remark 2.7. *See §5.3, §2.3 for some of the compelling reasons to introduce non-classical versions of orbifold morphisms and integral points.*

2.7 Examples of Orbifolds on $\mathbb{P}^1$

We shall illustrate these definitions with two examples of integral points over two orbifold structures on $\mathbb{P}^1$, supported on 2 (resp. 3) points, with infinite (resp. finite) multiplicities.

In both cases, we shall choose $k = \mathbb{Q}$, $S = p_1, \ldots, p_s$ for distinct primes p_j, so that $\mathcal{O}_{\mathbb{Q},S} = \mathbb{Z}[\frac{1}{p_1}, \ldots, \frac{1}{p_s}]$.

- **$\mathbb{P}^1$ minus two or three points:** Assume now that $\Delta = \{0, \infty\}$ reduced (i.e.: with infinite multiplicities. An element of $\mathbb{P}^1(\mathbb{Q})$ is of the form $\pm\frac{a}{b}$, with a, b nonnegative coprime integers, not both zero. The 'arithmetic surface' $\pi : \mathbb{P}^1_{\mathbb{Z}} \to Spec(\mathbb{Z})$ has fibre $\mathbb{P}^1_{\mathbb{F}_p}$ (the projective line over the finite field $\mathbb{F}_p$) over each $p \in Spec(\mathbb{Z})$. We associate to $\frac{a}{b}$ the section of this projection which sends each p to the mod p-reduction of $\frac{a}{b}$. The 2 points of Δ give similarly two sections $\{0\}$ and $\{\infty\}$ of this projection. The section $\frac{a}{b}$ meets the section $\{0\}$ exactly at the p's dividing a, and meets the section $\{\infty\}$ at the p's dividing b.

  The section $\frac{a}{b}$ will thus be contained in the arithmetic surface $(X \setminus \Delta)_{\mathbb{Z}}$ (that is: avoid the two sections $\{0\}$ and $\{\infty\}$) if and only if a and b are invertible in $\mathbb{Z}$, that is: if and only if $\pm\frac{a}{b} = \pm 1$, i.e., a unit of $\mathbb{Z}$.

  If instead of the ring $\mathbb{Z}$, we use the larger ring $\mathbb{Z}[\frac{1}{p_1}, \ldots, \frac{1}{p_s}] = \mathcal{O}_{\mathbb{Q},S}$, where $S = \{p_1, \ldots, p_s\} \subset Spec(\mathbb{Z})$, the set of sections $\frac{a}{b}$ meeting the sections $\{0\}$ and $\{\infty\}$ only over S are again exactly the units of $\mathcal{O}_{\mathbb{Q},S}$, that is, quotients $\frac{a}{b}$ of two coprime integers, both coprime with $p \notin S$.

  If we remove now the 3 points $0, 1, \infty$, the integral points for $\mathcal{O}_{\mathbb{Q},S}$ are the solutions of the 'S-unit equation' $a - b = c$, in which all three terms are S-units. Indeed, not only a and b should be S-units, but also $a - b$, since $\frac{a}{b}$ should not reduce to 1 modulo any p outside S. The 'classical' integral points are then the same as their 'non-classical' version. The situation is different for finite multiplicities, as we shall now see.

- **$\mathbb{P}^1$ with 3 orbifold points:** We consider $(\mathbb{P}^1, \Delta)$, where Δ consists of the 3 points $0, 1, \infty$, respectively, equipped with the integral finite multiplicities p, r, q, each at least 2.

  In other words: $\Delta = (1 - \frac{1}{p}).\{0\} + (1 - \frac{1}{r}).\{1\} + (1 - \frac{1}{q}).\{\infty\}$.

  We take here the simplest situation: $k = \mathbb{Q}, S = \emptyset$.

  Let us first describe the 'classical' integral points $x = \pm\frac{a}{c}$ of $(\mathbb{P}^1, \Delta)$, with a, c positive coprime integers, seen as a section of the arithmetic surface $\pi : \mathbb{P}^1_{\mathbb{Z}} \to Spec(\mathbb{Z})$. The section x meets the section 0 at the primes $\mathfrak{p}$ which divide a, with an intersection multiplicity equal to the exponent of $\mathfrak{p}$ in the prime decomposition of a. Similarly: the section x meets the section ∞ at the $\mathfrak{p}'s$ dividing c, with intersection multiplicity equal to the exponent of $\mathfrak{p}$ in the prime decomposition of c. The section x meets the section 1 at the primes dividing $x - 1 = \frac{a-c}{c}$, that is, those appearing in the prime decomposition of $(c - a)$, with exponents equal to the corresponding intersection multiplicities.

  There are now 2 different sets of orbifold integral points: the classical ones and the 'non-classical' ones.

- Description of the 'classical' integral points of $(\mathbb{P}^1, \Delta)$: for such an $x = \frac{a}{c}$, each of the exponents of a must be divisible by p. Thus: $a = \alpha^p$ for some positive integer α. Similarly: $c = \gamma^q$ (resp. $\pm(c - a) := b = \beta^r$), for some integers $\gamma > 0, \beta > 0$. In other words, the 'classical' integral points of $(\mathbb{P}^1, \Delta)$ over $\mathbb{Q}, S = \emptyset$ are (up to signs) the integral coprime solutions (α, β, γ) of the equation: $\alpha^p + \beta^q = \gamma^r$.

This is the construction used by Darmon-Granville in [25] to show the finiteness of solutions in coprime integers of the generalised Fermat equation $Ax^p + By^q = Cz^r$ (A, B, C become indeed S-units if we add to S the finite set consisting of the primes dividing ABC).

- Description of the integral points of $(\mathbb{P}^1, \Delta)$ (over $k = \mathbb{Q}, S = \emptyset$): a similar analysis shows that these are (up to signs, i.e.: units of $\mathbb{Z}$) solutions of the equation $a + b = c$ with: a a p-powerful integer, b a r-powerful integer and c a q-powerful integer, according to the:

Definition 2.8. *Let $k > 1$ be an integer. A positive integer m is said to be k-powerful if the k-th power of each prime dividing m still divides m, that is: if the k-th power of $rad(m)$ divides m, where $rad(m)$ (the 'radical of m') is the product (without multiplicities) of the primes dividing m. Exact k-th powers are k-powerful, but not conversely: $72 = 2^3.3^2$ is 2-powerful, but not a square.*

Nevertheless, by a result of Erdös–Szekeres, [27],§2, p. 101, the number of k-th powerful numbers less than a certain bound B is asymptotically, as $B \to +\infty$, of the form $C(k).B^{\frac{1}{k}}$ for a certain constant $C(k) > 1$, and so comparable to the number $B^{\frac{1}{k}}$ of exact k-th powers in the same range.

3   The Arithmetic of Orbifold Curves

3.1   *Projective Curves*

Let thus $C = X$ be a connected smooth projective curve defined over k. Its fundamental invariant is its genus $g \geq 0$, also equal to $h^0(C, K_C)$, the number of its (linearly independent) regular differentials, and also equal to $g = 1 + \frac{deg(K_C)}{2}$. The genus is also a topological invariant (the number of 'handles') of the set of complex points of C (and so purely 'geometric').

There are only 3 cases, according to the value of g, or equivalently to the sign of $deg(K_C)$:

- $g = 0$: if $C(k)$ is not empty, C is isomorphic to $\mathbb{P}^1$ over k, and so $C(k) \cong \mathbb{P}^1(k)$ is infinite. There always exists a quadratic extension k'/k such that $C(k') \neq \emptyset$.
- $g = 1$: after a finite extension k'/k (its degree depending on C), $C(k') \neq \emptyset$, and $C(k')$ is thus an elliptic curve with a group structure. A suitable

quadratic[10] extension k''/k' gives a point 'of infinite order' in the group $C(k'')$, and so $C(k'')$ is infinite.

- $g \geq 2$. Faltings' theorem (solving Mordell's conjecture) says that $C(k')$ is finite, however big k' is.
- **Conclusion:** C is potentially dense if and only if $deg(K_C) \leq 0$. Notice indeed that $deg(K_C) \leq 0$ if and only if $g \leq 1$.

3.2 Quasi-Projective Curves

These are just projective curves C with a non-empty finite set D removed. Here C and D are thus assumed to be defined over k (which means that D is preserved by the action of $Gal(\bar{\mathbb{Q}}/k)$).

The fundamental geometric invariant of the situation is now the sign of the log-canonical bundle $K_C + D$ (which replaces K_C when $D = 0$). The conclusion is exactly the same as in the proper case (by a theorem essentially due to C.L. Siegel).

- $deg(K_C + D) < 0$: the set of S'-integral points relative to D is Zariski dense for some k', S' sufficiently large. This case occurs only with $C = \mathbb{P}^1$, with 1 point deleted.
- $deg(K_C + D) = 0$: again, the set of S-integral points relative to D is Zariski dense for some k', S'. This case occurs only with $C = \mathbb{P}^1$, with 2 geometric points deleted.
- $deg(K_C + D) > 0$: the set of S-integral points relative to D is finite for any k', S'. This case occurs only with $C = \mathbb{P}^1$, with 3 or more points deleted, or if C is a curve of positive genus with at least 1 point deleted.

3.3 The Orbifold Mordell Conjecture

This is the one-dimensional special case of a more general conjecture to be formulated later. It relates the arithmetic of a curve orbifold pair (C, Δ) to the sign of its 'orbifold canonical bundle' $K_C + \Delta$, just as when $\Delta = 0$ or when $\Delta = D$, the (reduced) support of Δ.

Conjecture 3.1. *Let (C, Δ) be an orbifold pair defined over a number field k. Let k'/k be a finite extension, and S' a finite set of places of k'.*

---

[10]This is easily seen from a Weierstrass equation and the finiteness of torsion points of the group $C(k)$.

Then $(C, \Delta)(S', k')$ is finite for each (k', S') if and only if $deg(K_C + \Delta) > 0$. Equivalently: $(C, \Delta)(S', k')$ is infinite for some (k', S') if and only if $deg(K_C + \Delta) \leq 0$.

We have seen above that this conjecture is true when $\Delta = 0$ and when $\Delta = D$, its reduced support.

We shall see next that it is solved also when one considers the 'classical' (S', Δ) integral points $(C, \Delta)(S', k')^*$, but that it is open for $(C, \Delta)(S', k')$. By the former inclusion $(C, \Delta)(S', k')^* \subset (C, \Delta)(S', k')$, this shows that only the 'Mordell' case $deg(K_C + \Delta) > 0$ remains open. Notice that if $\Delta < \Delta'$ in the sense that $(\Delta' - \Delta)$ is an effective $\mathbb{Q}$-divisor, we have an inclusion $(C, \Delta')(S', k') \subset (C, \Delta)(S', k')$. It is thus sufficient to deal with the 'minimal' orbifold pairs (C, Δ) with $deg(K_C + \Delta) > 0$ listed below in order to solve the preceding conjecture.

Remark 3.2. *The 'minimal' cases with $deg(K_C + \Delta) > 0$ not solved by the preceding results are thus the following ones:*

- *C is elliptic, and $\Delta = (1 - \frac{1}{2}).\{a\}, a \in C(k)$.*
- *$C = \mathbb{P}^1$ and $s \geq 3$, where s is the cardinality of the support D of Δ. Let $(m_1 \leq m_2 \leq \ldots \leq m_s)$ be the corresponding multiplicities. We have thus: $\sum_j (1 - \frac{1}{m_j}) > 2$, or equivalently $\sum_j \frac{1}{m_j} < s - 2$. This gives the following possibilities, with $s = 3, 4, 5$ only:*
- *$s = 3$, and $(m_1, m_2, m_3) \in \{(2, 3, 7), (2, 4, 5), (3, 3, 4)\}$.*
- *$s = 4$, and $(m_1, \ldots, m_4) = \{2, 2, 2, 3\}$.*
- *$s = 5$ and $(m_1, \ldots, m_5) = \{2, 2, 2, 2, 2\}$.*

The 'orbifold Mordell Conjecture' thus reduces to showing finiteness of (S, Δ)-integral points for (S, Δ) in the above short list. Notice that its solution would imply in particular the finiteness of the infinite union of classical integral points for the orbifolds 'divisible' by Δ, which are the ones deduced from Δ by multiplying each of its multiplicities by an arbitrary positive integer (without changing the support). The orbifold conjecture thus looks much stronger than its 'classical' version.

Remark 3.3. *The complex function field version of the orbifold Mordell conjecture is solved in [13]. For function fields over finite fields, the solution is much more involved and more recent: see [32]. The hyperbolic version of the orbifold Mordell conjecture is also known (see §3.8).*

3.4 Solution of the Classical Version

This classical version is solved by Darmon-Granville in [25], the idea being to remove the orbifold divisor Δ by means of suitable ramified covers $\pi : C' \to C$ which are étale in the orbifold sense. We briefly sketch their arguments.

Definition 3.4. *Let $\pi : C' \to C$ be a surjective (hence finite) regular map defined over k between two smooth projective curves. Let $\Delta := \sum_j (1 - \frac{1}{m_j}).D_j$ be an orbifold divisor defined over k on C. We shall say that π is a 'classical' orbifold morphism if, for any j, and any $x' \in \pi^{-1}(D_j)$, the ramification order $e_{x'}$ of π at x' is a multiple of m_j.*

We shall say that π is 'classically' orbifold-étale over Δ if we have the equality $e_{x'} = m_j$ for any such x', j. This is easily seen to be equivalent to: $\pi^*(K_C + \Delta) = K_{C'}$.

The use of such covers is based on the following:

Proposition 3.5. *Let $\pi : C' \to C$, k, Δ be as in the previous definition, and let S be a finite set of places of k. Assume that π is classically orbifold-étale over Δ. We then have the following two properties:*

1. *$\pi(C'(k) \setminus R) \subset (C, \Delta)(S, k')^*$, R being the ramification of π.*
2. *There is a finite extension k'/k such that $\pi(C'(k')) \supset (C, \Delta)(S, k)$.*

Proof. The proof of Claim 1 is easy just by going through the definitions. By contrast, Claim 2 is an orbifold version of the theorem of Chevalley–Weil, which deals with the case $\Delta = 0$ in any dimension. Claim 2 is established, by reduction to this classical result, in [25], Proposition 3.2. □

The rest of the argument is purely geometric, by constructing suitable orbifold-étale covers.

- We first deal with the 'easy' case in which $deg(K_C + \Delta) \leq 0$. In this case $C = \mathbb{P}^1$. The proof just consists in producing a suitable orbifold-étale cover $\pi : C' \to \mathbb{P}^1$ over Δ and defined over $\bar{\mathbb{Q}}$, with C' either elliptic (if $deg(K_C + \Delta) = 0$), or $C' = \mathbb{P}^1$ (if $deg(K_C + \Delta) < 0$). This is classical (and easy, except in the case where $C = \mathbb{P}^1$, and Δ is supported on 3 points of multiplicities $(2, 3, 5)$, where the Klein icosahedral cover solves the problem). See [25], §6,7 and [3] for many more details. Only Claim 1 is needed here, together with the 'potential density' of rational and elliptic curves.
- The second case $deg(K_C + \Delta) > 0$ requires much more. First one needs an orbifold étale cover $\pi : C' \to C$ of (C, Δ). If C is elliptic, with $\Delta = (1 - \frac{1}{2}).a, a \in C(k)$, this is given by a cover C' of C which ramifies at order 2 only over a, by first taking a double étale cover (still elliptic) $\pi : C' \to C$ of C, and then a double cover of C' ramifying at order 2 over the two points of the inverse image of a in C'. Otherwise $C = \mathbb{P}^1$, and the only non-obvious cases are when $s = 3$ with 3 points $0, 1, \infty$ of multiplicities p, q, r with $\frac{1}{p} + \frac{1}{q} + \frac{1}{r} < 1$. The existence of such a cover C' follows from the existence of finite quotients $Q_{p,q,r}$ of $\pi_1(\mathbb{P}^1(\mathbb{C}) - \{0, 1, \infty\})$, which is a free group on two generators, and with Q a finite permutation group containing 3 elements A, B, C of respective orders p, q, r, with $C^{-1} = AB$ (see [37], 1.2.13, 1.2.15). Applying claim 2 of Proposition 3.5, we see that

$\pi(C'(k')) \supset (S, \Delta)(C)$. Since, by Faltings' theorem, $C(k')$ is finite, so is $(S, \Delta)(C)$.

Remark 3.6. *The reason why the Orbifold Mordell Conjecture cannot be proved by the same argument for 'non-classical' integral points is that (above orbifold version of) the Chevalley–Weil theorem does not apply to them: the lifting of integral Δ-points requires that the ramification orders divide (and not only be smaller than) the corresponding multiplicities. More precisely: contrary to what happens with the 'classical' integral points, the arithmetic ramification can occur anywhere geometrically for non-classical integral points. This is illustrated by the following simplest possible example. Let $(\mathbb{P}^1, \Delta)$ where Δ is supported on $\{0, \infty\}$, each of these two points being equipped with the multiplicity 2. The classical integral points over $\mathbb{Q}, S = \emptyset$, are thus simply the squares of non-zero integers up to sign, while the non-classical integral points are the non-zero 2-powerful numbers, which admit odd arithmetic ramification at any prime, and are not the squares of a ring of integer of the form $\mathcal{O}_{k,S}$ for any finitely generated extension of $\mathbb{Q}$.*

3.5 The abc-Conjecture

We state here its simplest form, for $k = \mathbb{Q}$ (a version for number fields has been given by Elkies):

Conjecture 3.7. *For each real $\varepsilon > 0$, there exists a constant $C_\varepsilon > 0$ such that for each triple (a, b, c) of positive coprime integers such that $a + b = c$, one has: $c \leq C_\varepsilon.rad(abc)^{1+\varepsilon}$. Recall that $rad(abc)$ is the product of the primes dividing abc.*

The rough meaning is that the exponents in the prime decompositions of a, b, c cannot be 'too' large.

- The *abc* conjecture can be interpreted geometrically in terms of the number of intersections counted without multiplicities of the section $x = \frac{a}{c}$ with the sections $0, 1, \infty$ on the arithmetic surface $\pi : \mathbb{P}^1_{\mathbb{Z}} \to Spec(\mathbb{Z})$. It simply says that the 'height', taken to exponent $(1 - \varepsilon)$, of x is bounded by the total number of intersection points (counted **without multiplicities**) of this section with the 3 sections $0, 1, \infty$.
- Let us visualise the *abc*-conjecture, using the sections $x, 0, 1, \infty$ of the arithmetic surface $\pi : \mathbb{P}^1_{\mathbb{Z}} \to Spec(\mathbb{Z})$. The section x only gives the intersection points of the section x with the 3 other sections, that is: $rad(a), rad(b), rad(c)$. To recover x, one needs additionally the arithmetic intersection numbers. The *abc*-conjecture claims they are 'small' (with a quantitative measure). The following exercise at least shows that they are finite in numbers, that is: the radicals of a, b, c determine $a, b, c = a + b$ 'up to a finite ambiguity' (Figure 3).

The two arithmetic sections $x = \frac{2^3}{3^1}$ and $x' = -\frac{2^2}{3^2}$ meet the sections (0), (1) and (∞) at the same points.

Intersections with: (0) ● (1) ● (∞) ●

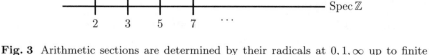

Fig. 3 Arithmetic sections are determined by their radicals at $0, 1, \infty$ up to finite ambiguity

Remark 3.8. *The abc-conjecture implies that there exists only a finite number of triples of coprime integers (a, b, c) such that $a + b = c$, and $rad(abc) \leq N$. This is a special case of the finiteness of solutions of the S-unit equation. It follows, for example, from the weak form of the abc-conjecture proved in [44]. This finiteness is due to K. Mahler, originally. See [28] and the references therein for more general statements. We illustrate below the case where $rad(abc) = 2.3.5 = 30$.*

Some of the solutions of the equation $2^x \pm 3^y = \pm 5^z$ are $(x, y, z) = (1, 1, 1), (2, 2, 1), (1, 3, 2), (4, 2, 2), (7, 1, 3)$. It is probably not easy to get a complete list of all solutions, even over $\mathbb{Z}$.

3.6 abc Implies Orbifold Mordell

Since this is shown in [26] when $\Delta = 0$, we only need to show this for the remaining 'minimal' cases listed in Remark 3.2. We start with $\mathbb{P}^1$ with 3 marked points.

- Let us show that abc implies the Mordell orbifold conjecture over $\mathbb{Q}$ for $(\mathbb{P}^1, \Delta)$ with Δ as in Example 2.7 above. Indeed: if a (resp. b, resp. c) is p-powerful (resp. q-powerful, resp. r-powerful), we have: $rad(a) \leq a^{\frac{1}{p}} \leq c^{\frac{1}{p}}$, and similarly $rad(b) \leq c^{\frac{1}{q}}$, $rad(c) \leq c^{\frac{1}{r}}$. We thus get: $rad(abc) \leq rad(a).rad(b).rad(c) \leq c^{\frac{1}{p}+\frac{1}{q}+\frac{1}{r}} \leq c^{1-\frac{1}{42}}$, since $\frac{1}{p}+\frac{1}{q}+\frac{1}{r} \leq 1-\frac{1}{42}$ for each of the minimal orbifolds listed in Example 2.7, the minimum being reached for the multiplicities $(2,3,7)$. The conjecture abc implies that: $c^{1-\frac{1}{42}} \geq rad(abc) \geq \frac{c^{1-\varepsilon}}{C_\varepsilon}$, for any $\varepsilon > 0$. Choosing $\varepsilon < \frac{1}{42}$ gives: $c^{\frac{1}{42}-\varepsilon} \leq C_\varepsilon$, and so the claimed finiteness.[11]
- The Orbifold Mordell conjecture can be deduced from the abc-conjecture also in the three remaining cases when either $C = \mathbb{P}^1$, and Δ is supported by 4 or 5 points on $\mathbb{P}^1$ with multiplicities $(2,2,2,3)$ and $(2,2,2,2,2)$, respectively, or when C is elliptic and Δ is supported on a single point with multiplicity 2. The derivation is, however, less direct less: one needs to apply a variant of the method used by N. Elkies in [26] to derive Faltings' theorem from the abc-conjecture. One can proceed as follows:[12]

- First step (the same thus as in [41]):

Let $f : C := \mathbb{P}^1 \to B := \mathbb{P}^1$ be a rational function $f = \frac{F}{G}$ of degree $d > 0$, quotient of polynomials F, G, defined over k, a number field. We shall use the notations of [26]. Let $P \in C(k)$, such that $f(P) \notin \{0, 1, \infty\}$. Let $H(P)$ (resp. H_P) be the height of P (resp. of $f(P)$). We denote by $N_0(f(P))$ the radical of $F(P)$. We have: $Log(H(f(P))) = d.Log(H(P)) + O(1)$.

Elkies shows that $Log(N_0(f(P))) \leq (\frac{k_0}{d}).Log(H(P)) + O(1)$, where k_0 is the cardinality (without multiplicities) of $f^{-1}(0)$. (The proof just consists in removing the ramifications on this fibre). One has then similar inequalities over the fibres of f over 1 and ∞ replacing f by $(f - 1)$ and $\frac{1}{f}$. From which he concludes (using the Riemann–Hurwitz formula) that $(k_0 + k_1 + k_\infty).Log(H(f(P))) \geq d.Log(N(f(P))) + O(1)$, with $N := N_0 + N_1 + N_\infty$, where N_1, N_∞ are defined as N_0, but considering the fibres over $1, \infty$ instead of 0.

---

[11] This observation has been communicated to me by Colliot-Thélène, who attributed it to P. Colmez.

[12] The referee informed me that this approach was already sketched in [1],§4.4, and treated completely in [43]. Abramovich's approach is based on Belyi maps and deals with all cases simultaneously. The proof given below is the same, but constructs Belyi maps explicitly in the three remaining cases mentioned above.

His argument easily extends to the case where C is equipped with an orbifold divisor Δ supported on the union of the fibres of f over $0, 1, \infty$. Let, for each point a_j in this union, m_j be its multiplicity in Δ, and t_j be the order of ramification of f at a_j. Define the number $d_0 := \sum_{a_j \in f^{-1}(0)} (m_j - 1).t_j$. Define similarly d_1, d_∞ for the fibres of f over 1 and ∞. Elkies argument then shows that: $k_0.Log(H(f(P))) \geq (d + d_0).Log(N_0(f(P)) + O(1)$. Adding the two other inequalities on the fibres of f over $1, \infty$, we get:

$$(k_0 + k_1 + k_\infty).Log(H(f(P))) \geq (d.Log(N(f(P)) + \delta + O(1),$$

where: $\delta = d_0.Log(N_0(f(P))) + d_1.Log(N_1(f(P))) + d_\infty.Log(N_\infty(f(P)))$

Assume now that f is unramified outside of the three fibres over $0, 1, \infty$. We then have: $(k_0 + k_1 + k_\infty) = d + 2$. Assume also that $min\{d_0, d_1, d_\infty\} \geq 3$.

We obtain: $(d + 2).Log(H(f(P))) \geq (d + 3).Log(N(f(P)))$, an inequality satisfied only for finitely many $P's \in k$, by the abc-conjecture. This implies Mordell orbifold for (C, Δ).

- Second step (construction of Belyi maps):

In order to show that this applies to $C = \mathbb{P}^1$, with Δ either of the form $(2, 2, 2, 3)$ or $(2, 2, 2, 2, 2)$, we consider $f : \mathbb{P}^1 \to \mathbb{P}^1$ defined by $f(x) := \frac{x^2(x-1)(x-w)}{ux-v}$. The fibre of f over 0 consists thus of 3 points, one double (0), two simple $(1, w)$. The fibre of f over ∞ consists of two points: the triple point ∞ and the single point $\frac{v}{u}$. We now fix 2 further points (distinct from the preceding ones): b, c, and notice that the equation: $x^2(x - 1)(x - w) = (ux + v) + (x - b)^2(x - c)(x - t)$ with unknowns u, v, w, t has a unique solution. This means that the fibre of f over 1 has 3 points: one double (b) and two simple ones: (c, t).

In order to deal with $\Delta = (2, 2, 2, 3)$, we attribute to the points $0, 1, b, \infty$, respectively, the multiplicities $2, 2, 3, 2$. An easy check shows that $d_1 = 4, d_0 = d_\infty = 3$.

In order to deal with $\Delta = (2, 2, 2, 2, 2)$, we attribute to all of the 5 points $0, 1, b, c, \infty$ the multiplicity 2. One again easily checks that $d_0 = d_1 = d_\infty = 3$.

The last remaining case is when C is elliptic, and $\Delta = (1 - \frac{1}{2}).\{a\}, a \in C(k)$. It can be reduced similarly to abc by composing the above map $f(x) := \frac{x^2(x-1)(x-w)}{ux-v}$ with the double cover $g : C \to \mathbb{P}^1$ so that its 4 ramification points are sent by f to $0, 1, b, c$, and a to ∞, equipping again each of these 5 points with the multiplicity 2.

This concludes the proof that abc implies orbifold Mordell.

3.7   Ramification of Belyi Maps

The question we would like to address here is whether the (non-classical) orbifold Mordell conjecture for one single orbifold pair $(\mathbb{P}^1, \Delta)$ of general type: $\mathbb{P}^1$ with the 3 marked points $0, 1, \infty$ of multiplicities $(3, 3, 4)$ (for example, one could choose $(2, 3, 7)$ or $(2, 4, 5)$ instead) implies Mordell Conjecture= Faltings' Theorem, for every curve defined over $\overline{\mathbb{Q}}$. One may of course raise this question for the other minimal orbifolds over $\mathbb{P}^1$ listed in Remark 3.2.

A positive answer to the following question implies this statement:

Question 3.9. *Let C be a curve defined over $\overline{\mathbb{Q}}$. Does there exist:*

1. *An unramified cover $u : \tilde{C} \to C$.*
2. *A Belyi map $\beta : \tilde{C} \to \mathbb{P}^1$ (unramified over the complement of $\{0, 1, \infty\}$) such that each of its ramification orders over 0 (resp. 1, resp. ∞) are at least 3, (resp. 3, resp. 4)?*

The usual construction of Belyi maps cannot produce Belyi maps such as in the preceding question. Assume indeed that g is already a Belyi map for C, but has some unramified point over each of $0, 1, \infty$. In order that $f \circ g$ be a Belyi map satisfying the condition 2 of 3.9, the map f itself should already be a Belyi map satisfying this very same condition. The Riemann–Hurwitz equality contradicts the existence of such an f.

Faltings' Theorem would follow from a positive answer to Question 3.9 and Orbifold Mordell. Indeed: fix k, a number field of definition of a given C, and let u, β answering positively the Question 3.9. Let k'/k be a finite extension such that $u(\tilde{C}(k')) \supset C(k)$ (using the Chevalley–Weil Theorem). Since β is an orbifold morphism to $(\mathbb{P}^1, \Delta)$, we get a map with uniformly finite fibres from $\tilde{C}(k')$ to $(\mathbb{P}^1, \Delta)(\mathcal{O}_{k'})$, the last set being finite by the Orbifold Mordell conjecture for any k'. We thus get the finiteness of $C(k)$.

Remark 3.10. *The Question 3.9 bears a certain similarity with the notion of universal curves introduced in [7] (although the étale covers there are over the universal curve). I thank A. Javanpeykar for bringing this reference to my knowledge.*

3.8   Link with Complex Hyperbolicity

Let C be a connected smooth projective curve C. By the Poincaré–Koebe uniformisation, there is a non-constant holomorphic map $h : \mathbb{C} \to C$ if and only if C is not uniformized by the unit disk $\mathbb{D} \subset \mathbb{C}$, that is: if $g(C) \leq 1$. Similarly, if C is defined over a number field k, the potential density of $C(k)$ holds if and only if there exists such a map h. It is very easy to check that this equivalence still holds for quasi-projective curves $(C - D)$, again by their uniformisation for the hyperbolic version.

We show in [18], using Nevanlinna's Second Main Theorem with truncation at order one, that the same thing is true for 'orbifold curves' (the notion of morphism $h : \mathbb{C} \to (C, \Delta)$ being defined as in Definition 2.5 in the two possible ways ('classical' and 'non-classical'). The orbifold Mordell Conjecture thus remains open only in its arithmetic version.

This link, initiated by S. Lang, will be studied in higher dimensions as well.

4   The Kodaira Dimension

4.1   The Iitaka Dimension of a Line Bundle

Since, for projective curves, the invariant $h^0(C, K_C) = g$ determines the qualitative arithmetic, it is natural to consider it also in higher dimensions. The invariant $h^0(X, K_X)$ is birational, but no longer preserved by étale covers in dimension 2 already, and one needs more information: the values $h^0(X, m.K_X) := p_m(X), m > 0$, the 'plurigenera' of Enriques. We shall even abstract more (in order to get a birational invariant preserved by étale covers), and only consider the asymptotic behaviour of the plurigenera as m goes to $+\infty$, for a given X. The notion actually makes sense, and is extremely useful, more generally, for arbitrary line bundles L, not only for $L = K_X$.

- Let X be a connected projective manifold of dimension n defined over a field k of characteristic 0. Let L be a line bundle on X. Let $h^0(X, L) \in \mathbb{N}$ be the k-dimension of its space $H^0(X, L)$ of sections. If $h^0(X, L) > 0$, let $\Phi_L = X \dashrightarrow \mathbb{P}(H^0(X, L)^*)$ be the rational map which sends a generic $x \in X$ to the hyperplane of $H^0(X, L)$ consisting of sections vanishing at x. We thus have: $0 \leq dim(\Phi_L(X)) \leq n$. We denote either with $m.L$ or with $L^{\otimes m}, m \in \mathbb{Z}$ the m-th power of L.

Definition 4.1. *We define* $\kappa(X, L) \in \{-\infty, 0, \ldots, n\}$ *as being* $-\infty$ *if* $h^0(X, mL) = 0, \forall m > 0$. *Otherwise,* $\kappa(X, L) := max_{m>0}\{dim(\Phi_{mL}(X))\}$.
An alternative definition, not immediately, equivalent is:

$$\kappa(X, L) := \overline{lim}_{m \to +\infty}\{\frac{Log h^0(X, m.L)}{Log m}\},$$

roughly meaning that $h^0(X, m.L)$ *grows like the* $\kappa(X, L)$-*th power of* m *as* m *goes to* $+\infty$.

Example 4.2.

- $\kappa(X, L) = -\infty$ if $L = \mathcal{O}_X(-D)$ for some effective divisor D. And also when X is an elliptic curve, if $c_1(L) = 0$, but L is not torsion in $Pic(X)$.
- $\kappa(X, L) = 0$ iff $h^0(X, mL) \leq 1, \forall m > 0$, with equality for some $m > 0$, for example, if L is torsion in $Pic(X)$.
- $\kappa(X, L) = n$ iff $mL = A + E$, for some $m > 0$, A ample and E effective. Then L is said to be 'big'.
- $\kappa(X, L) = d \in \{1, \ldots, n\}$ if $p : X \to Z$ be regular onto, with $d := dim(Z)$, and $L = p^*(A)$, $A \in Pic(Z)$, ample. Indeed, one has:
- $\kappa(X, p^*(M)) = \kappa(Z, M)$, for any line bundle M on Z.

The following theorem gives a weak analogue in general:

Theorem 4.3. *If $\kappa(X, L) = d \geq 0$, for any sufficiently large and divisible integer $m > 0$, the rational map $\Phi_{m.L}$ has connected fibres, its image $Z_m = Z$ has dimension d and its generic fibre X_z has $\kappa(X_z, L_{|X_z}) = 0$. Moreover, Z_m is birationally independent of $m > 0$ sufficiently large and divisible.*

If $d = n$, $\Phi_{mL}(X)$ is birational to X for m large enough.

Observe however that, in general, L will not be torsion on the general fibre of Φ_{mL}. Many more details and numerous examples can be found in [46].

The following Proposition gives an upper bound on $\kappa(X, L)$:

Proposition 4.4 ('Easy Additivity'). *Let $p : X \to Z$ be a fibration, and $L \in Pic(X)$. Let X_z be the general fibre of p. Then:*

$$\kappa(X, L) \leq \kappa(X_z, L_{|X_z}) + dim(Z).$$

4.2 The Kodaira Dimension κ

The fundamental case is when $L = K_X := det(\Omega_X^1)$, the canonical line bundle on X. One writes then: $\kappa(X) := \kappa(X, K_X)$.

The invariant $\kappa(X)$ enjoys several properties:

- It is birational, and preserved by finite étale covers.
- Additive for products: $\kappa(X := Y \times Z) = \kappa(Y) + \kappa(Z)$, since:

$$h^0(X, mK_X) = h^0(Y, mK_Y) \times h^0(Z, mK_Z), \forall m.$$

- In particular: $\kappa(X) = -\infty, \forall Z$, if $\kappa(Y) = -\infty$ (e.g.: $Y = \mathbb{P}^1$).
- Also: $\kappa(X) = \kappa(Z)$ if $\kappa(Y) = 0$.

4.3 First Examples: Curves and Hypersurfaces

For curves, $\kappa(X)$ tells (almost) everything, qualitatively, it indeed describes X, its topology, fundamental group, as well as hyperbolicity and arithmetic properties.

κ	g	X	$X(k)$
$-\infty$	$g = 0$	$\mathbb{P}^1$	Potentially dense
0	$g = 1$	$\mathbb{C}/\Lambda$	Potentially dense
1	$g \geq 2$	$\mathbb{D}/\Gamma$	Not potentially dense

The preceding trichotomy (according to the 'sign' of K_X: positive, zero or negative) still appears in the special case of smooth hypersurfaces in $\mathbb{P}_{n+1}$.

- **Hypersurfaces in $\mathbb{P}_{n+1}$.** Let $H_d \subset \mathbb{P}_{n+1}$ be a smooth hypersurface of degree d (defining by a homogeneous polynomial in $(n+2)$ variables of degree d). The adjunction formula shows that $K_{H_d} = \mathcal{O}(d - n + 2)_{|H_d}$. Thus K_{H_d} is ample if $d \geq (n+3)$, trivial if $d = (n+2)$ and anti-ample if $d \leq (n+1)$. We thus have, in particular: $\kappa(H_d) = n$ (resp. 0, resp. $-\infty$) if $d > n + 2$ (resp. $d = n+2$, resp. $d < n+2$).
- **Hypersurfaces in $\mathbb{P}_{n+1-k} \times \mathbb{P}_k$.** Let now $H := H_{d,d'}$ be a smooth hypersurface of bidegree (d, d') in this product (this means that $H \cap F$ is a hypersurface of degree d' (resp. d) when intersected with a generic $\mathbb{P}_{n+1-k} \times \{a'\}$ (resp. $\{a\} \times \mathbb{P}_k$). The adjunction formula now shows that $K_H = \mathcal{O}(d - (n+2-k), d' - (k+1))_{|H}$. One thus obtains that $\kappa(H) = -\infty$ if $d \leq n+1-k$, or if $d' \leq k$, that $\kappa(H) = 0$ if $d = n+2-k$ and $d' = k+1$, that $\kappa(H) = k$ if $d = n + 2 - k, d' \geq k+2$, that $\kappa(H) = n + 1 - k$ if $d > n + 2 - k, d' = k + 1$, and that $\kappa(H) = n$ if $d > n + 2 - k, d' > k + 1$.
- The smooth hypersurfaces in products of projective spaces show that arbitrary κ may occur, which are not determined simply by those of base and fibres.

4.4 The Iitaka–Moishezon Fibration

There are 3 fundamental cases (as for curves with $g = 0, 1, \geq 2$):

1. $\kappa(X) = -\infty$.
2. $\kappa(X) = 0$.
3. $\kappa(X) = n$. In this third case, X is said to be **'of general type'**.

Let us briefly comment on these 3 classes:

- $\kappa = n$ is a large class (as for curves), it contains the smooth hypersurfaces of degree at least $(n + 3)$ in $\mathbb{P}_{n+1}$. This is the reason for the term 'general type' introduced by B. Moishezon. They are conjectured to be Mordellic by S. Lang. Examples of manifolds of general type are quotients of bounded domains in $\mathbb{C}^n$ by discrete torsion-free groups of automorphisms, which are higher dimensional analogues of curves of genus greater than 1. But many manifolds of general type (such as hypersurfaces of dimension greater than 1) are simply connected.
- $\kappa = 0$ contains manifolds with trivial (or torsion) canonical bundle, the structure of which is partially understood by means of the Beauville–Bogomolov–Yau decomposition theorem. They are however classified only in dimension 2. Even in dimension 3, it is unknown whether or not there are finitely many deformation families.

We conjecture that the manifolds with $\kappa = 0$ are Potentially Dense. It is expected that on suitable mildly singular birational models their canonical bundle becomes torsion.

- $\kappa = -\infty$: this class contains products $\mathbb{P}^1 \times Z, \forall Z$. It is discussed below.

This class thus does not consist only of Potentially dense manifolds. We define below the more restricted class of 'rationally connected' manifolds, conjectured to be potentially dense, which permits to 'split' any manifold with $\kappa = -\infty$ by means of a single fibration into a rationally connected part (the fibres), and a part (conjecturally) with $\kappa \geq 0$ (the base).

- The structure of the intermediate cases when $1 \leq \kappa(X) = d \leq (n - 1)$ 'reduces' (to some extent) to the case of $\kappa = 0$ and lower dimension, by means of the following 'Iitaka–Moishezon fibration' J.

Proposition 4.5. *The map* $J := \Phi_{mK_X} : X \dashrightarrow Z := \Phi_{m.K_X}(X) = J(X)$, *for* $m > 0$ *suitably large and divisible is birationally well-defined, and may thus be assumed to be regular. Its generic fibres* X_z *are then smooth with* $\kappa(X_z) = 0$, *because* $\kappa(X_z, K_{X|X_z}) = 0$, *and* $K_{X|X_z} = K_{X_z}$ *(by the 'Adjunction formula').*
J is defined over k, if so is X

Example 4.6. *The fibration J is the projection onto the second (resp. first) factor when* $H_{d,d'} \subset \mathbb{P}_{n+1-k} \times \mathbb{P}_k$ *is a smooth hypersurface of bidegree* $(n + 2 - k, d')$ *(resp. $(d, k + 1)$) if* $d' > k + 1$ *(resp. $(d > n + 1 - k)$).*

When $\kappa(X) = 0$, Z is a point, and J does not give any information. In the other extreme case, where $\kappa(X) = n$, J embeds birationally X in the projective space $\mathbb{P}((H^0(X, m.K_X)^*))$, for appropriate $m > 0$. One thus 'reconstructs' X from its pluricanonical sections.

Caution In general, however, $\kappa(Z) \leq d := dim(Z) = \kappa(X)$ (and strict inequality may occur, as shown by Example 4.6, since the base of J is then a projective space). The fibration J thus does not in general decompose X in parts with $\kappa(X_z) = 0$ and $\kappa(Z) = dim(Z)$.

- Notice also that J is not defined when $\kappa(X) = -\infty$. This case $\kappa(X) = -\infty$ requires a completely different treatment, which we briefly describe below.

4.5 Rational Curves and $\kappa = -\infty$

In order not to overload the text with quotations, we have deleted them for this section. The results in this section are mainly due to Mori, Miyaoka–Mori, Campana, Kollár–Miyaoka–Mori, Graber–Harris–Starr.

Definition 4.7. *A 'rational curve' on X is the image of a regular nonconstant map:* $\mathbb{P}^1 \to X$. *We say that X is uniruled if it is covered by rational curves, or equivalently, if there exists a dominant rational map* $\mathbb{P}^1 \times T_{n-1} \dashrightarrow X$ *for some $(n-1)$ dimensional variety T_{n-1}.*

If X is uniruled : $\kappa(X) \leq \kappa(\mathbb{P}^1 \times T) = -\infty$. Thus $\kappa(X) = -\infty$. The converse is a central conjecture of birational geometry, known up to dimension 3:

Conjecture 4.8 ('Uniruledness Conjecture'). *If $\kappa(X) = -\infty$, X is uniruled.*

The decomposition of arbitrary X into parts with a 'birationally signed' canonical bundle depends on some or other form of this central conjecture.

4.6 Rational Connectedness and $\kappa^+ = -\infty$

Definition 4.9. *X is 'rationally connected' (RC for short) if any two generic points of X are joined by a rational curve.*

Example 4.10.

1. *Let $X = \mathbb{P}^1 \times C$, for C a projective curve of genus g: X is uniruled, but it is rationally connected if and only if $g = 0$.*
2. *Unirational manifolds (those dominated by $\mathbb{P}^n$) are RC.*
3. *Fano manifolds (those with $-K_X$ ample) are rationally connected.*
4. *Smooth hypersurfaces of degree at most $(n+1)$ in $\mathbb{P}^{n+1}$ are Fano.*
5. *Rationally connected manifolds are simply connected.*
6. *Although no rationally connected manifold is presently proved to be non-unirational, it is expected that this is the case for most rationally connected manifolds of dimension 3 or more. In particular, the (non) unirationality*

of the double cover of $\mathbb{P}_3$ ramified along a smooth sextic surface S_6 is an open problem.

Remark 4.11. *If X is defined over a field $k \subset \mathbb{C}$ and is uniruled (resp. rational, unirational, rationally connected over $\mathbb{C}$) it is not difficult to see that it has this property also over some finite extension of k.*

Theorem 4.12. *For any X, there is a unique fibration $r_X : X \to R_X$ such that:*

1. *its fibres are rationally connected, and:*
2. *R_X is not uniruled.*

It is called the 'rational quotient', or the 'MRC[13] of X.
  If X is defined over k, so is r_X.

The fibration r_X thus decomposes X into its antithetic parts: rationally connected (the fibres) and non-uniruled (the base R_X). The extreme cases are when $X = R_X$ (i.e.: X is not uniruled), and when R_X is a point (i.e.: X is rationally connected).

Remark that the uniruledness conjecture implies that $\kappa(R_X) \geq 0$. This leads to the following definition:

Definition 4.13. *Define, for any projective X:*

$$\kappa^+(X) := max\{\kappa(Y)|\exists \ dominant \ f : X \dashrightarrow Y\}$$

From Theorem 4.12, one gets:

Proposition 4.14. *Assume the Uniruledness Conjecture 4.8. The following are then equivalent:*

1. *X is rationally connected.*
2. *$\kappa^+(X) = -\infty$.*

Moreover, the 'rational quotient' is also the unique fibration $g : X \to Z$ on any X such that:

1. *$\kappa^+(X_z) = -\infty$ for the general fibre X_z of g, and:*
2. *$\kappa(Z) \geq 0$.*

Note that these conjectural characterisations of rational connectedness and of r do not rely on rational curves, but only on κ and its refinement κ^+. The rational quotient will also be constructed without mentioning rational curves, conditionally on conjecture $C_{n,m}$, in §6.5.

Remark 4.15. *We conjecture that manifolds with $\kappa^+ = -\infty$ are potentially dense. Thus so should be the rationally connected manifolds. Much more*

---

[13]Stands for 'maximally rationally connected'.

*generally, we conjecture that 'special manifolds' (defined later) are **exactly** the potentially dense manifolds.*

5 Surfaces

5.1 Classification of Surfaces

If S is a smooth projective surface, we have: $\kappa := \kappa(S) \in \{-\infty, 0, 1, 2\}$. The maps r and J permit to elucidate the structure of S when $\kappa(S) \neq 2$.

When $\kappa = -\infty$, the uniruledness conjecture is a classical result of Castelnuovo, and we thus get a non-trivial rational quotient $r : S \to R$, where R is either a curve C_q of genus $q = h^0(S, \Omega_S^1) > 0$, or a point (in which case S is rationally connected, and even rational).

When $\kappa = 1$, one has the Iitaka–Moishezon fibration $J : S \to B$, with smooth fibres elliptic, and B a curve. One says that S is an elliptic surface over B.

When $\kappa = 0$, a precise classification is known: S is covered by a blow-up of either an abelian surface or of a $K3$ surface, where $K3$-surfaces are defined by: $q = 0, K_S \cong \mathcal{O}_S$. They form a single deformation family containing the smooth quartics in $\mathbb{P}_3$.

One thus gets the 'Enriques–Kodaira–Shafarevich' classification, displayed in the table below (up to birational equivalence and finite étale covers), where C_q denotes a curve of genus q, $q := h^0(S, \Omega_S^1) = \frac{1}{2} b_1(X)$. We indicate the status of potential density for S defined over some large number field k. More details below.

κ	q	S(up to bir, étale $\cong$)	$S(k)$ potentially dense
$-\infty$	$q \geq 0$	$\mathbb{P}^1 \times C_q$	Yes iff $q \leq 1$
0	0	$K3$	Yes in many examples
0	2	$(\mathbb{C}^2/\Lambda)$	Yes, always
1	≥ 0	Elliptic over C_q	Yes in many examples if $q \leq 1$
2	≥ 0	No classification scheme	No, in all known examples

5.2 Remarks on Potential Density

Our guiding principle here consists of the following 3 facts, for X a smooth connected projective manifold defined over a number field k:

0. Potential density is a birational property.
1. Chevalley–Weil theorem: if $X' \to X$ is an étale covering, $X'(k)$ is potentially dense if $X(k)$ is (the converse is obvious).

2. Lang's conjecture:[14] if X is 'of general type', then $X(k)$ is **not** potentially dense.

By Faltings' theorem this holds for curves, but is open for surfaces.

Definition 5.1. *We say that X (defined over $\mathbb{C}$) is 'weakly special' if, for any finite étale cover $u : X' \to X$, there exists no dominant rational map $f : X' \dashrightarrow Z$, with Z of 'general type' and $\dim(Z) > 0$.*

Remark 5.2. *The 3 facts above imply that if X is not weakly special, $X(k)$ is not potentially dense. The following claims the converse also:*

Conjecture 5.3 ([30, Conjecture 1.2]). *A projective manifold X/k is potentially dense if and only if X is 'weakly special'.*

Remark 5.4. *This conjecture conflicts with other conjectures stated below[15] when $\dim(X) \geq 3$, but both conjectures agree for surfaces (because specialness and weak specialness coincide for them).*

Let us check the known cases of this conjecture for surfaces, according to $\kappa(S) = \kappa$, for S a surface defined over a number field k. Let $r : \tilde{S} \to S$ be any finite étale cover of S, and $\tilde{q}(S)$ the supremum (possibly infinite) of $q(\tilde{S})$ when $\tilde{S}$ ranges over all finite étale covers of S. For example, $\tilde{q}(S) = +\infty$ if some $\tilde{S}$ fibres over a curve of genus $g \geq 2$. Recall that a Theorem of Y.T. Siu shows that this happens if and only if some finite index subgroup of $\pi_1(S)$ admits a quotient which is a 'surface group' (i.e.: of the form $\pi_1(C)$ with $g(C) \geq 2$). Notice that $\tilde{q}(S) \geq 2$ and $\kappa(S) \neq 0, 2$ imply that some $\tilde{S}$ fibres over a curve of genus at least 2, and so that: $\tilde{q}(S) = +\infty$.

- $\kappa = 2$. If $\tilde{q}(S) \geq 2$, then S is Mordellic, by Faltings' Theorem (and Kawamata Theorem on the structure of ramified covers of Abelian varieties) showing that a subvariety of general type of an Abelian variety is Mordellic. If $\tilde{q}(S) = 0, 1$, S is Mordellic conditionally on Lang's conjecture.
- $\kappa = -\infty$. Then $S = \mathbb{P}^1 \times C_q$. Thus $S(k)$ is potentially dense if and only if so is C_q: The conjecture is true.
- $\kappa = 0$. Some $\tilde{S}$ is either an Abelian surface, or a $K3$ surface. Both are easily seen to be weakly special. If S is an Abelian surface, $S(k)$ is potentially dense, and the conjecture then holds.

The conjecture then claims that $K3$ surfaces are potentially dense. This is unknown in general, but known for $K3$ surfaces which are Kummer, or admit either an elliptic fibration, or an automorphism group of infinite order [6], the main idea of which is: if $f : S \to C$ is an elliptic fibration onto the

---

[14] Also attributed to E. Bombieri in the case of surfaces, although not in written form, even in [9].

[15] Where 'weak specialness' is replaced by 'specialness'.

curve C, and if S contains a rational or elliptic 'non-torsion multisection', then $S(k)$ is potentially dense.

A 'non-torsion multisection' is an irreducible curve $D \subset C$ such that $f(D) = C$, and moreover such that, over the generic point of C, the fibre of D has two points the difference of which is not torsion in the group of translations of this (elliptic) fibre.

It is shown in [6] (this is the hardest geometric part) that elliptic $K3$ surfaces always contain some rational or elliptic 'non-torsion' D.

- $\kappa = 1$. Let $f : S \to C$ be the (elliptic) Moishezon–Iitaka fibration. A major rôle is played by the 'multiple fibres' of f. Let indeed, for $s \in C$, $f^*(s) := (\sum_h t_h.F_h)$ be the scheme-theoretic fibre of f over s. Define: $m_s := gcd_h\{t_h\}$. This is the 'classical' multiplicity[16] of the fibre of f over s, and it is equal to 1, except for finitely many (possibly none) $s \in C$. We define now the 'orbifold base of f' to be the orbifold curve (C, Δ_f), with $\Delta_f := \sum_{s \in C}(1 - \frac{1}{m_s}).\{s\}$, a finite sum since $(1 - \frac{1}{m_s}) = 0$ iff $m_s = 1$.

In this situation, we now have the following (geometric):

Lemma 5.5. *An elliptic projective smooth surface S is weakly special if and only if $deg(K_C + \Delta_f) \leq 0$.*

Proof. The proof has two steps. First step: show that there exists[17] an 'orbifold-étale' cover $u : C' \to C$ over Δ_f. Then $K_{C'} = u^*(K_C + \Delta_f)$, so that $deg(K_{C'}) \leq 0$ iff $deg(K_C + \Delta_f) \leq 0$.

Second step: the (normalised) base-change $f' : S' := \widehat{S \times_C C'} \to C'$ has the property that $u : S' \to S$ is étale.

If $deg(K_C + \Delta_f) > 0$, $g(C') \geq 2$, and S is not weakly special in this case. Notice that Faltings' and Chevalley–Weil theorems imply that $S(k)$ is not potentially dense, and the conjecture is true unconditionally.

If $deg(K_C + \Delta_f) \leq 0$, C' is rational or elliptic, and since $f' : S' \to C'$ has no multiple fibre, there is an exact sequence of groups:

$$\pi_1(F'_s) \to \pi_1(S') \to \pi_1(C') \to \{1\}$$

which implies that no étale cover of S' has a fibration onto a curve C'' with $g(C'') \geq 2$ (since $\pi_1(C'')$ has the free group on 2 generators as a quotient, and is not solvable). $\qquad\square$

The Conjecture 5.3 is thus equivalent to the fact that $S(k')$ is dense when $deg(K_C + \Delta_f) \leq 0$, which is open, but verified on many examples.

---

[16]We shall introduce its 'non-classical' version in §5.3 below.

[17]Except in two quite simple cases of $\mathbb{P}^1$ with Δ supported on one or 2 points, which can be dealt with directly. We shall ignore these simple cases here.

5.3 Fibred Simply Connected Surfaces of General Type

We shall give here examples of smooth projective simply connected surfaces S of general type (defined over $\mathbb{Q}$) which are not potentially dense, conditionally on the Orbifold Mordell Conjecture.[18] Presently (July 2019) no such example is known unconditionally.[19]

Let $f : S \to C$ be a fibration (with connected fibres) from the smooth connected projective surface S onto the smooth projective curve C. We do not assume that the smooth fibres are elliptic.

Let $s \in C$, and $f^*(s) := \sum_h t_h.F_h$ be the scheme-theoretic fibre of f over s. We define two notions of multiplicity for this fibre:

- The 'classical' (or 'gcd') multiplicity $m_s^*(f) := gcd_h\{t_h\}$.
- The 'inf' multiplicity $m_s(f) := inf_h\{t_h\}$.

Of course, $m_s^*(f)$ divides $m_s(f)$, both are 1 except possibly on the finite set of singular fibres.

We now define two 'orbifold bases' of f:

- The 'classical' orbifold base (C, Δ_f^*), with $\Delta_f^* := \sum_{s \in C}(1 - \frac{1}{m_s^*(f)}).\{s\}$
- The orbifold base (C, Δ_f), with $\Delta_f := \sum_{s \in C}(1 - \frac{1}{m_s(f)}).\{s\}$

Remark 5.6.

1. *If f is an elliptic fibration, $\Delta_f = \Delta_f^*$. As we shall see, they may differ, but only if the smooth fibres of f have $g \geq 2$.*
2. *If (C, Δ_f^*) is of general type, there is always a base-change $v : C' \to C$, orbifold-étale over Δ_f^*, with $g(C') \geq 2$, such that the resulting normalised base-change $u : S' \to S$ is étale. Thus $\pi_1(S')$, which is a finite index subgroup of $\pi_1(S)$, maps onto $\pi_1(C')$, showing that $\pi_1(S)$ is a 'big' hyperbolic non-abelian group.*
3. *The map f induces natural group-morphisms $f_* : \pi_1(S) \to \pi_1(C, \Delta_f^*)$ and $\pi_1(C, \Delta_f) \to \pi_1(C, \Delta_f^*)$, but f_* does not lift to a natural group-morphism $\pi_1(S) \to \pi_1(C, \Delta_f)$. Here $\pi_1(C, \Delta_f^*)$ is the quotient of $\pi_1(S \setminus \Delta_f^*)$ by the normal subgroup generated by the m_j-th powers of a small loop winding once around D_j, this for any j if $\Delta_f^* := \sum_j(1 - \frac{1}{m_j}).\{a_j\}$.*

We shall now construct fibrations $f : S \to C$ with (non-classical) orbifold base (C, Δ_f) of general type with S simply connected.

---

[18]The particular case of $\mathbb{P}^1$ with $m \geq 5$ points of multiplicity 2 is sufficient.

[19]Unconditionally, quasi-projective examples are given in [24], and projective examples over $\mathbb{F}_q(t)$, inspired by the ones given here, are proposed in [32]. The Orbifold Mordell Conjecture over $\mathbb{C}(t)$ was previously established in [12].

Proposition 5.7 ([13]). *Let $f : S \to C$ be a fibration from the smooth projective connected surface S onto the projective curve C. Assume that $deg(K_C + \Delta_f) > 0$, and that S is simply connected. Then:*

1. *$\kappa(S) = 2$, the smooth fibres of f have $g \geq 2$.*
2. *There exist such fibrations defined over $\mathbb{Q}$. In this case:*
3. *If the orbifold Mordell conjecture is true, then $S(k)$ is contained in a finite number of fibres of f, for any number field k, and $S(\mathbb{Q})$ is not potentially dense.*

Proof. Claim 1 follows from an 'orbifold' version of the $C_{n,m}$ conjecture (see below). We shall give examples of claim 2 below. For Claim 3, it suffices to see that $f(S(k))$ is contained in $(S, \Delta_f)(C)$ (finite by the orbifold Mordell conjecture) for any k and a sufficiently large finite subset S of the places of k, determined by a 'model' of (C, Δ_f) over $\mathcal{O}_{S,k}$, such that (C, Δ_f) has good reduction outside of S. Let thus $x \in S(k)$, and t be a k-rational function which gives a local coordinate on C at $f(x)$. Let $\mathfrak{p}$ be a place of k outside S. Assume that $x \notin f^{-1}(s)$, if s is in the support of Δ_f. If the $\mathfrak{p}$ reduction of x belongs to the $\mathfrak{p}$ reduction $(F_h)_{\mathfrak{p}}$of some component F_h of $f^{-1}(s)$, let t_h be the multiplicity of F_h in $f^*(s)$. Then $t_h \geq m_s(f)$, by definition of $m_s(f)$. On the other hand, the arithmetic intersection number of $f(x)_{\mathfrak{p}}$ with $(s)_{\mathfrak{p}}$ is the product of t_h with the arithmetic intersection number of $(x)_{\mathfrak{p}}$ with $(F_h)_{\mathfrak{p}}$, and is thus a multiple of t_h, and thus at least $m_s(f)$. $\qquad\square$

Remark 5.8. *In the quasi-projective case, Corvaja–Zannier have given the first example of simply connected quasi-projective smooth surfaces with a non-Zariski dense set of integral points over any number field (see [24]). Their proof uses Schmidts' subspace theorem. Their examples (blow-ups of $\mathbb{P}_2$ on union of 4 lines, removing the strict transforms, not the total transform, of these lines, which permit to realise the simple-connectedness of the complement) are similar to the ones given in §8.7 below, using infinite multiplicities, instead of finite ones.*

Example 5.9. *We now give some examples of fibrations $f : S \to \mathbb{P}^1$ with orbifold base of general type, and S simply connected. Different examples where initially constructed in [13]. They are quite complicated, with fibres of high genus $g = 13$ (but relatively simple multiple fibres consisting of 5 rational curves meeting transversally in a single point, their multiplicities being $(2, 2, 2, 3, 3)$). In [45], L. Stoppino used former work of Namikawa–Ueno [38] to give much simpler explicit examples with fibres of (minimal possible) genus 2. In these examples, as in the examples produced in [13], the 'non-classical' multiple fibres have 'inf'-multiplicity 2. We describe here the simplest example of [41], to which we refer for more details, and in particular the (quite involved) description of the multiple fibres, which are trees of rational curves (and so are simply connected) (Figure 4).*

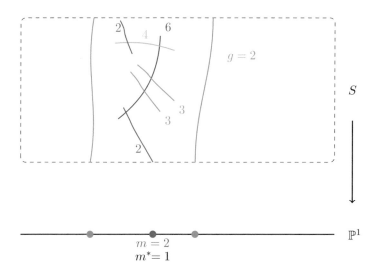

Fig. 4 A non-classical double fibre of genus 2

- Take the ramified 2-cover $\pi_0 : S_0 \to \mathbb{P}^1 \times \mathbb{P}^1$ of equation $y^2 = t(x^6 + t.x^3 + t^2)$ (with affine coordinates (t, x) on $\mathbb{P}^1 \times \mathbb{P}^1$). Resolve by $r : S \to S_0$ the singularities of S_0 to get an isotrivial fibration $f = q \circ \pi_0 \circ r : S \to \mathbb{P}^1$, where $q : \mathbb{P}^1 \times \mathbb{P}^1 \to \mathbb{P}^1$ is the first projection which sends (t, x) to t. The fibration has then smooth fibres of genus 2 and two simply connected fibres of 'inf'-multiplicity 2, over $t = 0, \infty$. More precisely, each of these fibres consists of 6 rational curves building a tree, their multiplicities being $(2, 6, 3, 3, 4, 2)$.
- The surface S so constructed is defined over $\mathbb{Q}$, and is rational. It is thus potentially dense. In order to get a fibration of general type, it is sufficient to make a generic cyclic base-change $u : \mathbb{P}^1 \to \mathbb{P}^1$ of degree $d \geq 3$ over the base of q, and to normalise. The resulting surface S' is then of general type, simply connected, defined over $\mathbb{Q}$, and the resulting fibration $f' : S' \to \mathbb{P}^1$ has $2d \geq 6$ 'non-classical' double fibres, and no 'classical' multiple fibre. The 'orbifold Mordell Conjecture' then implies that it is not potentially dense. This would provide the first non-potentially dense simply connected smooth surface defined over a number field.

5.4 Link with Hyperbolicity

A. In [18], Corollary 4, p. 208, based on Nevanlinna's second main theorem with truncation at level 1, it is shown that any entire curve $h : \mathbb{C} \to S$ has its image contained in a rational or elliptic component of some singular fibre of f, if S has a fibration on a curve such that its orbifold base is of general type. This is the exact hyperbolicity analog of the conjectural arithmetic statement of non-potential density.

B. It is shown in [10] that a complex projective surface S not of general type admits a holomorphic map $h : \mathbb{C}^2 \to S$ with dense image, if and only if S is weakly special (with the possible exception of non-elliptic and non-Kummer $K3$ surfaces). This leads to conjecture the equivalence of the following three properties:

1. S is weakly special
2. S admit a dense entire curve[20]
3. $S(k)$ is potentially dense (if S is defined over a number field k).

- **Insufficiencies of the 'weak specialness':** We shall see in §8.7 that from dimension 3 on, the property of 'weak specialness' is too weak to imply property 2 (and conjecturally also property 3) above. We shall replace it by the 'specialness' property, defined below.

6   Decomposition of Arbitrary X's

We have previously defined 3 classes of 'primitive' manifolds: those with $\kappa^+ = -\infty$, $\kappa = 0$, or with $\kappa = dim$ (i.e.: of general type), respective generalisations of rational, elliptic, and hyperbolic curves. We now decompose any higher dimensional X into 'twisted products' of manifolds of these 3 primitive types by a suitable sequence of canonical and (birationally) functorial fibrations. We first describe a decomposition by a canonically defined sequence of fibrations, which is however conditional in the uniruledness Conjecture 4.8. We next define a second decomposition by one single fibration which is unconditional and also birationally functorial (while the steps of the first are not). The abutments of both decompositions however agree (the first one existing only conditionally).

6.1   The $(J \circ r)^n$ Decomposition

Let X be arbitrarily be given, and let $r : X \to R_X$ be its 'rational quotient'. Assuming the 'uniruledness Conjecture' 4.8, one gets that $\kappa(R_X) \geq 0$, so that the Iitaka–Moishezon fibration $J : R_X \to J(R_X)$ is always birationally

---

[20]We **do not** conjecture the existence of a Zariski dense map $h : \mathbb{C}^2 \to S$ for any non-elliptic and non-Kummer $K3$ surface S.

defined.[21] The composite map: $J \circ r : X \to J(R_X)$ is thus defined for every X, and can be iterated. The following properties are easy:

1. $X = J(R_X)$ if and only if X is of general type. Thus:
2. Defining inductively the k-th iterate $(J \circ r)^k : X \to X_k = (J(R_{X_{k-1}}))$, with $X_0 := X$, we see that $d_k := dim(X_k)$ is decreasing. Next (by 1.), $d_{k+1} = d_k$ if and only if X_k is of general type.
3. In particular, $(J \circ r)^n : X \to X_n$ is a fibration over a manifold X_n of general type (possibly a point), with fibres towers of fibrations with fibres alternatively either rationally connected, or with $\kappa = 0$.

We call this map $c : (J \circ r)^n : X \to X_n$ the 'weak core map' of X. It has been constructed conditionally on Conjecture 4.8. We shall now give a (more general) unconditional construction.

The 'weak core map' however fails to be preserved even by finite étale covers (see Example 6.15). This is due to neglecting the multiple fibres of the fibrations J. This will be corrected later (see §8.1) by introducing 'orbifold bases' of fibrations.

The relevance to potential density will be explained in §8.4.

6.2   The $C_{n,m}$ Conjecture

Let $f : X \to Z$ be a fibration between complex projective manifolds, denote by X_z its generic (smooth) fibre.

Proposition 6.1 ('Easy Addition'). $\kappa(X) \le \kappa(X_z) + dim(Z)$.[22]

The following is a central conjecture of classification:

Conjecture 6.2 ('$C_{n,m}$-conjecture'). $\kappa(X) \ge \kappa(X_z) + \kappa(Z)$.

Theorem 6.3 (E. Viehweg). $\kappa(X) = \kappa(X_z) + dim(Z)$ when Z is of general type. In particular, if X_z is of general type, so is X.

We shall formulate an 'orbifold' version of this conjecture in §7.4. This orbifold version is known also when the 'orbifold base' of f is of general type.

Corollary 6.4. If $\kappa(X) = 0$, there is no rational fibration $f : X \to Z$, with Z of general type and $dim(Z) > 0$.

Indeed: $0 = \kappa(X) \ge \kappa(X_z) + dim(Z) > \kappa(X_z) \ge 0$ (the last inequality is easy).

---

[21] Note however that these maps are all almost holomorphic, that is: their indeterminacy loci do not dominate their images.

[22] This inequality is true for any line bundle, not only K_X.

6.3   A Decomposition Criterion

Let $\mathcal{C}$ be a class of complex (connected) projective manifolds, stable by birational equivalence. We denote by $\mathcal{C}^\perp$ the class of all (complex projective) manifolds X which do not admit any dominant rational fibration onto any $Z \in \mathcal{C}$. We call $\mathcal{C}^\perp$ the 'Kernel' of $\mathcal{C}$.

Definition 6.5. *We say that the class $\mathcal{C}$ is 'stable' if the following two properties E1 and E2 hold true.*

(E1) If $f : X \to Z$ is a surjective regular fibration with general (smooth) fibre $X_z \in \mathcal{C}$, and $Z \in \mathcal{C}$. Then $X \in \mathcal{C}$.

(E2) If a connected projective manifold Y is equipped with two (surjective) fibrations $h : Y \to Z, g : Y \to T$ such that $h : Y_t \to h(Y_t) \subset Z$ is birational for $t \in T$ generic, and if $Z \in \mathcal{C}$, then $Y_t \in \mathcal{C}$ for $t \in T$ generic. We abbreviate this property by saying that the general member of a Z-covering family of varieties is in $\mathcal{C}$ if $Z \in \mathcal{C}$.

Theorem 6.6. *Assume that $\mathcal{C}$ is stable. Then, for any complex projective X, there exists a unique fibration $\gamma_X : X \to C_X$ such that:*

1. *its general fibre $X_z \in \mathcal{C}^\perp$.*
2. *$C_X \in \mathcal{C}$.*

If X is defined over k, so is γ_X.

We call γ_X the $\mathcal{C}$-splitting of X.

The $\mathcal{C}$-splitting is functorial: any rational dominant fibration $f : X \to Z$ induces a unique rational fibration $\gamma_f : C_X \to C_Z$ such that $\gamma_Z \circ f = \gamma_f \circ \gamma_X$.

Proof. We proceed by induction on $n := dim(X)$, the assertion being true for $n = 0$ (in which case $X \in \mathcal{C} \cap \mathcal{C}^\perp$, by convention). Let $g : X \to Z$ be a rational fibration with $Z \in \mathcal{C}$, $d := dim(Z)$ being maximal with this property. If $d = 0$, we are finished since then $X \in \mathcal{C}^\perp$, by definition. Otherwise: $(n - d) > 0$, and so the proposition holds for X_z. By uniqueness of the map γ for X_z, Chow space theory shows the existence of fibration $\gamma_{X/Z} : X \to Y$ and $h : Y \to Z$ such that $h \circ \gamma_{X/Z} = g$, and such that the restriction $\gamma_z : X_z \to Y_z$ is γ_{X_z} (already inductively existing) for X_z, $z \in Z$ general. By property (E1), since we have: $Y_z \in \mathcal{C}$, and $Z \in \mathcal{C}$, we have $Y \in \mathcal{C}$. The maximality of $dim(Z)$ implies that $Y = Z$, the fibres X_z of g thus coincide with those of $\gamma_{X/Z}$, which are in $\mathcal{C}^\perp$. The map g thus enjoys the two claimed properties.

The uniqueness follows from (E2). Let indeed $k : X \to Y$ be a second fibration enjoying properties 1 and 2, with $dim(Y)$ maximal, thus $dim(Y) = dim(Z) = d$. Let $y \in Y$ be general, $X_y := k^{-1}(y)$, and $Z_y := g(X_y) \subset Z$. By property E2, $Z_y \in \mathcal{C}$. Since $X_y \in \mathcal{C}^\perp$, Z_y is a point. There thus exists a map $h : Y \to Z$ such that $g = h \circ k$. Since $dim(Y) = dim(Z)$, we have $Z = Y, g = k$ (birationally).

The functoriality follows from a similar argument: the fibres of γ_X, which are in $\mathcal{C}^\perp$, are mapped by $\gamma_Z \circ f$ to a covering family of subvarieties of $C_Z \in \mathcal{C}$;

they are thus points, by E2. This implies the claimed factorisation of $\gamma_Z \circ f$ through γ_X.   □

Remark 6.7. *The existence of γ_X follows from E1, the uniqueness from E2. The proof shows that the fibres of γ_X are the largest subvarieties of X in $\mathcal{C}^\perp$, and that $Z = C_X$ dominates any member of $\mathcal{C}$ dominated by X.*

Denoting with $\mathcal{P}$ the class of all (complex, connected) projective manifolds, it is tempting to write the content of the $\mathcal{C}$-splitting in the form of a short exact sequence $[\mathcal{C}^\perp] \to \mathcal{P} \to \mathcal{C}$, to mean that any $X \in \mathcal{P}$ is in a unique way an 'extension' of an element of $\mathcal{C}$ by a (deformation class) of $\mathcal{C}^\perp$, a fibration being seen as an 'extension' of its base by its general fibre.

We shall now apply this criterion in two situations.

6.4   The Weak Core Map

Proposition 6.8. *Let $\mathcal{C} := \mathcal{K}^{max}$ be the class of manifolds of general type. It is stable, i.e. enjoys the properties E1,E2 of Theorem 6.6.*

Proof. Property E1 follows directly from Theorem 6.3. Property E2 follows from the 'easy addition' property (6.1).   □

Let now $\mathcal{S}^w$ be the smallest class of complex projective manifolds containing those with $\kappa = 0, \kappa^+ = -\infty$, and stable by 'extensions' (i.e.: such that $X \in \mathcal{S}^w$ whenever there is a fibration $f : X \to Z$ with $Z \in \mathcal{S}^w$ and $X_z \in \mathcal{S}^w$).

Lemma 6.9. $\mathcal{S}^w \subset (\mathcal{K}^{max})^\perp$, *the class of manifolds not dominating any positive-dimensional manifold of general type.*

Proof. $(\mathcal{K}^{max})^\perp$ is clearly stable by extensions, and contains the manifolds with $\kappa^+ = -\infty$, by definition. It also contains those with $\kappa = 0$, by Corollary 6.4.   □

Corollary 6.10. *Let $c_X : X_n \to C_X$ be the 'weak core map' of an arbitrary n-dimensional $X = X_n$. Assume Conjecture 4.8, so that the map $(J \circ r)^n$ is defined. Then $c_X = (J \circ r)^n$, and $\mathcal{S}^w = (\mathcal{K}^{max})^\perp$.*

The weak core map is functorial: any fibration $f : X \to Z$ induces a (rational, dominant) map $c_f : C_X \to C_Z$.

Proof. Both maps have a base in $\mathcal{K}^{max}$ and general fibres in $(\mathcal{K}^{max})^\perp$, they thus coincide by uniqueness of the weak core. Applying this to any $X \in (\mathcal{K}^{max})^\perp$ shows that $X \in \mathcal{S}^w$. The functoriality is a special case of Theorem 6.6.   □

Remark 6.11. *Let us stress that the weak core map is defined unconditionally, contrary to $(J \circ r)^n$. Also, the map J is not functorial, and so the functoriality of $(J \circ r)^n$ does not follow directly from its construction.*

6.5   The κ-Rational Quotient

We show here how to construct the rational quotient map $r_X : X \to R_X$ without mentioning rational curves (but assuming $C_{n,m}$ and 4.8).

Let $\mathcal{K}_{\geq 0}$ be the class of projective manifolds X with $\kappa(X) \geq 0$. The class $(\mathcal{K}_{\geq 0})^{\perp}$ thus consists, by definition, of all manifolds with $\kappa^+ = -\infty$.

Lemma 6.12. *Assume Conjecture $C_{n,m}$. The class $\mathcal{K}_{\geq 0}$ then enjoys properties E1, E2 of Theorem 6.6.*

Proof. Property E1 follows directly from $C_{n,m}$, property E2 is shown as for the class $\mathcal{K}^{max}$ (by 'easy addition'). $\square$

Applying Theorem 6.6 and the same argument as in Corollary 6.10, we get:

Proposition 6.13. *Assume conjecture $C_{n,m}$. For any X, there is a unique fibration $\rho_X : X \to R(X)$ such that:*

1. *$\kappa^+(X_z) = -\infty$ for its general fibre X_z, and:*
2. *$\kappa(R(X)) \geq 0$.*

We call ρ_X the 'κ-rational quotient' of X.

Remark 6.14. *We cannot however here show that ρ_X coincides with the 'true' rational quotient $r_X : X \to R_X$, because we do not know whether all manifolds with $\kappa^+ = -\infty$ are rationally connected. We can only show (assuming $C_{n,m}$) that we have a factorisation $\varphi : R_X \to R(X)$ such that $\rho_X = \varphi \circ r_X$. The fibres of ρ_X are indeed not uniruled with $\kappa^+ = -\infty$. The Conjecture 4.8 thus implies that $\rho_X = r_X$.*

6.6   The Weak Core Is Not Preserved by étale Covers

This is shown by the following (simplest possible) example. This implies (among other things) that it is inappropriate for the description of $X(k')$. We shall replace it later with the 'true' core map, which takes into account the multiple fibres of fibrations, and is preserved by finite étale covers.

Example 6.15. *Let C be a hyperelliptic curve of genus $g \geq 2$, $h : C \to \mathbb{P}^1 := C/ <\tau>$ be the double cover induced by the hyperelliptic involution τ of C. Let E be an elliptic curve, and t a translation of order 2 on E. Let $S' := E \times C$, and $\iota := t \times \tau$ the fixed-point free involution on S'. Let $u : S' \to S$ be the quotient by ι.*

The projections $J : S \to \mathbb{P}^1 := C/ <\tau>$ (resp. $J' : S' \to C$) are the Iitaka fibrations of S, S', and $J \circ u = h \circ J'$. The weak core map $c_S := (J \circ r)^2 : S \to C_S$ of S maps S to a point, but $c_{S'} = (J \circ r)^2 : S' \to C_{S'} = C$ is simply the fibration $J' : S' \to C$, since $g(C') \geq 2$. The natural map $c_u : C_{S'} \to C_S$ thus does not preserve the dimension.

The surface S has an 'orbifold quotient' of general type, revealed on its double cover S', but may be seen directly on S if one considers the 'orbifold base' of J, which is indeed of general type.

The 'orbifold base' of J consists of the base $B = \mathbb{P}^1 := C/ <\tau>$ of J, in which the points p_j over which the fibre is multiple (here double) are equipped with the multiplicity (2, here) of the corresponding fibre. The points p_j are here obviously the $2g + 2$ points images of the hyperelliptic points of C. We obtain thus the 'orbifold base' (B, Δ) with $\Delta = \sum_{j=1}^{j=2g+2}(1 - \frac{1}{2}).\{j\}$, in such a way that $h^*(K_B + \Delta) = K_C$, by the ramification formula. Which indeed shows that the orbifold curve (B, Δ) is of general type.

A second way to see this quotient of general type is to consider not only the line bundle $J^*(K_{\mathbb{P}^1})$, but its saturation L_J in Ω_S^1, which has $\kappa = 1$ (See Example 7.8). As we shall see in Theorem 7.6, the two aspects (orbifold base, saturation of $f^*(K_S)$) actually coincide.

- The failure of the weak core map will be corrected by the introduction of 'orbifold base' of fibrations, as in the preceding example. One has then, however, to work in the larger category of 'orbifold pairs'. Even if one only wants to decompose projective manifolds without orbifold structures, these will appear, as in the preceding example, in general when considering the Moishezon–Iitaka fibration. For surfaces, this can be dealt with by suitable étale covers, but no longer in dimension 3 or more (see Example in §8.7 below).

7 Special Manifolds

7.1 Definition, First Examples and Properties

From now on, X_n is a smooth and connected complex projective manifold[23] of dimension n. Our exposition here is very sketchy. Details can be found in [11] and [13].

---

[23]Or compact Kähler, more generally.

Definition 7.1. X is special[24] if $\kappa(X, L) < p$ for any line bundle $L \subset \Omega_X^p$, and for any $p > 0$.

Example 7.2.

1. If X is a curve, the unique $p > 0$ to consider is $p = 1$, and so $L = K_X = \Omega_X^1$. A curve is thus special if either rational or elliptic.
2. If X is rationally connected, it is special (since it satisfies the much stronger vanishing: $h^0(X, \otimes^m \Omega_X^1) = 0, \forall m > 0$). This generalises rational curves.
3. If $\kappa(X) = 0$, X is special. (See 7.11 below). This generalises elliptic curves. Much more is expected to be true: $\kappa(X, L) \leq 0$ for any $L \subset \otimes^m(\Omega_X^1), \forall m > 0$, L of rank 1, if $\kappa(X) = 0$.
4. If X is of general type, it is not special, using $L = K_X = \Omega_X^n$.
5. More generally: if there is a fibration $f : X \dashrightarrow Z_p$, with $p = dim(Z) > 0$, and if Z is of general type, then X is not special (take $L = f^*(K_Z) = f^*(\Omega_Z^p) \subset \Omega_X^p$), then $\kappa(X, L) = \kappa(Z, K_Z) = p$, contradicting the specialness of X).
6. Being special is preserved by birational equivalence and finite étale covers. Thus 'special' implies 'weakly special'. The converse holds for curves and surfaces, but no longer for threefolds (see §8.7 below). See Theorem 7.4 for a characterisation of specialness in this direction.
7. The Kodaira dimension does not characterise (non-)specialness (except for $k = 0, n$): if $n \geq 1, k \in \{-\infty, 1, \ldots, (n-1)\}$, there exist both special and non-special manifolds with $dim = n$, $\kappa = k$.

   Non-special examples are given by obvious products.

   'Special' examples are given, if $k \geq 0$, by smooth divisors X in $\mathbb{P}^{n-k+1} \times \mathbb{P}^k$ of bidegree $(n - k + 2, k + 2)$.
8. If $h : \mathbb{C}^n \dashrightarrow X$ is a meromorphic (possibly transcendental) non-degenerate map, X is special. 'Non-degenerate' means that it has non-vanishing Jacobian generically. This is an orbifold version of a result of Kobayashi–Ochiai.
9. If S is a smooth projective weakly special surface, it is special. When $\kappa(S) = -\infty, 0$, it is easy from the classification and 7.11. When $\kappa(S) = 1$, this follows from Lemma 5.5.

   Special surfaces thus have a very simple characterisation: $\kappa(S) \leq 1$, and $\tilde{q}(S) \leq 2$. Specialness is preserved by deformation (and even diffeomorphism) for surfaces.

   We conjecture that specialness is preserved by deformations and specialisation of smooth (compact Kähler) manifolds.

---

[24]The name is inspired from Moishezon's definition of 'general type', and supposed to convey the idea that these manifolds are in a precise sense 'antithetic' to those of general type, as will be amply illustrated below.

Remark 7.3. *One could replace the condition $\kappa(X,L) < p$ by the stronger condition $\nu(X,L) < p$ for any rank-one $L \subset \Omega_X^p$, where $\nu(X,L) \geq \kappa(X,L)$ is the numerical dimension of L. It is an open question whether one obtains the same class of manifolds. It has been shown by C. Mourougane and S. Boucksom that $\nu(X,L) \leq p, \forall p, L, X$, strengthening Bogomolov's theorem. Notice however that it may happen that $\kappa(X,L) = -\infty$ if $\nu(X,L) = p$ for $L \subset \Omega_X^p$, as observed by Brunella on surfaces covered by the bidisk. The situation is similar to the one considered in the next §7.2.*

7.2 The Birational Stability of the Cotangent Bundle

Let X be a complex connected projective manifold.

The canonical algebra $K(X) := \oplus_{m \geq 0} H^0(X, m.K_X)$, and so also $\kappa(X)$ are not (birationally) functorial in the sense that a dominant rational map $f : X \to Z$ does not induce any natural (injective) morphism of algebras $f^* : K(Z) \to K(X)$, or inequality $\kappa(X) \geq \kappa(Z)$ when $dim(X) > dim(Z)$.

The 'cotangent algebra' $\Omega(X) := \oplus_{m \geq 0} H^0(X, \otimes^m \Omega_X^1)$ is, by contrast, obviously functorial, as well as $\kappa^{++}(X) := max\{\kappa(X,L)|L \subset (\otimes^m \Omega_X^1)$ coherent of rank 1, $\forall m > 0\}$. We obviously have: $\kappa^{++} \geq \kappa^+ \geq \kappa$, where κ^+ is defined in 4.13, and also obviously functorial.

One can show[25] that $\kappa^{++}(X) = \kappa^{++}(R_X)$, where $r_X : X \to R_X$ is the rational quotient of X (the same holds easily for κ^+). This permits to reduce the study of κ^{++} to the case when K_X is pseudo-effective (i.e.: X not uniruled). Assuming Conjecture 4.8, one even reduces the study of κ^{++} to the case when $\kappa(X) \geq 0$.

A stronger version is obtained by replacing $\kappa(X,L)$ by its 'numerical' version $\nu(X,L) \in \{-\infty, 0, 1, \ldots, dim(X)\}$ (as defined by N. Nakayama):

$$\nu(X,L) := inf\{k \in \mathbb{Z}|\overline{lim}_{m \to +\infty}\left(\frac{h^0(X, mL + A)}{m^k}\right) > 0\},$$

where A is a sufficiently ample line bundle on X, for example: $K_X + (2n + 2).H, H$ any ample line bundle on X. We have: $\nu(X,L) \geq \kappa(X,L)$ for any line bundle L on X.

We defined (in [17]) $\nu^+(X)$ just as $\kappa^{++}(X)$, just replacing there κ by ν, and showed that $\nu^+(X) = \nu(X, K_X)$ when K_X is pseudo-effective. This is the 'birational stability' of the cotangent bundle: the positivity of its line subsheaves is controlled by the canonical bundle (and similarly for its tensor powers) when X is not uniruled.

---

[25] Using arguments in [14].

If we now assume the conjecture that $\nu(X, K_X) = \kappa(X)$ for any X such that K_X is pseudo-effective, we obtain in this case: $\nu^+(X) = \nu(X) = \kappa^{++}(X) = \kappa(X)$, and $\nu^+(X) = \kappa^{++}(X) = \kappa(R_X)$ for any X. A particularly important case is when $\kappa(X) = 0$, in which case the conjecture is that $\nu(X) = 0$, implying that $\nu^+(X) = \kappa^{++}(X) = 0$, a statement considerably stronger than the proved specialness.

Another consequence of the conjecture $\nu(X) = \kappa(X)$ for X non-uniruled were that $\kappa(X) \geq \kappa(Z)$ for any dominant rational map $f : X \to Z$ between non-uniruled manifolds: apply the equality $\nu^{++} = \nu$ of [17] to X and Z together with the equalities $\nu^{++} = \kappa^+$ implied by the conjecture, and the obvious inequality $\kappa^+(X) \geq \kappa^+(Z)$.

Similar results and conjectures hold for smooth orbifold pairs (X, Δ) as well (see [17], [14]). When Δ is reduced, one just has to consider $\Omega_X^1(Log(\Delta))$ in place of Ω_X^1.

Let us finally observe that the rate of growth of the spaces of sections of the symmetric powers of the cotangent bundle is in general unrelated to the 'Kodaira' dimension, as shown by the smooth hypersurfaces of the projective spaces (since their cotangent bundles are known to be non-pseudo-effective).

7.3   Specialness as Opposed to Base Orbifolds of General Type

The following is due to F. Bogomolov:

Theorem 7.4 ([4]). *Let X be projective smooth, and $L \subset \Omega_X^p$ a line bundle. Then:*

1. $\kappa(X, L) \leq p$.
2. *If $\kappa(X, L) = p$, there exists a fibration $f : X \dashrightarrow Z_p$ such that $L = f^*(K_Z)$ generically[26] on X.*

Line bundles as in 2) are called 'Bogomolov sheaves'.

Remark 7.5.

1. *Bogomolov sheaves are thus 'maximally big' line subsheaves of $\Omega_X^\bullet$. And X is special if $\Omega_X^\bullet$ does not contain such maximally big line subsheaves.*
2. *There are many examples of Bogomolov sheaves $L = f^*(K_Z) \subset \Omega_X^\bullet$, generically over Z, and such that $\kappa(Z) = -\infty$. This is due to the multiple fibres of f, encoded in the 'orbifold base of f'. Hence the geometric characterisation of 'specialness' is given in 7.7.*

---

[26]I.e.: on a nonempty Zariski open subset.

Theorem 7.6. *Let $f : X \to Z_p = Z$ be a fibration.[27] Let $L := f^*(K_Z)^{sat} \subset \Omega^p_X$ be the saturation[28] of $f^*(K_Z)$ in Ω^p_X.*
   Then: $\kappa(X, L) = \kappa(Z, K_Z + \Delta_f)$.[29]

Thus Δ_f encodes the difference between $f^*(K_Z)$ and its saturation: $\kappa(X, f^*(K_Z)^{sat}) - \kappa(X, f^*(K_Z)) = \kappa(Z, K_Z + \Delta_f) - \kappa(Z, K_Z)$. **This fails** for the 'classical' orbifold base of f, and is the main reason for the introduction of this 'non-classical' notion.

We thus get a geometric characterisation of 'specialness':

Corollary 7.7. *X is special if and only if, for any fibration $f : X \dashrightarrow Z$, the orbifold base of any of its 'neat models' is **not** of general type.*

Of course, this implies that (but turns out to be much stronger than) the non-existence of fibrations $f : X \dashrightarrow Z$ with Z of general type.

Example 7.8. *Let us give the concrete meaning of the saturation in a simple example: let $f : S \to C$ be a fibration of the surface S onto the curve C, with an irreducible smooth fibre $F = f^{-1}(s)$ of multiplicity $t > 1$, thus given in local analytic coordinates (x, y) on S by: $f(x, y) = u := x^t$. Then $\Delta_f = (1 - \frac{1}{t}).\{s\} + \ldots$ near s in C.*
   Thus $f^*(K_C) = f^*(du) = t.x^{t-1}.dx$ near s, while: $f^*(K_Z + \Delta_f) = f^*\left(\frac{du}{u^{(1-\frac{1}{t})}}\right) = t.dx$, which is indeed the saturation of $f^*(du)$ in Ω^1_S.

7.4   The Orbifold Version of the $C_{n,m}$ Conjecture

Conjecture 7.9 (Conjecture $C^{orb}_{n,m}$). *Let $f : X \to Z$ be a fibration, with generic fibre X_z. Then $\kappa(X) \geq \kappa(X_z) + \kappa(Z, \Delta_f)$.*

Without Δ_f, this conjecture is due to S. Iitaka. More general versions[30] exist. The special case where (Z, Δ_f) is of general type is known:

Theorem 7.10 (Viehweg). *In the situation of 7.9, if $\kappa(Z, \Delta_f) = dim(Z)$, we have: $\kappa(X) = \kappa(X_z) + dim(Z)$.*

---

[27]Recall that we sometimes indicate with a subscript the dimension of a complex manifold, writing thus X_n, Z_p. Here Z is thus p-dimensional.

[28]This is the largest subsheaf of Ω^p_X containing $f^*(K_Z)$, generically equal to it.

[29](Z, Δ_f) is here the (non-classical) orbifold base of f on any suitable birational 'neat model' of f.

[30]One can, for example, consider an orbifold pair (X, Δ) instead of X, and increase accordingly the orbifold base divisor.

This result is due to Viehweg when $\Delta_f = 0$. The proof extends with some adaptations to cover this more general case. The range of applicability is considerably extended by the adjunction of the orbifold term.

Corollary 7.11. X *is special if* $\kappa(X) = 0$.

Proof. $0 = \kappa(X) = \kappa(X_z) + dim(Z) \geq dim(Z)$ since $\kappa(X_z) \geq 0$.   □

This is one of the basic examples of special manifolds, generalising elliptic curves.

8   The Core Map

8.1   A Splitting Criterion

We briefly explain that one can extend Theorem 6.6 to the orbifold category (Figure 5).

Let $\mathcal{C}$ be a class of (smooth projective) orbifold pairs.[31] We define the class $\mathcal{C}^\perp$ of smooth orbifolds admitting no dominant fibration such that a neat model of its orbifold base belongs to $\mathcal{C}$.

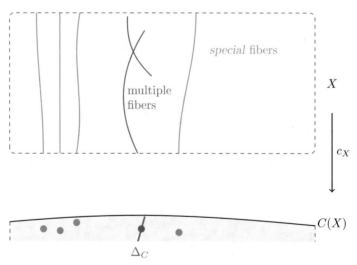

orbifold base $\big(C(X), \Delta_C\big)$ of general type

Fig. 5 The core map

---

[31] Also stable by birational equivalence (in a suitable sense, not defined here).

If we assume that the class $\mathcal{C}$ possesses the properties E1, E2, then we have a $\mathcal{C}$-splitting theorem entirely similar to 6.6 with the same proof.

We shall apply this to the following 2 cases, already considered when $\Delta = 0$. For them, property E2 is elementary, proved as when $\Delta = 0$.

1. $\mathcal{C}$ is the class of orbifold pairs of general type. Property E1 follows from the orbifold version 6.3 of Viehweg's Theorem 6.3. This leads to the 'core map' described in Theorem 8.1 below.
2. $\mathcal{C}$ is the class of orbifold pairs with $\kappa \geq 0$. Property E1 is conditional in $C_{n,m}^{orb}$. This gives the 'κ-rational quotient' of Proposition 8.7.

8.2   The Core Map

Theorem 8.1. *For any X, there is a unique fibration $c_X : X \to C_X$, called the 'core of X', such that:*

1. *Its general fibres are special.*
2. *Its 'orbifold base' (C_X, Δ_{c_X}) is of general type.*

Functoriality: any dominant $g : Y \dashrightarrow X$ induces $c_g : C_Y \to C_X$ with $c_X \circ g = c_g \circ c_Y$.

If X is defined over k, so is c_X by its uniqueness.

The proof works by induction on $dim(X)$, using Theorem 7.10, in a way entirely similar to the proof of Theorem 6.6.

- We use the same notation $c_X : X \to C_X$ for both the core map and the weak core map. From now on we shall **only** consider the 'true' core map (of Theorem 8.1), this should thus not lead to any confusion.

Let us first note that the 'true' core map corrects the failure of its weak version:

Corollary 8.2. *If $u : X' \to X$ is étale finite, $c_u : C_{X'} \to C_X$ is generically finite, (ramified, but orbifold-étale).*
In particular: if X is special, so is X'.

Indeed: we can assume that X' is Galois over X, by uniqueness of the core map of X', it is defined by a Bogomolov subsheaf which is preserved by the Galois group, and thus descends to X as a Bogomolov subsheaf, since X' is étale over X.

Corollary 8.3. *If X is special, it is weakly special.*

Indeed: any finite étale cover X' of X is still special, and thus does not fibre over any positive-dimensional manifold of general type.

Example 7.2.9 shows that for surfaces, these two properties are equivalent, this is however no longer true in dimensions 3 or more (see §8.7).

8.3 The Conjectures for Arbitrary Projective Manifolds

We formulate here our main conjecture without using orbifold notions. Its solution (if any) will however require the orbifold version in §8.6 below.

Conjecture 8.4.

1. If X is special, $\pi_1(X)$ is almost abelian.
2. Being special is preserved by deformations and specialisations of smooth manifolds.
3. X is special if and only if it contains a dense entire curve.
4. Let $c_X : X \to C(X)$ its core map. There exists a complex projective subvariety $W \subsetneq C_X$ such that any entire curve $h : \mathbb{C} \to X$ has image either contained in $c_X^{-1}(W)$, or in some fibre of c_X.

    If X is defined over a number field k:
5. $X(k)$ is potentially dense if and only if X is special.
6. Let $c_X : X \to C(X)$ its core map. There exists a complex projective subvariety $W \subsetneq C_X$ such that, for any finite extension k'/k, $c_X(X(k')) \cap U$, is finite, $U := (C_X \setminus W)$. The smallest such $W \subset C_X$ has to be defined over k. Let $U := X \setminus W$.

Moreover, there exists k' such that for any $k'' \supset k'$, $X(k'')$ is Zariski dense in each fibre of c_X lying over $c_X(X(k'')) \cap U$.

8.4 The $c = (j \circ r)^n$ Decomposition of the Core

The 'orbifold version' of the 'decomposition' $(J \circ r)^n$ of the 'weak core map' mentioned in Remark 6.1 coincides with the core. We give a very succinct description, here.

Theorem 8.5. Let $c_X : X \to C_X$ be the core map of a smooth connected projective manifold of dimension n. Assume the orbifold version[32] $C_{n,m}^{orb}$ of conjecture $C_{n,m}$ given in 7.9. Then $c_X = (j \circ r)^n$, where r, j are the fibrations defined below.

Let (X, Δ) be a smooth orbifold pair.

- **The orbifold Iitaka fibration** j: This is just the Iitaka fibration of the $\mathbb{Q}$-line bundle $(K_X + \Delta)$ on X if $\kappa(X, K_X + \Delta) \geq 0$. It induces a fibration $j : (X, \Delta) \to (J, \Delta_{j,\Delta})$ with $dim(J) = \kappa(X, \Delta)$ and $\kappa(X_z, \Delta_{|X_z}) = 0$, if X_z is the generic smooth fibre of j.

---

[32] One needs the version for an orbifold pair (X, Δ), not just for X.

- **The 'κ-rational quotient' r :**

Definition 8.6. *We say that $\kappa^+(X, \Delta) = -\infty$ if some/any neat orbifold base $(Z, \Delta_{f,\Delta})$ of any fibration $f : (X, \Delta) \dashrightarrow Z$ has $\kappa(Z, \Delta_{f,\Delta}) = -\infty$.*

When $\Delta = 0$, this is equivalent (under the 'uniruledness conjecture') to X being rationally connected. We conjecture (see next subsection) that this is still true for orbifolds (with the usual definition of rational connectedness, replacing rational curves by 'orbifold (or Δ)-rational curves', as defined in Definition 2.5). Similarly to 4.12, we have:

Proposition 8.7. *Assume $C_{n,m}^{orb}$ as stated in 7.9. Any smooth (X, Δ) admits a unique fibration $r : (X, \Delta) \to (R, \Delta_{r,\Delta})$ such that:*

1. *$\kappa^+(X_z, \Delta_{|X_z}) = -\infty$ for the generic fibre X_z of r.*
2. *$\kappa(R, \Delta_{r,\Delta}) \geq 0$.*
   r is called the 'κ-rational quotient' of (X, Δ).

Corollary 8.8. *X is special if and only if it has a birational model which is a tower of neat fibrations with orbifold fibres having either $\kappa^+ = -\infty$, or $\kappa = 0$.*

Notice that 'orbifold divisors' will in general appear when encoding multiple fibres, as shown by Example 6.15.

Remark 8.9. *It is sometimes said that the 'building blocks' for the construction of arbitrary manifolds are (terminal or canonical) varieties with canonical bundles either anti-ample (i.e.: Fano), or numerically trivial, or ample. The birational version being: rationally connected, $\kappa = 0$, or of general type, respectively. We show here that these 'building blocks' need to be chosen in the larger category of orbifold pairs.*

8.5 Rationally Connected Orbifolds and $\kappa^+ = -\infty$

Definition 8.10. *Let (X, Δ) be a smooth orbifold pair, with X complex projective. We say that (X, Δ) is rationally connected if any two generic points of X are contained in an orbifold rational curve[33] $h : \mathbb{P}^1 \to (X, \Delta)$.*

Remark 8.11. *One may expect that, just as when $\Delta = 0$, the above properties are equivalent to the 'chain-connected' version, and also to the fact that any finite subset of $X \setminus \Delta$ is contained in a single irreducible orbifold rational curve.*

---

[33] As defined in 2.5.

Conjecture 8.12. *Let (X, Δ) be a smooth orbifold pair with X projective. The following are equivalent:*

1. (X, Δ) *is rationally connected.*
2. $\kappa^+(X, \Delta) = -\infty$.
3. $h^0(X, [Sym^m(\wedge^p)](\Omega^1(X, \Delta))) = 0, \forall m > 0, p > 0$.

We refer to [14], §2.7, for the definition of the integral parts of orbifold tensors $[Sym^m(\wedge^p)](\Omega^1(X, \Delta))$ appearing in 8.12.3, and more details on this notion. This conjecture is solved (see [33]) in dimension 2 when Δ is reduced (i.e.: with multiplicities infinite).

8.6 The Orbifold Version of the Conjectures

Conjecture 8.13. *Let (X, Δ) be a smooth projective orbifold pair.*

1. *Assume first that (X, Δ) is of general type, then, there exists a Zariski closed subset $W \subsetneq X$ such that:*
1. *H. Any orbifold entire curve[34] $h : \mathbb{C} \to (X, \Delta)$ has image contained in W.*
1. *A. If (X, Δ) is defined over k, for any model over $k', S' \subset Spec(\mathcal{O}_{k'})$, the set of (S', k') integral points of (X, Δ) contained in $X \setminus W$ is finite.*
2. *Assume that either $\kappa(X, \Delta) = 0$ or that $\kappa^+(X, \Delta) = -\infty$. Then:*
2. *H. There exists an orbifold entire curve $h : \mathbb{C} \to (X, \Delta)$ with dense image in X.*
2. *A. There exists k', S' such that the (S', k') integral points of (X, Δ) are Zariski dense in X.*

The decomposition $c = (j \circ r)^n$ of the core and conjectures 8.13 (essentially) imply the main conjectures 8.4. Here 'essentially' means that two further properties are still needed: the (orbifold) birational invariance of Mordellicity and potential density, together with the fact that if the generic orbifold fibres and the orbifold base of a fibration $f : (X, \Delta) \to (Z, \Delta_{f, \Delta})$ are potentially dense, then so is (X, Δ), when everything is defined over $\overline{\mathbb{Q}}$.

8.7 Examples of Weakly Special, But Non-special Threefolds

From dimension 3 on, the two notions differ, due to the existence of smooth and simply connected 'orbifold surfaces' of general type.

---

[34]See Definition 2.5 and subsequent lines.

Lemma 8.14. *Let $F : X_3 \to S_2$ be an elliptic fibration from a simply connected smooth projective threefold X onto a smooth surface S with $\kappa(S) \leq 1$.*

Assume that all fibres of F have dimension 1, and that the orbifold base (S, Δ_F) of f is smooth of general type (i.e.: $\kappa(S, K_S + \Delta_F) = 2$). Then:

1. *X is weakly special, but not special: its 'core map' is F.*
2. *There exists such fibrations defined over $\mathbb{Q}$.*

Proof. Let us prove the first claim: since $\kappa(S, K_S + \Delta_f) = 2$, X is not special, and F is the core map of X. In order to show that X is weakly special, it is sufficient (because X is simply connected) to see that there is no fibration $g : X \dashrightarrow Z$ with Z of general type, and $p := dim(Z) > 0$. Indeed since g had then to factorise through F, we had either $p = 2$ and $Z = S$, or $p = 1$, and Z simply connected hence $Z = \mathbb{P}^1$. Contradiction since both S and $\mathbb{P}^1$ are not of general type.

We now prove the existence of such X's as in 8.14. The following construction follows and extends slightly the one given in [8]. The recipe to construct X needs two 'ingredients':

1. A projective elliptic surface $f : T \to \mathbb{P}^1$ with one simply connected fibre $T_1 := f^{-1}(1)$, and a multiple smooth fibre $T_0 = f^{-1}(0)$ of multiplicity $m > 1$. One can obtain such a surface from a Halphen pencil,[35] which allows to get examples defined over $\mathbb{Q}$ (Special cases of Halphen pencils of index $m > 0$ are obtained by blowing up 9 points of a smooth cubic C in Weierstrass form in $\mathbb{P}^2$, whose sum is m-torsion on C; see [20] for details).
2. A surface $g : S \to \mathbb{P}^1$ with $\kappa(S) \leq 1$ and smooth fibre $S_0 = g^{-1}(0)$ such that $\pi_1(S - S_0) = \{1\}$. This can be constructed from any simply connected surface S' with $\kappa(S') \leq 1$, by choosing on S' a base-point free ample linear system defined by a smooth ample divisor $D' \subset S'$, and a second generic member D'' of this linear system which meets transversally D' at $d := (D')^2$ distinct points, and such that, moreover, $\kappa(S', K'_S + (1 - \frac{1}{m}).D') = 2$.

For example, $S' = \mathbb{P}_2$, and D', D'' two generic quartic curves satisfy these conditions.

One then blows up all points of $D' \cap D''$ to obtain S, and $g : S \to \mathbb{P}^1$ is the map defined by the pencil generated by D', D''. One takes for $D = S_0$ the strict transform of D' in S. The simple-connectedness of $(S - D)$ is a consequence of a version of Lefschetz theorem.

We now choose $X_3 := S \times_{\mathbb{P}^1} T$, and $F : X \to S$ the first projection.

We show that the orbifold base (S, D_F) of $F : X \to S$ is of general type. Indeed: $F^*(D) = m.F^{-1}(D)$, since $D = g^{-1}(0)$, and $f^{-1}(0) = m.T_0$.

---

[35] The use of Halphen pencils has been suggested to me by I. Dolgachev. It permits to avoid the transcendental Logarithmic Transformations of Kodaira.

Thus $D_F \geq (1 - \frac{1}{m}).D$, and an easy computation shows that $\kappa(S, (1 - \frac{1}{m}).D) = \kappa(S', (1 - \frac{1}{m}).D') = 2$, since $K_S = b^*(K_{S'}) + E$, while $D = b^*(D') - E$, if $b : S \to S'$ is the blow-up and E its exceptional divisor.

And so: $K_S + (1 - \frac{1}{m}).D = b^*(K_{S'} + (1 - \frac{1}{m}).D') + \frac{1}{m}.E$ □

Remark 8.15. *The Conjecture 5.3 of [30], conjecture 1.2, claims that any X such as in 8.14 is potentially dense, while the Conjecture 8.4.(4) above claims it is not. Vojta's conjectural 'arithmetic second main theorem' implies also that such an X is not potentially dense (using the core map). The hyperbolic analogue claims that there are no Zariski dense entire curves on such an X, and this is proved for some examples in [16].*

9   Entire Curves on Special Manifolds

Recall that an entire curve in a complex manifold M is just a non-constant holomorphic map $h : \mathbb{C} \to M$. Algebraic entire curves are simply rational curves, and entire curves are thus seen as transcendental analogues of rational curves. The following observations indicate that they can serve as testing ground for arithmetic geometry.

9.1   Entire Curves and Sequences of k-Rational Points

Let X be complex projective smooth, defined over a number field k.

In [49], an analogy and dictionary between entire curves and infinite sequences in $X(k)$ are described. Assuming the Conjecture 8.4, this becomes an equivalence.

Proposition 9.1. *Assume Conjecture 8.4. The following properties are then equivalent:*

1. *There is an entire curve $h : \mathbb{C} \to X$.*
2. *$X(k')$ is infinite for some finite extension k'/k.*
3. *X contains a positive-dimensional special subvariety.*

Proof. Assume that $X(k')$ is infinite. Let Z be the Zariski closure of $X(k')$. Since $Z(k')$ is Zariski dense in Z (or in any of its resolutions), Z is special, and thus admits a Zariski dense entire curve, and X has thus also an entire curve.

Assume conversely that there is an entire curve $h : \mathbb{C} \to X$. Let Z be the Zariski closure of $h(\mathbb{C})$, and $Z' \to Z$ a resolution of singularities. Then h lifts to a Zariski dense entire curve in Z'. If Z, and so Z' is defined over

k, Z' is thus special, and $Z'(k')$ is Zariski dense in Z', and so infinite (since $dim(Z) > 0$). Thus so is $X(k')$. In the general case, let Y be a resolution of singularities of the smallest closed irreducible projective subset of X defined over k and containing Z. Assume Y is not special, and let $c : Y \to C$ be its core map (defined over k). Then $c \circ h(\mathbb{C})$ is contained in a strict algebraic subset $W \subset C$ defined over k. Contradiction. Thus Y is special, and $Y(k')$ is Zariski dense in Y, hence infinite.

The equivalence with 3 has been shown in the course of the proof.   □

This motivates the study of the relationship between the distribution of entire curves on projective (and more generally compact Kähler) manifolds X and their core map.

9.2 Specialness and Entire Curves

Some variants of Conjecture 8.4 are:

Conjecture 9.2. *The following are equivalent, for X compact Kähler smooth:*

1. *X is special.*
2. *The Kobayashi pseudodistance[36] d_X of X vanishes identically.*
2'. *The infinitesimal Kobayashi pseudometric d_X^* vanishes on TX.*
3. *Any 2 points of X are joined by an entire curve.*
3'. *Any 2 points of X are joined by a chain of entire curves.*
4. *Any countable subset of X is contained in some entire curve.*
5. *There exists a Zariski dense entire curve on X.*
5'. *There exists a metrically dense entire curve on X.*

Remark 9.3.

1. *Special manifolds are seen as generalisations of rationally connected manifolds, rational curves replaced by entire curves.*
2. *Special manifold are **not** conjectured to be all $\mathbb{C}^n$-dominable (i.e.: to admit a non-degenerate meromorphic map $H : \mathbb{C}^n \dashrightarrow X$). See §9.6.*

We shall mention some partial results, extracted from [19]. Although much efforts have been devoted to the Green–Griffiths–Lang conjecture (asserting that there are no Zariski dense entire curves if X is of general type), the results below seem to be the first ones in the opposite direction: produce

---

[36]Defined as the largest pseudodistance δ on X such that $h^*(\delta) \leq d_{\mathbb{D}}$, for any holomorphic map $h : \mathbb{D} \to X$, where $d_{\mathbb{D}}$ is the Poincaré distance on the unit disk. See [34] for this notion and its infinitesimal version.

dense entire curves on X if it is special, beyond the obvious cases where X is either (uni)rational or Abelian/Kummer.

9.3 Special Surfaces

From surface classification, approximability of $K3$ surfaces by Kummer ones, a classical result by Mori–Mikai, and [10], one gets:

Proposition 9.4 ([19]). *Let S be a special compact Kähler surface. Then:*

1. d_S *vanishes on S.*
2. S *is $\mathbb{C}^2$-dominable unless possibly when S is a K3-surface which is non-elliptic and non-Kummer.*
3. *If S is projective, any 2 points are connected by a chain of 2 elliptic curves.*
4. *If S is not projective, it contains a Zariski dense entire curve.*

The interesting remaining cases are thus $K3$-surfaces either of algebraic dimension zero or projective 'general'. It is far from clear whether the later ones should be expected to be $\mathbb{C}^2$-dominable.

9.4 Rationally Connected Manifolds

Theorem 9.5 ([19]). *Let X be projective, smooth, rationally connected. Let $A \subset X$ be algebraic of codimension at least 2, and let $N \subset X$ be a countable subset of $X \setminus A$. There exists $h : \mathbb{C} \to X \setminus A$ holomorphic such that $N \subset h(\mathbb{C})$.*

A simplified version of the main step of the proof is the following:

Lemma 9.6. *Let $f : \mathbb{P}^1 \to X$ be a very free rational curve going through $x_1, \ldots, x_m$, let $R > 0$ and $\varepsilon > 0$. If x_{m+1} is given, there exists a very free rational curve $g : \mathbb{P}^1 \to X$ going through $x_1, \ldots, x_{m+1}$ and such that $d(f(z), g(z)) \leq \varepsilon$ if $|z| \leq R$, if d is any Hermitian metric on X.*

The proof rests on the 'comb-smoothing' technique of [35]. The lemma consists in joining $x_{n+1} := h(1)$ and $f(\infty) := h(0)$ by a very free rational curve $h : \mathbb{P}^1 \to X$, and approximating sufficiently closely the 'comb' $f(\mathbb{P}^1) \cup h(\mathbb{P}^1)$ by a family of rational curves g_ε which go through $x_1, \ldots, x_{n+1}$.

The rest of the proof consists in constructing inductively on m a sequence of very free rational curves f_n going through the m-first points $x_1, \ldots, x_n$ of the set N, in such a way that they converge uniformly on the disks of radii m.

Stronger versions are proved in [19], to which we refer. For example, the following analog of the 'Weak Approximation Property'[37] on rationally connected manifolds can be immediately derived from the proof of Theorem 9.5, the fact that blown up rationally connected manifolds are still rationally connected, and the Weierstrass products of entire functions:

Corollary 9.7. *Let X be rationally connected smooth. Let $M \subset X$ be a countable set, and for each $m \in M$, let a jet j_m of finite order k_m of holomorphic function from $\mathbb{C}$ to X at m. There then exists an entire function $h : \mathbb{C} \to X$ going through each $m \in M$, and whose k_m-jet at m is j_m.*

The following 'orbifold version' follows from Theorem 9.5:

Example 9.8. *Let $S \subset \mathbb{P}_3$ be a smooth sextic surface. There exists a dense entire curve $h : \mathbb{C} \to \mathbb{P}_3$ which is tangent to S at each intersection point of $h(\mathbb{C})$ with S. Indeed: the double cover $\pi : X \to \mathbb{P}_3$ ramified along S is smooth Fano, hence rationally connected. Any (dense) entire curve $h : \mathbb{C} \to X$ projects to $\mathbb{P}_3$ tangentially along S.*

We do not show the preceding statement directly on $\mathbb{P}_3$ without applying Theorem 9.5 on the double cover, by lack of an orbifold comb-smoothing technique on the Fano Orbifold Pair $(\mathbb{P}_3, S_6)$. Notice that it is unknown whether X is unirational or not.

The following singular version can be obtained, using the MMP for surfaces, [50], and applies to prove Proposition 9.11 below.

Theorem 9.9 ([19]). *Let S be a normal projective surface with only quotient singularities. Assume there exists on S a non-zero $\mathbb{Q}$-effective divisor Δ such that (S, Δ) is Log-terminal and $-K_S = \Delta$. If $F \subset S$ is a finite set containing the singular locus of S, then $S \setminus F$ contains a dense entire curve.*

9.5   Manifolds with $c_1 = 0$

The second fundamental class of special manifolds are those with $\kappa = 0$, in particular those with $c_1 = 0$. They decompose after an étale cover as products with factors belonging to three subclasses: tori, hyperkähler and Calabi–Yau.

- **Complex tori** are easy to deal with: they admit dense affine entire curves, for Abelian varieties, one can do more: construct entire curves (no longer affine) going through any given countable set.

---

[37] This analogy was pointed to us by P. Corvaja, who also noticed that in arithmetic geometry, the WAP implies the Hilbert Property, an implication also implicit in the proof of Theorem 10.3.

By S.T. Yau's solution of Calabi's conjecture, a compact Kähler manifold with $c_1 = c_2 = 0$ is covered by a complex compact torus, and thus satisfies all statements of Conjecture 9.2.

- **Hyperkähler manifolds.** If X is compact Kähler and has no complex analytic compact subvariety (except points and itself), then any entire curve on X is obviously Zariski dense. Since such manifolds have algebraic dimension zero, they are special, and the existence of a Zariski dense entire curve should follow from Conjecture 9.2.

- Examples of such manifolds are:

  1. General deformations of $Hilb^m(K3)$, for any $m > 0$ (by [47], [48]). These contain at least an entire curve (by [47]), which is thus Zariski dense.
  2. All compact Kähler threefolds without subvariety (because they are simple compact tori, by [15]), and thus contain dense entire curves.

Remark 9.10.

1. *Conversely, we conjecture that any compact Kähler manifold without subvariety is either Hyperkähler or simple tori.*
2. *It was interesting to get some information about the 'size' of the entire curves constructed in the general deformations of $Hilb^m(K3)$'s (as measured, for example, by the Hausdorff dimension of their topological closures).*
3. *A much more difficult case is the one of compact Kähler manifolds without subvariety through their general point. These have in particular algebraic dimension zero. And we conjecture that they are either covered by a torus, or have a holomorphic 2-form which is symplectic generically. The solution of this conjecture in dimension 3 implies that any compact Kähler 3-fold with algebraic dimension zero contains a Zariski dense entire curve. See [19].*

- **Calabi–Yau manifolds** are much harder to deal with.

A class for which Conjecture 9.2 can be solved is:

Proposition 9.11. *Any elliptic Calabi–Yau Threefold contains dense entire curves.*

The proof combines Theorem 9.9, [29], [39] and [10] when $c_2 \neq 0$, and follows from Yau's solution of Calabi's conjecture when $c_2 = 0$.

9.6 Remarks on $\mathbb{C}^n$-dominability and Uniform Rationality

We do not expect the $\mathbb{C}^n$-dominability of special n-dimensional manifolds for the following reasons:

1. The algebraic version of $\mathbb{C}^n$-dominability is unirationality. And it is expected that most rationally connected manifolds should be non-unirational from dimension 3 on, starting with the double covers of $\mathbb{P}_3$ branched over a smooth sextic, or standard conic bundles over $\mathbb{P}_2$ with smooth discriminant of large degree.
2. Non-elliptic and non-Kummer $K3$ surfaces are covered by countably many different families of elliptic curves. However, these families might be (and are presumably) parametrised by hyperbolic curves.

The following questions concern the relations between unirationality and $\mathbb{C}^n$-dominability for rationally connected manifolds:

Question 9.12.

1. Are there $\mathbb{C}^n$-dominable rationally connected manifolds which are not unirational?
2. Special case: X is a smooth model of $X_0 = A/G$, where A is an abelian variety, and G a finite group acting holomorphically on A. If X is rationally connected, is it unirational?

The answer is positive in the few cases where it is known. Note also that these examples provide an interesting testing ground for the problem of 'uniform rationality'. Recall (see [5])

Definition 9.13. *A smooth rational n-fold X is said to be 'uniformly rational' if any point of X has a Zariski open neighbourhood algebraically isomorphic to a Zariski open set of $\mathbb{C}^n$.*

When rational, the smooth models of quotients A/G, obtained by blowing up A at the points of non-trivial isotropy, may fail to be uniformly rational at the points of some of the exceptional divisors. For example:

Question 9.14. *Let X be the Ueno threefold, smooth model of $E^3/\mathbb{Z}_4$ obtained by blowing up each point of E^3 of non-trivial isotropy, where $E := \mathbb{C}/\mathbb{Z}[i], i$ a primitive 4-th root of unity, is the Gauss elliptic curve, and $\mathbb{Z}_4$ acts by multiplication by i^k simultaneously on each factor. This manifold is unirational [21], and even rational [22]. Is it uniformly rational? Note that no explicit rational parametrisation of X is known. A similar question can be raised for the similar example $F^3/\mathbb{Z}_6$, where $F := \mathbb{C}/\mathbb{Z}[j], j$ a primitive 6-th root of unity, for which an explicit parametrisation is known.*

10   The Nevanlinna Version of the Hilbert Property

10.1   The Hilbert Property and Its Nevanlinna Version

Definition 10.1 ([23, §2.2]). *Let X/k be a (smooth) projective variety defined over a number field k. Then X is said to have the 'Weak Hilbert Property' over k (WHP for short)[38] if $(X(k) \setminus \cup_j Y_j(k))$ is Zariski dense in X, for any finite set of covers $\pi_j : Y_j \to X$ defined over k, each ramified over a non-empty divisor D_j of X.*

Note that $X(k)$ being Zariski dense, X has to be special, and its fundamental group almost abelian, by Conjecture 8.4.

In [23], Corvaja–Zannier propose an analytic version of the WHP in the following form [23, §2.4]:

Question-Conjecture 10.1 Let X be a special compact Kähler[39] manifold. For any finite cover $\pi : Y \to X$ ramified over a non-empty divisor, with Y irreducible, there exists a dense entire curve $h : \mathbb{C} \to X$ which does not lift to an entire curve $h' : \mathbb{C} \to Y$ (i.e.: such that $\pi \circ h' = h$). We write $NHP(X)$ if X possesses this property, and say that X has NHP (for Nevanlinna–Hilbert Property).

Notice that these NHP properties are preserved by finite étale covers and smooth blow-ups.

A simple tool in checking the non-liftability is the following:

Proposition 10.2 ([19]). *Let $h : \mathbb{C} \to X$ be an entire curve and H an hypersurface of X such that there exists a regular point $a \in H$ in which $h(\mathbb{C})$ and H intersect with order of contact t.*

Let $\pi : X_1 \to X$ be a finite Galois covering with branch locus containing H, such that π ramifies at order $s \geq 2$ over H at a. Then h cannot be lifted to an entire curve $\tilde{h} : \mathbb{C} \to X_1$ if t does not divide s.

Thus, if $h(\mathbb{C})$ meets H transversally at a, h does not lift to Y.

Proof. If π is Galois, it ramifies at order s at any point of Y over $a \in H$. Since $h(\mathbb{C})$ intersect at order s at a, if it lifted to Y, its order of contact with H were a multiple of s.   □

---

[38]The classical Hilbert property does not require the covers $Y_j \to X$ to be ramified. By the Chevalley–Weil Theorem X is then algebraically simply connected.

[39]In [23], X is supposed to be complex projective and to contain a Zariski dense entire curve. We extend their expectation to the compact Kähler case, and replace the dense entire curve by the specialness of X.

10.2   Rationally Connected and Abelian Manifolds

We have the following stronger form for rationally connected manifolds, in which a *fixed* entire curve h does not lift to *any* Galois[40] ramified cover $\pi : Y \to X$:

Theorem 10.3 ([19]). *Let X be a rationally connected complex projective manifold or a complex compact torus.*
    Then there exists an entire curve $f : \mathbb{C} \to X$ such that:

1. *The image $f(\mathbb{C})$ is dense.*
2. *f cannot be lifted to any ramified Galois covering $\tau : X' \to X$.*

Proof. Combine (stronger forms proved in [19] of) Theorem 9.5 with Proposition 10.2. The Abelian case is obtained similarly.   □

10.3   Special Surfaces

Theorem 10.4 ([19]). *Let $f : S \to B$ an elliptic surface with $\pi_1(S)$ is almost abelian (or equivalently: S is special). For any irreducible cover $\pi : Y \to X$ ramifying over a non-empty divisor $R \subset S$, there exists a dense entire curve $h : \mathbb{C} \to S$ which does not lift to Y.*

Proof. Assume that $R \subset S$ meets a regular point of some reduced component of some fibre of f. From [10], one gets a submersive map $H : \mathbb{C}^2 \to S$ whose image contains all smooth fibres of f, and the regular part of the component of the fibre of f which meets R. This produces an entire curve $h : \mathbb{C} \to \mathbb{C}^2$ which meets transversally $H^*(R)$. We refer to [19] for the reduction to this particular case.   □

Remark 10.5. *The above result together with the simpler case of special surfaces S with $\kappa(S) = -\infty$ solves the Conjecture 10.1 for special surfaces except for $K3$ surfaces which are neither Kummer nor elliptic.*

11   The Kobayashi Pseudodistance

We explain here how to get from the core map a conjectural (qualitative) description of the Kobayashi pseudodistance of any complex projective (or compact Kähler) manifold X, using the notion of orbifold Kobayashi pseudodistance.

---

[40]The Galois assumption can be removed using more delicate arguments.

- Recall first that if M is a complex manifold, its Kobayashi pseudodistance d_M is the largest pseudodistance δ on M such that $h^*(\delta) \leq p_{\mathbb{D}}$, for any holomorphic map $h : \mathbb{D} \to M$, where $p_{\mathbb{D}}$ is the Poincaré metric on the unit disc $\mathbb{D} \subset \mathbb{C}$. It enjoys the following properties:

  1. $d_{\mathbb{D}} = p_{\mathbb{D}}$.
  2. It is distance decreasing: $f^*(d_N) \leq d_M, \forall f : M \to N$ holomorphic.
  3. It is preserved by $Aut(M)$.
  4. $d_{\mathbb{C}} \equiv 0$.
  5. It is continuous wrt the metric topology on any M.
  6. $d_{M|\overline{h(\mathbb{C})}} \equiv 0$ for $h : \mathbb{C} \to M$ holomorphic, $\overline{E}$ the metric closure.
  7. $d_M \equiv 0$ if $M = \mathbb{P}^n$, or $M = $ a complex torus.
  8. If M is compact, d_M is a distance iff M does not contain any entire curve (Brody Theorem).

We thus see that there is a close relationship between d_M and the distribution of entire curves on M. In particular, $d_M \equiv 0$ if there exists a dense entire curve on M, or if any two points in a dense subset of M can be joined by a connected chain of entire curves. The reverse implications are however widely open, even for $K3$ surfaces M, for which d_M is known to vanish identically.

Entirely similarly to the case when $\Delta = 0$, we define the Kobayashi pseudodistance in the orbifold setting. Let thus (X, Δ) be a smooth orbifold pair with X compact Kähler and $\Delta := \sum_j (1 - \frac{1}{m_j}).D_j$ an orbifold divisor with SNC support $D := (\cup_j D_j)$.

Recall that $Hol(\mathbb{D}, (X, \Delta))$ (resp. $Hol^*(\mathbb{D}, (X, \Delta))$) denotes the set of orbifold (resp. classical orbifold) morphisms from the unit disk $\mathbb{D}$ to (X, Δ) as defined in 2.5.

Definition 11.1. *The Kobayashi (resp. The Classical Kobayashi) Pseudodistance $d_{(X,\Delta)}$ (resp. $d^*_{(X,\Delta)}$) of the orbifold (X, Δ) is the largest pseudodistance δ on X such that $\delta \leq h^*(d_{\mathbb{D}}), \forall h \in Hol(\mathbb{D}, (X, \Delta))$ (resp. $\forall h \in Hol^*(\mathbb{D}, (X, \Delta))$). We thus have: $d_{(X,\Delta)} \leq d^*_{(X,\Delta)}$, but have equality if $\Delta = 0$ or if $\Delta = Supp(\Delta)$ (projective and quasi-projective cases, in which cases we recover d_X and $d_{X \setminus D}$, respectively). For orbifold curves, these pseudodistances agree, but no longer for orbifold surfaces in general (see [18], Theorem 2, and [40], Theorem 3.17).*

We shall not use the 'classical' version here (except in the proof of 11.8, for $X = \mathbb{D}$. The example given in [40] however suggests the following:

Question 11.2.

1. *Is there a continuous function $c : X \times X \to [0, 1]$, positive outside of $A \times A$, for some Zariski closed subset $A \subsetneq X$, such that $d_{(X,\Delta)}(x, y) = c(x, y).d^*_{(X,\Delta)}(x, y), \forall (x, y) \in X \times X$?*
2. *Assume that (X, Δ) is smooth. If $A \subset X$ is Zariski closed of codimension at least 2 in X, is $d_{(X,\Delta)|X^*} = d_{(X^*,\Delta^*)}$, where $X^* := X \setminus A$, and $\Delta^* :=$*

$\Delta \cap X^*$? When $\Delta = 0$ and when $\Delta = Supp(\Delta)$, this is true by [34], Theorem 2.3.19.

Recall the general notion of orbifold morphism between orbifold pairs:

Definition 11.3. Let (X, Δ) and (Y, Δ_Y) be orbifold pairs, Y smooth (or $\mathbb{Q}$-factorial) and $h : X \to Y$ be a holomorphic map such that $h(X)$ is not contained in $Supp(\Delta_Y)$. Then $h : (X, \Delta) \to (Y, \Delta_Y)$ is an orbifold morphism if, for each irreducible divisor $F \subset Y$, and each irreducible divisor $E \subset X$ such that $h(E) \subset F$, one has: $m_\Delta(E) \geq t_{E,F}.m_{\Delta_Y}(F)$, where: $m_{Delta}(E)$ is the multiplicity of E in Δ (and similarly for $m_{\Delta_Y}(F)$, while $t_{E,F}$ is the scheme-theoretic multiplicity of E in $h^*(F)$ (i.e.: $h^*(F) = t_{E,F}.E + R$, where r does not contain E in its support).

Clearly, orbifold morphisms can be composed. We have the following obvious functoriality property: $h_*(Hol(\mathbb{D}, (X, \Delta)) \subset Hol(\mathbb{D}, (Y, \Delta_Y))$ if h is an orbifold morphism, and so also the usual distance decreasing property: $h^*(d_{(Y, \Delta_Y)}) \leq d_{(X, \Delta)}$.

We shall need the following birational invariance property also:

Proposition 11.4. Let X be smooth, and $A \subset X$ a Zariski closed subset of codimension at least 2. Let $X^* := (X \setminus A)$, and let $\mu : X' \to X$ be a bimeromorphic holomorphic map which is isomorphic over X^*. Let E be the exceptional divisor of μ, and let Δ' be an orbifold divisor on X' supported on E. Then $d_{(X', \Delta')} = \mu^*(d_X)$ (whatever large and possibly infinite are the multiplicities on the components of Δ').

Proof. From [34], Theorem 2.3.19, we know that $d_{X^*} = d_{X|X^*}$. We identify X^* with its inverse image in X', and extend by continuity d_{X^*} to X' and X, with the same (abusive) notation. On the other hand, we also have: $d_{X'} \leq d_{(X', \Delta')} \leq d_{X^*}$ on X'. This implies the claim, since $\mu^*(d_{X^*}) = d_{X^*}$ (where the LHS is on X, and the RHS on X'). □

Theorem 11.5. Let $f : X \to Z$ be a fibration, with X a connected complex compact manifold. Let $f' : X' \to Z'$ be a bimeromorphic 'neat model' of f, where $\mu : X' \to X$ is bimeromorphic. Let (Z', Δ') be the (smooth) orbifold base of f'. Then:

1. $f^*(d_{(Z', \Delta')}) \leq d_{X'} = \mu^*(d_X)$.
2. $f^*(d_{(Z', \Delta')}) = d_{X'}$ if $d_{X_z} \equiv 0$, for a dense set of fibres X_z of f.

Corollary 11.6. Let $c : X \to C_X$ be the core map of some compact Kähler manifold X. Then: $d_X = c^*(d_{(C_X, \Delta_c)})$.

Assume Conjecture 8.4, and Conjecture 11.7 below. Then: $d_{(C_X, \Delta_c)}$ is a metric on a non-empty Zariski open subset $C_X \setminus W$ of C_X.

The following is simply an orbifold version of the strong Lang's generic hyperbolicity conjecture for manifolds of general type.

Conjecture 11.7. *Let (Z, Δ) be a smooth orbifold pair of general type. There exists a strict Zariski closed subset $W \subsetneq Z$ such that $d_{(Z,\Delta)}$ is a metric on $Z \setminus W$. Moreover, the smallest such W is defined over k if so is Z.*

Proof (of Theorem 11.5). Since $f' : X' \to (Z', \Delta')$ is a neat model of f, we have the following properties: there exist two Zariski closed subsets $B \subset Z'$ and $A \subset X$, A contained in the indeterminacy locus of μ^{-1}, such that: $\mu((f')^{-1}(B)) \subset A$, and $f' : X^* := X' \setminus (f')^{-1}(B) \to (Z', \Delta')$ has equidimensional fibres and is an orbifold morphism. If we equip the components of the exceptional divisor E of μ with sufficiently large multiplicities, we get an orbifold divisor $\Delta_{X'}$ on X' such that all of $f' : (X', \Delta_{X'}) \to (Z', \Delta')$ becomes an orbifold morphism. We thus get, from the definition of orbifold Kobayashi pseudometrics, the inequality: $(f')^*(d_{(Z^*, \Delta^*)}) \le d_{X^*}$. We can thus conclude from the continuity of these pseudometrics, and Proposition 11.4 that $(f')^*(d_{Z', \Delta'}) \le d_{X^*} = d_{X'}$. $\square$

Let us now prove the reverse inequality when the fibres all have a vanishing Kobayashi pseudometric (which is the case if a dense subset of them have this property, by the continuity of the Kobayashi pseudometric). We may, and shall, assume here that $X' = X$ and $f' = f$, we then write $(Z', \Delta') = (Z, \Delta)$ to simplify notations. Notice that, due to Proposition 11.4 and the preceding argument, it will be sufficient to show that $(f')^*(d_{(Z^*, \Delta^*)}) \le d_{X^*}$.

Proposition 11.8. *Let $g : M \to \mathbb{D}$ be a proper fibration from a complex manifold to the unit disk. Assume that $d_{M_z} \equiv 0$ for all fibres of g, and that Δ_g is supported on a finite set of $\mathbb{D}$. Then $d_M = g^*(d_{\mathbb{D}, \Delta_g})$.*

Let us first show that the inequality 2 of Theorem 11.5 follows from Proposition 11.8.

Let $h_i, i = 0, \ldots, N, a_i, b_i$ be a Kobayashi chain in X^* joining two points $a, b \in X$, that is: a sequence of holomorphic disks $h_i : \mathbb{D} \to X$, together with points $a_i, b_i \in \mathbb{D}$ such that $h_0(a_0) = a$, $h_N(b_N) = b$, and $h_i(b_i) = h_{i+1}(a_{i+1})$ for $i = 0, \ldots, (N - 1)$. From the choice of A, B, X^*, Z^*, we deduce that $g_i := f_*(h_i) := f \circ h_i \in Hol(\mathbb{D}, (Z^*, \Delta^*))$. From Proposition 11.8 we deduce that the Kobayashi lengths of the chains $\{h_i, a_i, b_i\}$ and $\{g_i, a_i, b_i\}$, given by $\sum_i d_X(h_i(a_i), h_i(b_i))$ and $\sum_i d_{(Z^*, \Delta^*)}(g_i(a_i), g_i(b_i))$ coincide. Taking the infimum (on either side) for given $a, b \in X$ (or $a', b' \in Z^*$) gives the claimed equality.

We now prove Proposition 11.8. It will be the consequence of the following three lemmas:

Lemma 11.9. *Let $g : M \to N$ be a surjective holomorphic map with connected fibres between two connected complex manifolds. Assume that g has everywhere local sections and that the fibres of g all have zero Kobayashi pseudometric. Then $d_M = g^*(d_N)$.*

Proof (of Theorem 11.5). The Kobayashi lengths on M of any arc joining a, b in M and of its image by g on N coincide, using local sections and the vanishing of d along the fibres of g. $\square$

Lemma 11.10. *If $\Delta := \sum_i (1 - \frac{1}{m_i}).\{a_i\}$ is a finitely supported orbifold divisor on $\mathbb{D}$, there is a finite unfolding $u : C \to \mathbb{D}$ from a complex smooth curve C which ramifies at order m_i over each point lying over a_i, this for any i, and unramified over the complement of the $a'_i s$.*

Proof (of Theorem 11.5). The fundamental group of the complement $\mathbb{D}^*$ of the $a'_i s$ is a free group F_N on N generators generated by small loops γ_i winding once around a_i, for each $i = 1, \ldots, N$, if N is the cardinality of the $a'_i s$. There is thus a natural surjective group morphism of F_N onto $\oplus_i \mathbb{Z}_{m_i}$ which induces a finite Galois cover $C^* \to \mathbb{D}^*$ which can be partially compactified over the $a'_i s$ so as to give the claimed unfolding. $\square$

The Kobayashi pseudodistance $d_{(\mathbb{D}, \Delta_u)}$ is obtained by integrating the Kobayashi–Royden infinitesimal pseudometric $d^R_{(\mathbb{D}, \Delta_u)}$, and similarly for $d^*_{(\mathbb{D}, \Delta_u)}$ and $d^{*,R}_{(\mathbb{D}, \Delta_u)}$, which are computed explicitly in [40] . By [40], Theorems 3.9, 3.13, we have: $d^{*,R}_{(\mathbb{D}, \Delta_u)} = d^R_{(\mathbb{D}, \Delta_u)}$, and $d^R_C = u^*(d^{*,R}_{(\mathbb{D}, \Delta_u)})$.

Let $g_C : M_C \to C$ be the (desingularised) base change of $g : M \to \mathbb{D}$. It has everywhere local sections (by the definition of the (non-classical) orbifold base). We thus have: $d^R_{M_C} = g_C^*(d^R_C)$.

Let $v : M_C \to M$ be the natural projection; we thus have:

$$v^*(d^R_M) \leq d^R_{M_C} = g^*(d^R_C) = g^*(u^*(d^R_{(\mathbb{D}, \Delta_u)})) = v^*(f^*(d^R_{(\mathbb{D}, \Delta_u)})).$$

Thus: $d^R_M = f^*(d^R_{(\mathbb{D}, \Delta_u)})$, and also the claim: $d_M = f^*(d_{(\mathbb{D}, \Delta_u)})$.

References

1. Dan Abramovich. Birational geometry for number theorists. In *Arithmetic geometry*, volume 8 of *Clay Math. Proc.*, pages 335–373. Amer. Math. Soc., Providence, RI, 2009.
2. Dan Abramovich and Anthony Várilly-Alvarado. Campana points, Vojta's conjecture, and level structures on semistable abelian varieties. *J. Théor. Nombres Bordeaux*, 30(2):525–532, 2018.
3. Frits Beukers. The Diophantine equation $Ax^p + By^q = Cz^r$. *Duke Math. J.*, 91(1):61–88, 1998.
4. Fedor Bogomolov. Holomorphic tensors and vector bundles on projective manifolds. *Izv. Akad. Nauk SSSR Ser. Mat.*, 42(6):1227–1287, 1439, 1978.
5. Fedor Bogomolov and Christian Böhning. On uniformly rational varieties. In *Topology, geometry, integrable systems, and mathematical physics*, volume 234 of *Amer. Math. Soc. Transl. Ser. 2*, pages 33–48. Amer. Math. Soc., Providence, RI, 2014.

6. Fedor Bogomolov and Yuri Tschinkel. Density of rational points on elliptic $K3$ surfaces. *Asian J. Math.*, 4(2):351–368, 2000.

7. Fedor Bogomolov and Yuri Tschinkel. Unramified correspondences. In *Algebraic number theory and algebraic geometry*, volume 300 of *Contemp. Math.*, pages 17–25. Amer. Math. Soc., Providence, RI, 2002.

8. Fedor Bogomolov and Yuri Tschinkel. Special elliptic fibrations. In *The Fano Conference*, pages 223–234. Univ. Torino, Turin, 2004.

9. Enrico Bombieri and Walter Gubler. *Heights in Diophantine geometry*, volume 4 of *New Mathematical Monographs*. Cambridge University Press, Cambridge, 2006.

10. Gregery T. Buzzard and Steven S. Y. Lu. Algebraic surfaces holomorphically dominable by $\mathbf{C}^2$. *Invent. Math.*, 139(3):617–659, 2000.

11. Frédéric Campana. Orbifolds, special varieties and classification theory. *Ann. Inst. Fourier (Grenoble)*, 54(3):499–630, 2004.

12. Frédéric Campana. Fibres multiples sur les surfaces: aspects géométriques, hyperboliques et arithmétiques. *Manuscripta Math.*, 117(4):429–461, 2005.

13. Frédéric Campana. Orbifoldes géométriques spéciales et classification biméromorphe des variétés kählériennes compactes. *J. Inst. Math. Jussieu*, 10(4):809–934, 2011.

14. Frédéric Campana. Orbifold slope-rational connectedness, 2016. arXiv 1607.07829.

15. Frédéric Campana, Jean-Pierre Demailly, and Misha Verbitsky. Compact Kähler 3-manifolds without nontrivial subvarieties. *Algebr. Geom.*, 1(2):131–139, 2014.

16. Frédéric Campana and Mihai Păun. Variétés faiblement spéciales à courbes entières dégénérées. *Compos. Math.*, 143(1):95–111, 2007.

17. Frédéric Campana and Mihai Păun. Foliations with positive slopes and birational stability of orbifold cotangent bundles. *Publ. Math. Inst. Hautes Études Sci.*, 129:1–49, 2019.

18. Frédéric Campana and Jörg Winkelmann. A Brody theorem for orbifolds. *Manuscripta Math.*, 128(2):195–212, 2009.

19. Frédéric Campana and Jörg Winkelmann. Dense entire curves in rationally connected manifolds, 2019. arXiv 1905.01104.

20. Serge Cantat and Igor Dolgachev. Rational surfaces with a large group of automorphisms. *J. Amer. Math. Soc.*, 25(3):863–905, 2012.

21. Fabrizio Catanese, Keiji Oguiso, and Tuyen Trung Truong. Unirationality of Ueno-Campana's threefold. *Manuscripta Math.*, 145(3-4):399–406, 2014.

22. Jean-Louis Colliot-Thélène. Rationalité d'un fibré en coniques. *Manuscripta Math.*, 147(3-4):305–310, 2015.

23. Pietro Corvaja and Umberto Zannier. Integral points, divisibility between values of polynomials and entire curves on surfaces. *Adv. Math.*, 225(2):1095–1118, 2010.

24. Pietro Corvaja and Umberto Zannier. On the Hilbert property and the fundamental group of algebraic varieties. *Math. Z.*, 286(1-2):579–602, 2017.

25. Henri Darmon and Andrew Granville. On the equations $z^m = F(x, y)$ and $Ax^p + By^q = Cz^r$. *Bull. London Math. Soc.*, 27(6):513–543, 1995.

26. Noam D. Elkies. abc implies Mordell. *Int. Math. Res. Not.*, 1991(7):99–109, 1991.

27. Pál Erdős and George Szekeres. Über die Anzahl der Abelschen Gruppen gegebener Ordnung und über ein verwandtes zahlentheoretisches Problem. *Acta Litt. Sci. Szeged*, 7:95–102, 1934.

28. J.-H. Evertse. On equations in S-units and the Thue-Mahler equation. *Invent. Math.*, 75(3):561–584, 1984.

29. Antonella Grassi. On minimal models of elliptic threefolds. *Math. Ann.*, 290(2):287–301, 1991.

30. Joe Harris and Yuri Tschinkel. Rational points on quartics. *Duke Math. J.*, 104(3):477–500, 2000.

31. Marc Hindry and Joseph H. Silverman. *Diophantine geometry*, volume 201 of *Graduate Texts in Mathematics*. Springer-Verlag, New York, 2000. An introduction.

32. Stefan Kebekus, Jorge Vitório Pereira, and Arne Smeets. Brauer-manin failure for a simply connected fourfold over a global function field, via orbifold Mordell, 2019. arXiv 1905.02795.

33. Seán Keel and James McKernan. Rational curves on quasi-projective surfaces. *Mem. Amer. Math. Soc.*, 140(669):viii+153, 1999.

34. Shoshichi Kobayashi. Intrinsic distances, measures and geometric function theory. *Bull. Amer. Math. Soc.*, 82(3):357–416, 1976.

35. János Kollár, Yoichi Miyaoka, and Shigefumi Mori. Rationally connected varieties. *J. Algebraic Geom.*, 1(3):429–448, 1992.

36. Serge Lang. Hyperbolic and Diophantine analysis. *Bull. Amer. Math. Soc. (N.S.)*, 14(2):159–205, 1986.

37. Makoto Namba. *Branched coverings and algebraic functions*, volume 161 of *Pitman Research Notes in Mathematics Series*. Longman Scientific & Technical, Harlow; John Wiley & Sons, Inc., New York, 1987.

38. Yukihiko Namikawa and Kenji Ueno. The complete classification of fibres in pencils of curves of genus two. *Manuscripta Math.*, 9:143–186, 1973.

39. Keiji Oguiso. On algebraic fiber space structures on a Calabi-Yau 3-fold. *Internat. J. Math.*, 4(3):439–465, 1993. With an appendix by Noboru Nakayama.

40. Erwan Rousseau. Hyperbolicity of geometric orbifolds. *Trans. Amer. Math. Soc.*, 362(7):3799–3826, 2010.

41. Jean-Pierre Serre. *Topics in Galois theory*, volume 1 of *Research Notes in Mathematics*. Jones and Bartlett Publishers, Boston, MA, 1992.

42. I. R. Shafarevich. *Basic algebraic geometry*. Springer-Verlag, Berlin-New York, study edition, 1977. Translated from the Russian by K. A. Hirsch, Revised printing of Grundlehren der mathematischen Wissenschaften, Vol. 213, 1974.

43. Arne Smeets. Insufficiency of the étale Brauer-Manin obstruction: towards a simply connected example. *Amer. J. Math.*, 139(2):417–431, 2017.

44. C. L. Stewart and Kunrui Yu. On the *abc* conjecture. II. *Duke Math. J.*, 108(1):169–181, 2001.

45. Lidia Stoppino. A note on fibrations of Campana general type on surfaces, 2010. arXiv 1002.2789.

46. Kenji Ueno. *Classification theory of algebraic varieties and compact complex spaces*. Lecture Notes in Mathematics, Vol. 439. Springer-Verlag, Berlin-New York, 1975. Notes written in collaboration with P. Cherenack.

47. Misha Verbitsky. Trianalytic subvarieties of the Hilbert scheme of points on a $K3$ surface. *Geom. Funct. Anal.*, 8(4):732–782, 1998.

48. Misha Verbitsky. Ergodic complex structures on hyperkähler manifolds. *Acta Math.*, 215(1):161–182, 2015. See also the Erratum in arXiv: 1708.05802.

49. Paul Vojta. Diophantine approximation and Nevanlinna theory. In *Arithmetic geometry*, volume 2009 of *Lecture Notes in Math.*, pages 111–224. Springer, Berlin, 2011.

50. Chenyang Xu. Strong rational connectedness of surfaces. *J. Reine Angew. Math.*, 665:189–205, 2012.

The Lang–Vojta Conjectures on Projective Pseudo-Hyperbolic Varieties

Ariyan Javanpeykar

2010 Mathematics Subject Classification 14G99 (11G35, 14G05, 32Q45)

1 Introduction

These notes grew out of a mini-course given from May 13th to May 17th at UQÀM in Montréal during a workshop on Diophantine Approximation and Value Distribution Theory.

1.1 *What Is in These Notes?*

We start with an overview of Lang–Vojta's conjectures on pseudo-hyperbolic *projective* varieties. These conjectures relate various different notions of hyperbolicity. We start with Brody hyperbolicity and discuss conjecturally related notions of hyperbolicity in arithmetic geometry and algebraic geometry in subsequent sections. We slowly work our way towards the most general version of Lang–Vojta's conjectures and provide a summary of all the conjectures in Section 12.

After having explained the main conjectures with the case of curves and closed subvarieties of abelian varieties as our guiding principle, we collect recent advances on Lang–Vojta's conjectures and present these in a unified manner. These results are concerned with endomorphisms of hyperbolic varieties, moduli spaces of maps into a hyperbolic variety, and also the

A. Javanpeykar (✉)
Institut für Mathematik, Johannes Gutenberg-Universität Mainz, Staudingerweg 9, 55099 Mainz, Germany
e-mail: peykar@uni-mainz.de

© Springer Nature Switzerland AG 2020
M.-H. Nicole (ed.), *Arithmetic Geometry of Logarithmic Pairs and Hyperbolicity of Moduli Spaces*, CRM Short Courses, https://doi.org/10.1007/978-3-030-49864-1_3

behaviour of hyperbolicity in families of varieties. The results presented in these sections are proven in [15, 49, 50, 55, 56].

We also present results on the Shafarevich conjecture for smooth hypersurfaces obtained in joint work with Daniel Litt [52]. These are motivated by Lawrence–Venkatesh's recent breakthrough on the non-density of integral points on the moduli space of hypersurfaces [63], and are in accordance with Lang–Vojta's conjecture for *affine* varieties. Our results in this section are proven using methods from Hodge theory, and are loosely related to Bakker–Tsimerman's chapter in this book [12].

In the final section we sketch a proof of the fact that being groupless is a Zariski-countable open condition, and thus in particular stable under generization. To prove this, we follow [55] and introduce a non-archimedean notion of hyperbolicity. We then state a non-archimedean analogue of the Lang–Vojta conjectures which we prove under suitable assumptions. These results suffice to prove that grouplessness is stable under generization.

1.2   Anything New in These Notes?

The main contribution of these notes is the systematic presentation and comparison between different notions of hyperbolicity, and their "pseudofications". As it is intended to be a broad-audience introduction to the Lang–Vojta conjectures, it contains all definitions and well-known relations between these. Also, Lang–Vojta's original conjectures are often only stated for varieties over $\overline{\mathbb{Q}}$, and we propose natural extensions of their conjectures to varieties over arbitrary algebraically closed fields of characteristic zero. We also define for each notion appearing in the conjecture the relevant "exceptional locus" (which Lang only does for some notions of hyperbolicity in [62]).

The final version of Lang–Vojta's conjecture as stated in Section 12 does not appear anywhere in the literature explicitly. Furthermore, the section on groupless varieties (Section 4) contains simple proofs that do not appear explicitly elsewhere. Also, we have included a thorough discussion of the a priori difference between being arithmetically hyperbolic and Mordellic for a projective variety in Section 7. This difference is not addressed anywhere else in the literature.

1.3   Rational Points over Function Fields

We have not included any discussion of rational points on projective varieties over function fields of smooth connected curves over a field k, and unfortunately ignore the relation to Lang–Vojta's conjecture throughout these notes.

1.4 Other Relevant Literature

Lang stated his conjectures in [62]; see also [23, Conjecture XV.4.3] and [1, §0.3]. In [85, Conj. 4.3] Vojta extended this conjecture to quasi-projective varieties. In [62] Lang "pseudofied" the notion of Brody hyperbolicity. Here he was inspired by Kiernan–Kobayashi's extension of the notion of Kobayashi hyperbolicity introduced in [58].

There are several beautiful surveys of the Green–Griffiths and Lang–Vojta conjectures. We mention [24–26, 31, 38, 86].

The first striking consequence of Lang–Vojta's conjecture was obtained by Caporaso–Harris–Mazur [19]. Their results were further investigated by Abramovich, Ascher–Turchet, Hassett, and Voloch; see [1–4, 9, 42].

Campana's conjectures provide a complement to Lang–Vojta's conjectures, and first appeared in [17, 18]; see also Campana's chapter in this book [16]. In a nutshell, the "opposite" of being pseudo-hyperbolic (in any sense of the word "hyperbolic") is conjecturally captured by Campana's notion of a "special" variety.

Conventions. Throughout these notes, we will let k be an algebraically closed field of characteristic zero. If X is a locally finite type scheme over $\mathbb{C}$, we let X^{an} be the associated complex-analytic space [39, Expose XII]. If K is a field, then a variety over K is a finite type separated K-scheme.

If X is a variety over a field K and L/K is a field extension, then $X_L := X \times_{\mathrm{Spec}\,K} \mathrm{Spec}\,L$ will denote the base-change of $X \to \mathrm{Spec}\,K$ along $\mathrm{Spec}\,L \to \mathrm{Spec}\,K$. More generally, if $R \to R'$ is an extension of rings and X is a scheme over R, we let $X_{R'}$ denote $X \times_{\mathrm{Spec}\,R} \mathrm{Spec}\,R'$.

If K is a number field and S is a finite set of finite places of K, then $\mathcal{O}_{K,S}$ will denote the ring of S-integers of K.

2 Brody Hyperbolicity

We start with the classical notion of Brody hyperbolicity for complex varieties.

Definition 2.1. A complex-analytic space X is *Brody hyperbolic* if every holomorphic map $\mathbb{C} \to X$ is constant. A locally finite type scheme X over $\mathbb{C}$ is *Brody hyperbolic* if X^{an} is Brody hyperbolic.

If X is a complex-analytic space, then a non-constant holomorphic map $\mathbb{C} \to X$ is commonly referred to as an entire curve in X. Thus, to say that X is Brody hyperbolic is to say that X has no entire curves.

We recall that a complex-analytic space X is *Kobayashi hyperbolic* if Kobayashi's pseudo-metric on X is a metric [59]. It is a fundamental result of Brody that a *compact* complex-analytic space X is Brody hyperbolic if and only if it is Kobayashi hyperbolic; see [59, Theorem 3.6.3].

Remark 2.2 (Descending Brody Hyperbolicity). Let $X \to Y$ be a proper étale (hence finite) morphism of varieties over $\mathbb{C}$. It is not hard to show that X is Brody hyperbolic if and only if Y is Brody hyperbolic. (It is crucial that $X \to Y$ is finite **and** étale.)

Fundamental results in complex analysis lead to the following classification of Brody hyperbolic projective curves.

Theorem 2.3 (Liouville, Riemann, Schwarz, Picard). *Let X be a smooth projective connected curve over $\mathbb{C}$. Then X is Brody hyperbolic if and only if* $\mathrm{genus}(X) \geq 2$.

More generally, a smooth quasi-projective connected curve X over $\mathbb{C}$ is Brody hyperbolic if and only if X is not isomorphic to $\mathbb{P}^1_{\mathbb{C}}$, $\mathbb{A}^1_{\mathbb{C}}$, $\mathbb{A}^1_{\mathbb{C}} \setminus \{0\}$, nor a smooth proper connected genus one curve over $\mathbb{C}$.

Remark 2.4. It is implicit in Theorem 2.3 that elliptic curves are not Brody hyperbolic. More generally, a non-trivial abelian variety A of dimension g over $\mathbb{C}$ is not Brody hyperbolic, as its associated complex-analytic space is uniformized by $\mathbb{C}^g$. Since A even has a dense entire curve, one can consider A to be as far as possible from being Brody hyperbolic. We mention that Campana conjectured that a projective variety has a dense entire curve if and only if it is "special". We refer the reader to Campana's article in this book for a further discussion of Campana's conjecture [16].

By Remark 2.4, an obvious obstruction to a projective variety X over $\mathbb{C}$ being Brody hyperbolic is that it contains an abelian variety. The theorem of Bloch–Ochiai–Kawamata says that this is the only obstruction if X can be embedded into an abelian variety (see [57]).

Theorem 2.5 (Bloch–Ochiai–Kawamata). *Let X be a closed subvariety of an abelian variety A over $\mathbb{C}$. Then X is Brody hyperbolic if and only if IT does not contain the translate of a positive-dimensional abelian subvariety of A.*

Throughout these notes, we mostly focus on closed subvarieties of abelian varieties, as in this case the results concerning Lang–Vojta's conjectures are complete; see Section 13 for details.

The theorem of Bloch–Ochiai–Kawamata has been pushed further by work of Noguchi–Winkelmann–Yamanoi; see [76–78, 87, 88]. Other examples of Brody hyperbolic varieties can be constructed as quotients of bounded domains, as we explain now.

Remark 2.6 (Bounded Domains). Let D be a bounded domain in the affine space $\mathbb{C}^N$, and let X be a reduced connected locally finite type scheme over $\mathbb{C}$. Then, any holomorphic map $X^{\mathrm{an}} \to D$ is constant; see [55,

Remark 2.9] for a detailed proof. In particular, the complex-analytic space D is Brody hyperbolic (take $X = \mathbb{A}_\mathbb{C}^1$).

It follows from Remark 2.6 that a (good) quotient of a bounded domain is Brody hyperbolic. This observation applies to locally symmetric varieties, Shimura varieties, and thus moduli spaces of abelian varieties. We conclude this section by recording the fact that the moduli space of abelian varieties (defined appropriately) is a Brody hyperbolic variety.

Example 2.7. Let $g \geq 1$ and let $N \geq 3$ be integers. Then, the (fine) moduli space of g-dimensional principally polarized abelian varieties with level N structure is a smooth quasi-projective variety over $\mathbb{C}$ which is Brody hyperbolic. Indeed, its universal cover is biholomorphic to a bounded domain in $\mathbb{C}^{g(g+1)/2}$, so that we can apply Remark 2.6. (As the coarse moduli space of elliptic curves is given by the j-line $\mathbb{A}_\mathbb{C}^1$, we see that it is not Brody hyperbolic. This is the reason for which we consider the moduli space of abelian varieties with level structure.)

3 Mordellic Varieties

What should correspond to being Brody hyperbolic in arithmetic geometry? Lang was the first to propose that a "Mordellic" projective variety over $\overline{\mathbb{Q}}$ should be Brody hyperbolic (over the complex numbers). Roughly speaking, a projective variety over $\overline{\mathbb{Q}}$ is Mordellic if it has only finitely many rational points in any fixed number field. To make this more precise, one has to choose models (see Definition 3.1 below). Conversely, a projective variety over a number field which is Brody hyperbolic (over the complex numbers) should be Mordellic. In this section we will present this conjecture of Lang.

Throughout this section, we let k be an algebraically closed field of characteristic zero. We first clarify what is meant with a model.

Definition 3.1. Let X be a finite type separated scheme over k and let $A \subset k$ be a subring. A *model for X over A* is a pair $(\mathcal{X}, \phi)$ with $\mathcal{X} \to \operatorname{Spec} A$ a finite type separated scheme and $\phi : \mathcal{X}_k \xrightarrow{\sim} X$ an isomorphism of schemes over k. We will often omit ϕ from our notation.

Remark 3.2. What constitutes the data of a model for X over A? To explain this, let X be an affine variety over $\mathbb{C}$, say $X = \operatorname{Spec} R$. Note that the coordinate ring R of X is a finite type $\mathbb{C}$-algebra. Suppose that X is given by the zero locus of polynomials $f_1, \ldots, f_r$ with coefficients in a subring A, so that $R \cong \mathbb{C}[x_1, \ldots, x_n]/(f_1, \ldots, f_r)$. Then $\mathcal{R} := A[x_1, \ldots, x_n]/(f_1, \ldots, f_r) \subset R$ is a finitely generated A-algebra and $\mathcal{R} \otimes_A \mathbb{C} = R$. That is, if $\mathcal{X} = \operatorname{Spec} \mathcal{R}$, then $\mathcal{X}$ is a model for X over A. We will be interested in studying A-valued points on $\mathcal{X}$. We follow common notation and let $\mathcal{X}(A)$ denote the set $\mathrm{Hom}_A(\mathrm{Spec}\,A, \mathcal{X})$. Note that $\mathcal{X}(A)$ is the set of solutions in A of the polynomial system of equations $f_1 = \ldots = f_r = 0$.

With the notion of model now clarified, we are ready to define what it means for a proper variety to be Mordellic. We leave the more general definition for non-proper varieties to the end of this section.

Definition 3.3. A proper scheme X over k is *Mordellic over k* (or: *has-only-finitely-many-rational-points over k*) if, for every finitely generated subfield $K \subset k$ and every (proper) model $\mathcal{X}$ over K, the set $\mathcal{X}(K) := \mathrm{Hom}_K(\mathrm{Spec}\,K, \mathcal{X})$ is finite.

Remark 3.4 (Independence of Models). We point out that the finiteness property required for a projective variety to be Mordellic can also be tested on a fixed model. That is, a proper scheme X over k is Mordellic over k if and only if there is a finitely generated subfield $K \subset k$ and a proper model $\mathcal{X}$ for X over K such that for all finitely generated subfields $L \subset k$ containing K, the set $\mathcal{X}(L) := \mathrm{Hom}_K(\mathrm{Spec}\,L, \mathcal{X})$ is finite.

We note that Mordellicity (just like Brody hyperbolicity) descends along finite étale morphisms (Remark 2.2).

Remark 3.5 (Descending Mordellicity). Let $X \to Y$ be a finite étale morphism of projective varieties over k. Then it follows from the Chevalley–Weil theorem that X is Mordellic over k if and only if Y is Mordellic over k; see Theorem 7.9 for a proof (of a more general result).

It is clear that $\mathbb{P}^1_k$ is not Mordellic, as $\mathbb{P}^1(\mathbb{Q})$ is dense. A deep theorem of Faltings leads to the following classification of projective Mordellic curves. If $k = \overline{\mathbb{Q}}$, then this theorem is proven in [32]. The statement below is proven in [33] (see also [82]).

Theorem 3.6 (Faltings). *Let X be a smooth projective connected curve over k. Then X is Mordellic over k if and only if $\mathrm{genus}(X) \geq 2$.*

Recall that abelian varieties are very far from being Brody hyperbolic (Remark 2.4). The following remark says that abelian varieties are also very far from being Mordellic.

Remark 3.7. It is not at all obvious that a smooth projective connected curve of genus one over $\overline{\mathbb{Q}}$ is not Mordellic. Indeed, it is not an obvious fact that an elliptic curve over a number field K has positive rank over some finite field extension of K, although this is certainly true and can be proven in many different ways. In fact, by a theorem of Frey–Jarden [36] (see also [49, §3.1] or [41, §3]), if A is an abelian variety over k, then there is a finitely generated subfield $K \subset k$ and an abelian variety $\mathcal{A}$ over K with $\mathcal{A}_k \cong A$ such that $\mathcal{A}(K)$ is dense in A. This theorem is not hard to prove when k is uncountable but requires non-trivial arguments otherwise. Thus, if $\dim A \neq 0$, then one can consider the abelian variety A to be as far as possible from being Mordellic. This statement is to be compared with the conclusion of Remark 2.4.

By Remark 3.7, an obvious obstruction to a projective variety X over k being Mordellic is that it contains an abelian variety. The following theorem of Faltings says that this is the only obstruction if X can be embedded into an abelian variety; see [34].

Theorem 3.8 (Faltings). *Let X be a closed subvariety of an abelian variety A over k. Then X is Mordellic over k if and only if X does not contain the translate of a positive-dimensional abelian subvariety of A.*

There are strong similarities between the statements in the previous section and the current section. These similarities (and a healthy dose of optimism) lead to the first version of the Lang–Vojta conjecture. To state this conjecture, let us say that a variety X over k is *strongly-Brody hyperbolic* over k if, for every subfield $k_0 \subset k$, every model $\mathcal{X}$ for X over k_0, and every embedding $k_0 \to \mathbb{C}$, the variety $\mathcal{X}_\mathbb{C}$ is Brody hyperbolic.

Conjecture 3.9 (Weak Lang–Vojta, I). *Let X be an integral projective variety over k. Then X is Mordellic over k if and only if X is strongly-Brody hyperbolic over k.*

As stated, this conjecture does not predict that, if X is a projective Brody hyperbolic variety over $\mathbb{C}$, then every conjugate of X is Brody hyperbolic. We state this conjecture separately.

Conjecture 3.10 (Conjugates of Brody Hyperbolic Varieties). *If X is an integral variety over k. Then X is strongly-Brody hyperbolic over k if and only if there is a subfield $k_0 \subset k$, a model $\mathcal{X}$ for X over k_0, and an embedding $k_0 \to \mathbb{C}$ such that the variety $\mathcal{X}_\mathbb{C}$ is Brody hyperbolic.*

Concretely, Conjecture 3.10 says that, if X is a Brody hyperbolic variety over $\mathbb{C}$ and σ is a field automorphism of $\mathbb{C}$, then the σ-conjugate X^σ of X is again Brody hyperbolic.

We briefly discuss the notion of Mordellicity for quasi-projective (not necessarily proper) schemes. We will also comment on this more general notion in Section 7. This notion appears in this generality (to our knowledge) for the first time in Vojta's paper [86], and it is also studied in [56]. It is intimately related to the notion of "arithmetic hyperbolicity" [49, 53]; see Section 7 for a discussion.

In the non-proper case, it is natural to study integral points rather than rational points. Vojta noticed in [86] that, in fact, it is more natural to study "near-integral points". Below we make this more precise.

Definition 3.11. Let $X \to S$ be a morphism of schemes with S integral. We define $X(S)^{(1)}$ to be the set of P in $X(K(S))$ such that, for every point s in S of codimension one, the point P lies in the image of $X(\mathcal{O}_{S,s}) \to X(K(S))$.

Vojta refers to the points in $X(S)^{(1)}$ as "near-integral" S-points. We point out that on an affine variety, there is no difference between the finiteness of integral points and "near-integral" points; see Section 7.

Definition 3.12 (Quasi-Projective Mordellic Varieties). A variety X over k is *Mordellic over* k if, for every $\mathbb{Z}$-finitely generated subring $A \subset k$ and every model $\mathcal{X}$ for X over A, the set $\mathcal{X}(A)^{(1)}$ of near-integral A-points is finite.

The study of near-integral points might seem unnatural at first. To convince the reader that this notion is slightly more natural than the notion of integral point, we include the following remark.

Remark 3.13 (Why "Near-Integral" Points?). Consider a proper scheme $\mathcal{X}$ over $\mathbb{Z}$ with generic fibre $X := \mathcal{X}_{\mathbb{Q}}$. Let K be a finitely generated field of characteristic zero and let $A \subset K$ be a regular $\mathbb{Z}$-finitely generated subring. Then, the set of K-rational points $X(K)$ **equals** the set of near-integral A-points of $\mathcal{X}$. On the other hand, if K has transcendence degree at least one over $\mathbb{Q}$, then it is not necessarily true that every K-point of X is an A-point of $\mathcal{X}$. Thus, studying K-rational points on the proper variety X over $\mathbb{Q}$ is equivalent to studying near-integral points of the proper scheme $\mathcal{X}$ over $\mathbb{Z}$.

With this definition at hand, we are able to state Faltings's finiteness theorem for abelian varieties over number rings as a statement about the Mordellicity of the appropriate moduli space. The analogous statement on its Brody hyperbolicity is Example 2.7.

Theorem 3.14 (Faltings, Shafarevich's Conjecture for Principally Polarized Abelian Varieties). *Let k be an algebraically closed field of characteristic zero. Let $g \geq 1$ and let $N \geq 3$ be integers. Then, the (fine) moduli space $\mathcal{A}_{g,k}^{[N]}$ of g-dimensional principally polarized abelian varieties with level N structure is a smooth quasi-projective Mordellic variety over k.*

Example 2.7 and Theorem 3.14 suggest that there might also be an analogue of Lang–Vojta's conjecture for quasi-projective schemes. It seems reasonable to suspect that an affine variety over k is Mordellic over k if and only if it is strongly-Brody hyperbolic over k; see for example [46] for a discussion of Lang's conjectures in the affine case. However, stating a reasonable conjecture for quasi-projective varieties requires some care, and would take us astray from our current objective. We refer the interested reader to articles of Ascher–Turchet and Campana in this book [8, 16] for a related discussion, and the book by Vojta [85].

Remark 3.15 (From Shafarevich to Mordell). Let us briefly explain how Faltings shows that Theorem 3.14 implies Faltings's finiteness theorem for curves (Theorem 3.6). Let X be a smooth projective connected curve of genus at least two over k. By a construction of Kodaira [69], there is a finite étale morphism $Y \to X$, an integer $g \geq 1$, and a non-constant morphism $Y \to \mathcal{A}_{g,k}^{[3]}$. Since $\mathcal{A}_{g,k}^{[3]}$ is Mordellic over k and $Y \to \mathcal{A}_{g,k}^{[3]}$ has finite fibres, it follows that Y is Mordellic over k. As Mordellicity descends along finite étale morphisms (Remark 3.5), we conclude that X is Mordellic, as required.

4   Groupless Varieties

To study Lang–Vojta's conjectures, it is natural to study varieties which do not "contain" any algebraic groups. Indeed, as we have explained in Remark 2.4 (resp. Remark 3.7), a Brody hyperbolic variety (resp. a Mordellic variety) does not admit any non-trivial morphisms from an abelian variety. For projective varieties, it turns out that this is equivalent to not admitting a non-constant map from any connected algebraic group (see Lemma 4.4 below).

As before, we let k be an algebraically closed field of characteristic zero. We start with the following definition.

Definition 4.1. A variety X over k is *groupless* if every morphism $\mathbb{G}_{m,k} \to X$ (of varieties over k) is constant, and for every abelian variety A over k, every morphism $A \to X$ is constant.

Remark 4.2. We claim that, for proper varieties, the notion of grouplessness can be tested on morphisms (or even rational maps) from abelian varieties. That is, a proper variety X over k is groupless if and only if, for every abelian variety A over k, every rational map $A \dashrightarrow X$ is constant. To show this, first note that a morphism $\mathbb{G}_{m,k} \to X$ extends to a morphism $\mathbb{P}_k^1 \to X$ and that $\mathbb{P}_k^1$ is surjected upon by an elliptic curve. Therefore, if every morphism from an abelian variety is constant, then X is groupless and has no rational curves. Now, if X is proper over k and has no rational curves, every rational map $A \dashrightarrow X$ with A an abelian variety extends to a morphism (see [50, Lemma 3.5]). Thus, if every morphism $A \to X$ is constant with A an abelian variety, we conclude that every rational map $A \dashrightarrow X$ is constant. This proves the claim. We also conclude that a proper variety is groupless if and only if it is "algebraically hyperbolic" in Lang's sense [62, p. 176].

Remark 4.3 (Lang's Algebraic Exceptional Set). For X a proper variety over k, Lang defines the *algebraic exceptional set* $\mathrm{Exc}_{alg}(X)$ *of* X to be the union of all non-constant rational images of abelian varieties in X. With Lang's terminology at hand, as is explained in Remark 4.2, a proper variety X over k is groupless over k if and only if $\mathrm{Exc}_{alg}(X)$ is empty.

Let us clear up why we refer to this property as groupless.

Lemma 4.4 (Why Call This Groupless?). *A variety X over k is groupless if and only if for all finite type connected group schemes G over k, every morphism $G \to X$ is constant.*

Proof. This follows from Chevalley's structure theorem for algebraic groups over the algebraically closed field k of characteristic zero. A detailed proof is given in [50, Lemma 2.5]. $\square$

The notion of grouplessness is well-studied, and sometimes referred to as "algebraic hyperbolicity" or "algebraic Lang hyperbolicity"; see [43], [62, page 176], [59, Remark 3.2.24], or [60, Definition 3.4]. We will only use the term "algebraically hyperbolic" for the notion introduced by Demailly in [29] (see also [15, 50, 56]). The term "groupless" was first used in [50, Definition 2.1] and [55, Definition 3.1].

Example 4.5. A zero-dimensional variety is groupless. Note that $\mathbb{P}^1_k$, $\mathbb{A}^1_k$, $\mathbb{A}^1_k \setminus \{0\}$ and smooth proper genus one curves over k are not groupless.

Much like Brody hyperbolicity and Mordellicity, grouplessness descends along finite étale morphisms. We include a sketch of the proof of this simple fact.

Lemma 4.6 (Descending Grouplessness). *Let $X \to Y$ be a finite étale morphism of varieties over k. Then X is groupless over k if and only if Y is groupless over k.*

Proof. If Y is groupless, then X is obviously groupless. Therefore, to prove the lemma, we may assume that X is groupless. Let G be $\mathbb{G}_{m,k}$ or an abelian variety over k. Let $G \to Y$ be a morphism. Consider the pull-back $G' := G \times_Y X$ of $G \to Y$ along $X \to Y$. Then, as k is algebraically closed and of characteristic zero, each connected component of G' is (or: can be endowed with the structure of) an algebraic group isomorphic to $\mathbb{G}_{m,k}$ or an abelian variety over k. Therefore, the morphism $G' \to X$ is constant. This implies that $G \to Y$ is constant. $\square$

We include an elementary proof of the fact that the classification of one-dimensional groupless varieties is the same as that of one-dimensional Brody hyperbolic curves.

Lemma 4.7. *A smooth quasi-projective connected curve X over k is groupless over k if and only if X is not isomorphic to $\mathbb{P}^1_k$, $\mathbb{A}^1_k$, $\mathbb{A}^1_k \setminus \{0\}$, nor a smooth proper connected curve of genus one over k.*

Proof. If X is groupless, then X is not isomorphic to $\mathbb{P}^1_k$, $\mathbb{A}^1_k$, $\mathbb{A}^1_k \setminus \{0\}$, nor a smooth proper connected curve of genus one over k; see Example 4.5. Thus to prove the lemma, we may (and do) assume that X is not isomorphic to either of these curves. Let $Y \to X$ be a finite étale cover of X such that the smooth projective model $\overline{Y}$ of Y is of genus at least two. (It is clear that such a cover exists when $X = \mathbb{G}_{m,k} \setminus \{1\}$ or $X = E \setminus \{0\}$ with E an elliptic curve over k. This is enough to conclude that such a cover always exists.) By Lemma 4.6, the variety X is groupless if and only if Y is groupless. Thus, it suffices to show that $\overline{Y}$ is groupless. To do so, assume that we have a morphism $\mathbb{G}_{m,k} \to \overline{Y}$. By Riemann–Hurwitz, this morphism is constant, as $\overline{Y}$ has genus at least two. Now, let A be an abelian variety over k and let $A \to \overline{Y}$ be a morphism. To show that this morphism is constant, we compose $A \to \overline{Y}$ with the Jacobian map $\overline{Y} \to \operatorname{Jac}(\overline{Y})$ (after choosing some point on $\overline{Y}$). If the morphism $A \to \overline{Y}$ is non-constant, then it is surjective. Since a morphism of abelian varieties is a homomorphism (up to translation of the origin), this induces a group structure on the genus > 1 curve $\overline{Y}$. However, as the automorphism group of (the positive-dimensional variety) $\overline{Y}$ is finite, the curve $\overline{Y}$ cannot be endowed with the structure of an algebraic group. This shows that $A \to \overline{Y}$ is constant, and concludes the proof. □

Bloch–Ochiai–Kawatama's theorem (Theorem 2.5) and Faltings's analogous theorem for rational points on closed subvarieties of abelian varieties (Theorem 3.8) characterize "hyperbolic" subvarieties of abelian varieties. It turns out that this characterization also holds for groupless varieties, as we explain now.

If X is a closed subvariety of an abelian variety A over k, we define the special locus $\operatorname{Sp}(X)$ of X to be the union of the translates of positive-dimensional abelian subvarieties of A contained in X.

Lemma 4.8. *Let X be a closed integral subvariety of an abelian variety A over k. Then X is groupless over k if and only if $\operatorname{Sp}(X)$ is empty.*

Proof. Clearly, if X is groupless over k, then X does not contain the translate of a positive-dimensional abelian subvariety of A, so that $\operatorname{Sp}(X)$ is empty. Conversely, assume that X does not contain the translate of a non-zero abelian subvariety of A. Let us show that X is groupless. Since the Albanese variety of $\mathbb{P}^1_k$ is trivial, any map $\mathbb{G}_{m,k} \to X$ is constant. Thus, to conclude the proof, we have to show that all morphisms $A' \to X$ are constant, where A' is an abelian variety over k. To do so, note that the image of $A' \to X$ in A is the translate of an abelian subvariety of A, as morphisms of abelian varieties are homomorphisms up to translation. This means that the image of $A' \to X$ is the translate of an abelian subvariety, hence a point (by our assumption). □

Remark 4.9. Let A be a simple abelian surface. Let $X = A \setminus \{0\}$. Then X is groupless. This remark might seem misplaced, but it shows that "grouplessness" as defined above does not capture the non-hyperbolicity of a quasi-projective variety. The "correct" definition in the quasi-projective case is discussed in Section 6 (and is also discussed in [56, 86]).

Although grouplessness does not capture the non-hyperbolicity of quasi-projective varieties (Remark 4.9), Lang conjectured that grouplessness is equivalent to being Mordellic and to being Brody hyperbolic (up to choosing a model over $\mathbb{C}$) for *projective* varieties. This brings us to the second form of Lang–Vojta's conjecture.

Conjecture 4.10 (Weak Lang–Vojta, II). *Let X be an integral projective variety over k. Then the following are equivalent.*

(1) The projective variety X is Mordellic over k.
(2) The variety X is strongly-Brody hyperbolic over k.
(3) The variety X is groupless over k.

5   Varieties of General Type

In this section we discuss the role of varieties of general type in Lang–Vojta's conjecture. Recall that a line bundle L on a smooth projective variety S over k is big if there is an ample line bundle A and an effective divisor D such that $L \cong A \otimes \mathcal{O}_S(D)$; see [64, 65]. We follow standard terminology and say that an integral proper variety X over k is of general type if it has a desingularization $X' \to X$ with X' a smooth projective integral variety over k such that the canonical bundle $\omega_{X'/k}$ is a big line bundle. For example, if $\omega_{X'/k}$ is ample, then it is big. Moreover, we will say that a proper variety X over a field k is of general type if, for every irreducible component Y of X, the reduced closed subscheme Y_{red} is of general type.

Varieties of general type are well-studied; see [64, 65]. For the sake of clarity, we briefly collect some statements. Our aim is to emphasize the similarities with the properties presented in the earlier sections.

For example, much like Brody hyperbolicity, Mordellicity, and grouplessness, the property of being of general type descends along finite étale morphisms. That is, if $X \to Y$ is a finite étale morphism of proper schemes over k, then X is of general type if and only if Y is of general type. Moreover, a simple computation of the degree of the canonical bundle of a curve implies that, if X is a smooth projective connected curve over k, then X is of general type if and only if genus$(X) \geq 2$.

Kawamata and Ueno classified which closed subvarieties of an abelian variety are of general type. To state their result, for A an abelian variety over k and X a closed subvariety of A, recall that the special locus $\mathrm{Sp}(X)$ of X is the union of translates of positive-dimensional abelian subvarieties of A contained in X. Note that Bloch–Ochiai–Kawamata's theorem (Theorem 2.5) can be stated as saying that a closed subvariety X of an abelian variety A over $\mathbb{C}$ is Brody hyperbolic if and only if $\mathrm{Sp}(X)$ is empty. Similarly, Faltings's theorem (Theorem 3.8) can be stated as saying that a closed subvariety of an abelian variety A over k is Mordellic if and only if $\mathrm{Sp}(X)$ is empty. The latter is also equivalent to saying that X is groupless over k by Lemma 4.8. The theorem of Kawamata–Ueno now reads as follows.

Theorem 5.1 (Kawamata–Ueno). *Let A be an abelian variety and let X be a closed integral subvariety of A. Then $\mathrm{Sp}(X)$ is a closed subset of X, and X is of general type if and only if $\mathrm{Sp}(X) \neq X$.*

Note that being of general type and being groupless are not equivalent. This is not a surprise, as the notion of general type is a birational invariant, whereas the blow-up of a smooth groupless surface along a point is no longer groupless. The conjectural relation between varieties of general type and the three notions (Brody hyperbolicity, Mordellicity, and grouplessness) introduced above is as follows.

Conjecture 5.2 (Weak Lang–Vojta, III). *Let X be an integral projective variety over k. Then the following are equivalent.*

(1) The projective variety X is Mordellic over k.
(2) The variety X is strongly-Brody hyperbolic over k.
(3) Every integral subvariety of X is of general type.
(4) The variety X is groupless over k.

Note that the notion of general type is a birational invariant, but hyperbolicity is not. What should (conjecturally) correspond to being of general type? The highly optimistic conjectural answer is that being of general type should correspond to being "pseudo"-Brody hyperbolic, "pseudo"-Mordellic, and "pseudo"-groupless. The definitions of these notions are essentially the same as given above, the only difference being that one has to allow for an "exceptional locus". In the following sections we will make this more precise.

6   Pseudo-Grouplessness

Let k be an algebraically closed field of characteristic zero. Roughly speaking, a projective variety X over k is groupless if it admits no non-trivial morphisms from a connected algebraic group. Conjecturally, a projective variety X over k is groupless if and only if every subvariety of X is of general type. To see what should correspond to being of general type, we will require the more general notion of pseudo-grouplessness.

Definition 6.1. Let X be a variety over k and let $\Delta \subset X$ be a closed subset. We say that X is *groupless modulo* Δ (*over* k) if, for every finite type connected group scheme G over k and every dense open subscheme $U \subset G$ with $\mathrm{codim}(G \setminus U) \geq 2$, every non-constant morphism $U \to X$ factors over Δ.

Hyperbolicity modulo a subset was first introduced by Kiernan–Kobayashi [58], and is thoroughly studied in Kobayashi's book [59]. As we will see below, it is quite natural to extend the study of hyperbolic varieties to the study of varieties which are hyperbolic modulo a subset.

For proper schemes, the notion of "groupless modulo the empty set" coincides with the notion of grouplessness introduced before (and studied in [49, 50, 55]). For the reader's convenience, we include a detailed proof of this.

Lemma 6.2. *Let X be a proper scheme over k. Then the following are equivalent.*

(1) The scheme X is groupless modulo the empty subscheme $\emptyset$ over k.
(2) The scheme X is groupless.
(3) For every finite type connected group scheme G over k and every dense open subscheme $V \subset G$, every morphism $V \to X$ is constant.

Proof. It is clear that (1) implies (2). To show that (2) implies (3), let G be a finite type connected group scheme over k, let $V \subset G$ be a dense open subscheme, and let $f : V \to X$ be a morphism of schemes over k. Then, as X is proper over k, there is an open subscheme $U \subset G$ containing V with $\mathrm{codim}(G \setminus U) \geq 2$ such that the morphism $f : V \to X$ extends to a morphism $f' : U \to X$. Since X is groupless and proper, it does not contain any rational curves. Therefore, as the variety underlying G is smooth over k [81, Tag 047N], it follows from [50, Lemma 3.5] (see also [27, Corollary 1.44]) that the morphism $f' : U \to X$ extends (uniquely) to a morphism $f'' : G \to X$. Since X is groupless, the morphism f'' is constant. This implies that f is constant. Finally, it is clear (from the definitions) that (3) implies (1).   □

Definition 6.3. A variety X is *pseudo-groupless* (*over k*) if there is a proper closed subset $\Delta \subsetneq X$ such that X is groupless modulo Δ.

The word "pseudo" in this definition refers to the fact that the non-hyperbolicity of the variety is concentrated in a proper closed subset. Note that a variety X is pseudo-groupless if and only if every irreducible component of X is pseudo-groupless.

Example 6.4. Let C be smooth projective connected curve of genus at least two and let X be the blow-up of $C \times C$ in a point. Then X is not groupless. However, its "non-grouplessness" is contained in the exceptional locus Δ of the blow-up $X \to C \times C$. Thus, as X is groupless modulo Δ, it follows that X is pseudo-groupless.

Let us briefly say that an open subset U of an integral variety V is *big* if $\mathrm{codim}(V \setminus U)$ is at least two. Now, the reader might wonder why we test pseudo-grouplessness on maps whose domain is a big open subset of some algebraic group. The example to keep in mind here is the blow-up of a simple abelian surface in its origin. In fact, as we test pseudo-grouplessness on big open subsets of abelian varieties (and not merely maps from abelian varieties), such blow-ups are *not* pseudo-groupless. Also, roughly speaking, one should consider big open subsets of abelian varieties as far as possible from being hyperbolic, in any sense of the word "hyperbolic". For example, much like how abelian varieties admit a dense entire curve (Remark 2.4), a big open subset of an abelian variety admits a dense entire curve. This is proven using Sard's theorem in [86]. Thus, big open subsets of abelian varieties are also as far as possible from being Brody hyperbolic.

We now show that the statement of Lemma 4.6 also holds in the "pseudo" setting, i.e., we show that pseudo-grouplessness descends along finite étale morphisms. As we have mentioned before, this descent property also holds for general type varieties.

Lemma 6.5. *Let $f : X \to Y$ be a finite étale morphism of varieties over k. Then X is pseudo-groupless over k if and only if Y is pseudo-groupless over k.*

Proof. We adapt the arguments in the proof of [55, Proposition 2.13]. First, if Y is groupless modulo a proper closed subset $\Delta_Y \subset Y$, then clearly X is groupless modulo the proper closed subset $f^{-1}(\Delta_Y)$. Now, assume that X is groupless modulo a proper closed subset $\Delta_X \subsetneq X$. Let G be a finite type connected (smooth quasi-projective) group scheme over k, let $U \subset G$ be a dense open subscheme with $\mathrm{codim}(G \setminus U) \geq 2$ and let $\phi : U \to Y$ be a morphism which does not factor over $f(\Delta_X)$. The pull-back of $G \to Y$ along the finite étale morphism $f : X \to Y$ induces a finite étale morphism $V := U \times_Y X \to U$. Since U is smooth over k, by purity of the branch locus [39, Théorème X.3.1], the finite étale morphism $V \to U$ extends (uniquely) to a finite étale morphism $G' \to G$. Note that every connected component G'' of G' has the structure of a finite type connected group scheme over k (and with this structure the morphism $G'' \to G$ is a homomorphism). Now, since smooth morphisms are codimension-preserving, we see that $\mathrm{codim}(G'' \setminus V) \geq 2$. As the morphism $V \to X$ does not factor over $f^{-1}(f(\Delta_X))$, it does not factor over Δ_X, and is thus constant (as X is groupless modulo Δ_X). This implies that the morphism $U \to Y$ is constant, as required.   □

Remark 6.6 (Birational Invariance). Let X and Y be proper schemes over k. Assume that X is birational to Y. Then X is pseudo-groupless over k if and only if Y is pseudo-groupless over k. This is proven in [56]. Thus, as pseudo-grouplessness is a birational invariant among proper varieties, this notion is more natural to study from a birational perspective than grouplessness.

Remark 6.7. Contrary to a hyperbolic proper variety, a proper pseudo-groupless variety could have rational curves. For example, the blow-up of the product of two smooth curves of genus two in a point (as in Example 6.4) contains precisely one rational curve. However, a pseudo-groupless proper variety is not covered by rational curves, i.e., it is non-uniruled, as all rational curves are contained in a proper closed subset (by definition).

Remark 6.8. Let X be a proper scheme over k and let $\Delta \subset X$ be a closed subset. It follows from the valuative criterion of properness that X is groupless modulo Δ if and only if, for every finite type connected group scheme G over k and every dense open subscheme $U \subset G$, any non-constant morphism $U \to X$ factors over Δ.

Recall that Lemma 4.4 says that the grouplessness of a proper variety entails that there are no non-constant morphisms from *any* connected algebraic group. One of the main results of [56] is the analogue of Lemma 4.4 for pseudo-groupless varieties. The proof of this result (see Theorem 6.9 below) relies on the structure theory of algebraic groups.

Theorem 6.9. *If X is a proper scheme X over k and Δ is a closed subset of X, then X is groupless modulo Δ over k if and only if, for every abelian variety A over k and every open subscheme $U \subset A$ with $\mathrm{codim}(A \setminus U) \geq 2$, every non-constant morphism of varieties $U \to X$ factors over Δ.*

Theorem 6.9 says that the pseudo-grouplessness of a proper variety can be tested on morphisms from big open subsets of abelian varieties (or on rational maps from abelian varieties). A similar, but different, statement holds for affine varieties. Indeed, if X is an affine variety over k, then X is groupless modulo $\Delta \subset X$ if and only if every non-constant morphism $\mathbb{G}_{m,k} \to X$ factors over Δ.

Lang conjectured that a projective variety is pseudo-groupless if and only if it is of general type. Note that, by the birational invariance of these two notions, this conjecture can be reduced to the case of smooth projective varieties by Hironaka's resolution of singularities.

Conjecture 6.10 (Strong Lang–Vojta, I). *Let X be an integral projective variety over k. Then X is pseudo-groupless over k if and only if X is of general type over k.*

Note that this conjecture predicts more than the equivalence of (3) and (4) in Conjecture 5.2. Also, even though it is stated for projective varieties, one could as well formulate the conjecture for proper varieties (or even proper algebraic spaces). The resulting "more general" conjecture actually follows from the above conjecture.

Example 6.11. By Kawamata–Ueno's theorem (Theorem 5.1) and Lemma 4.8, the Strong Lang–Vojta conjecture holds for closed subvarieties of abelian varieties.

Remark 6.12. If X is a proper pseudo-groupless surface, then X is of general type (see [56] for a proof). For higher-dimensional varieties, Conjecture 6.10 predicts a similar statement, but this is not even known for threefolds. However, assuming the Abundance Conjecture and certain conjectures on Calabi–Yau varieties, one can show that every proper pseudo-groupless variety is of general type (i.e., (1) $\implies$ (2) in Conjecture 6.10). Regarding the implication (2) $\implies$ (1), not much is known beyond the one-dimensional case. For example, if X is a proper surface of general type, then

Conjecture 6.10 implies that there should be a proper closed subset $\Delta \subset X$ such that every rational curve $C \subset X$ is contained in Δ. Such statements are known to hold for certain surfaces of general type by the work of Bogomolov and McQuillan; see [30, 71].

7 Pseudo-Mordellicity and Pseudo-Arithmetic Hyperbolicity

In the previous section, we introduced pseudo-grouplessness and stated Lang–Vojta's conjecture that a projective variety is of general type if and only if it is pseudo-groupless. In this section, we explain what the "pseudo" analogue is of the notion of Mordellicity, and explain Lang–Vojta's conjecture that a projective variety is of general type if and only if it is pseudo-Mordellic.

7.1 Pseudo-Arithmetic Hyperbolicity

As we have said before, Lang coined the term "Mordellic". We will now introduce the related (and a priori different) notion of arithmetic hyperbolicity (as defined in [49, 52, 53]); see also [83, §2], and [10, 11]. In Section 3 we ignored that the extension of the notion of Mordellicity over $\overline{\mathbb{Q}}$ to arbitrary algebraically closed fields can actually be done in two a priori different ways. We discuss both notions now and give them *different* names. We refer the reader to Section 3 for our conventions regarding models of varieties, and we continue to let k denote an algebraically closed field of characteristic zero.

Definition 7.1. Let X be a variety over k and let Δ be a closed subset of X. We say that X is *arithmetically hyperbolic modulo Δ over k* if, for every $\mathbb{Z}$-finitely generated subring A and every model $\mathcal{X}$ for X over A, we have that every positive-dimensional irreducible component of the Zariski closure of $\mathcal{X}(A)$ in X is contained in Δ.

Definition 7.2. A variety X over k is *pseudo-arithmetically hyperbolic over k* if there is a proper closed subset $\Delta \subset X$ such that X is arithmetically hyperbolic modulo Δ over k.

Remark 7.3. A variety X over k is arithmetically hyperbolic over k (as defined in [49] and [53, §4]) if and only if X is arithmetically hyperbolic over k modulo the empty subscheme.

Lemma 7.4 (Independence of Model). *Let X be a variety over k and let Δ be a closed subset of k. Then the following are equivalent.*

(1) The finite type scheme X over k is arithmetically hyperbolic modulo Δ.

(2) There is a $\mathbb{Z}$-finitely generated subring $A \subset k$, there is a model $\mathcal{X}$ for X over A, and there is a model $\mathcal{D} \subset \mathcal{X}$ for $\Delta \subset X$ over A such that, for every $\mathbb{Z}$-finitely generated subring $B \subset k$ containing A, the set

$$\mathcal{X}(B) \setminus \mathcal{D}(B)$$

is finite.

Proof. This follows from standard spreading out arguments. These type of arguments are used in [53] to prove more general statements in which the objects are algebraic stacks.   □

Remark 7.5. We unravel what the notion of arithmetic hyperbolicity modulo Δ entails for affine varieties. To do so, let X be an affine variety over k, and let Δ be a proper closed subset of X. Choose the following data.

- integers $n, \delta, m \geq 1$;
- polynomials $f_1, \ldots, f_n \in k[x_1, \ldots, x_m]$;
- polynomials $d_1, \ldots, d_\delta \in k[x_1, \ldots, x_m]$;
- an isomorphism

$$X \cong \mathrm{Spec}(k[x_1, \ldots, x_m]/(f_1, \ldots, f_n));$$

- an isomorphism

$$\Delta \cong \mathrm{Spec}(k[x_1, \ldots, x_m]/(d_1, \ldots, d_\delta)).$$

Let A_0 be the $\mathbb{Z}$-finitely generated subring of k generated by the (finitely many) coefficients of the polynomials $f_1, \ldots, f_n, d_1, \ldots, d_\delta$. Now, the following statements are equivalent.

(1) The variety X is arithmetically hyperbolic modulo Δ over k.
(2) For every $\mathbb{Z}$-finitely generated subring $A \subset k$ containing A_0, the set

$$\{a \in A^m \mid f_1(a) = \ldots = f_n(a) = 0\} \setminus \{a \in A^m \mid d_1(a) = \ldots = d_\delta(a) = 0\}$$

is finite.

Thus, roughly speaking, one could say that an algebraic variety over k is arithmetically hyperbolic modulo Δ over k if "X minus Δ" has only finitely many A-valued points, for any choice of finitely generated subring $A \subset k$.

7.2   Pseudo-Mordellicity

The reader might have noticed a possibly confusing change in terminology. Why do we not refer to the above notion as being "Mordellic modulo Δ"? The precise reason brings us to a subtle point in the study of integral points valued in higher-dimensional rings (contrary to those valued in $\mathcal{O}_{K,S}$ with S a finite set of finite places of a number field K). To explain this subtle point, let us first define what it means to be pseudo-Mordellic. For this definition, we will require the notion of "near-integral" point (Definition 3.11).

Definition 7.6. Let X be a variety over k and let Δ be a closed subset of X. We say that X is *Mordellic modulo* Δ *over* k if, for every $\mathbb{Z}$-finitely generated subring A and every model $\mathcal{X}$ for X over A, we have that every positive-dimensional irreducible component of the Zariski closure of $\mathcal{X}(A)^{(1)}$ in X is contained in Δ, where $\mathcal{X}(A)^{(1)}$ is defined in Definition 3.11.

Remark 7.7. Let X be a proper scheme over k and let Δ be a closed subset of X. Then, by the valuative criterion of properness, the proper scheme X is Mordellic modulo Δ if, for every finitely generated subfield $K \subset k$ and every proper model $\mathcal{X}$ over K, the set $\mathcal{X}(K) \setminus \Delta$ is finite.

Definition 7.8. A variety X over k is *pseudo-Mordellic over* k if there is a proper closed subset $\Delta \subset X$ such that X is Mordellic modulo Δ over k.

Note that X is Mordellic over k (as defined in Section 3) if and only if X is Mordellic modulo the empty subset. It is also clear from the definitions that, if X is Mordellic modulo Δ over k, then X is arithmetically hyperbolic modulo Δ over k. In particular, a pseudo-Mordellic variety is pseudo-arithmetically hyperbolic and a Mordellic variety is arithmetically hyperbolic. Indeed, roughly speaking, to say that a variety is arithmetically hyperbolic is to say that any set of integral points on it is finite, and to say that a variety is Mordellic is to say that any set of "near-integral" points on it is finite. The latter sets are a priori bigger. However, there is no difference between these two sets when $k = \overline{\mathbb{Q}}$. That is, a variety X over $\overline{\mathbb{Q}}$ is arithmetically hyperbolic modulo Δ if and only if it is Mordellic modulo Δ over $\overline{\mathbb{Q}}$.

Following the exposition in the previous sections, let us prove the fact that pseudo-arithmetic hyperbolicity (resp. pseudo-Mordellicity) descends along finite étale morphisms of varieties.

Theorem 7.9 (Chevalley–Weil). *Let* $f : X \to Y$ *be a finite étale surjective morphism of varieties over* k. *Let* $\Delta \subset X$ *be a closed subset. If* X *is Mordellic modulo* Δ *over* k *(resp. arithmetically hyperbolic modulo* Δ *over* k*), then* Y *is Mordellic modulo* $f(\Delta)$ *over* k *(resp. arithmetically hyperbolic modulo* $f(\Delta)$ *over* k*).*

Proof. We assume that X is Mordellic modulo Δ, and show that Y is Mordellic modulo $f(\Delta)$. (The statement concerning arithmetic hyperbolicity is proven similarly.)

Let $A \subset k$ be a regular $\mathbb{Z}$-finitely generated subring, let $\mathcal{X}$ be a model for X over A, let $\mathcal{Y}$ be a model for Y over A, and let $F : \mathcal{X} \to \mathcal{Y}$ be a finite étale surjective morphism such that $F_k = f$. Assume for a contradiction that Y is not Mordellic modulo $f(\Delta)$. Then, replacing A by a larger regular $\mathbb{Z}$-finitely generated subring of k if necessary, for $i = 1, 2, \ldots$, we may choose pairwise distinct elements a_i of $\mathcal{Y}(A)^{(1)}$ whose closure in Y is an irreducible positive-dimensional subvariety $R \subset Y$ such that $R \not\subset f(\Delta)$. For every $i = 1, 2, \ldots$, choose a dense open subscheme U_i of Spec A whose complement in Spec A has codimension at least two and such that a_i defines a morphism $a_i : U_i \to \mathcal{X}$. Consider $V_i := U_i \times_{\mathcal{Y}, F} \mathcal{X} \to \mathcal{X}$, and note that $V_i \to U_i$ is finite étale. By Zariski–Nagata purity of the branch locus [39, Théorème X.3.1], the morphism $V_i \to U_i$ extends to a finite étale morphism Spec $B_i \to A$. By Hermite's finiteness theorem, as the degree of B_i over A is bounded by $\deg(f)$, replacing a_i by an infinite subset if necessary, we may and do assume that $B := B_1 \cong B_2 \cong B_3 \cong \ldots$. Now, the $b_i : V_i \to \mathcal{X}$ define elements in $\mathcal{X}(B)^{(1)}$. Let S be their closure in X. Note that $R \subset S$. In particular, $S \not\subset \Delta$. This contradicts the fact that X is Mordellic modulo Δ. Thus, we conclude that Y is Mordellic modulo $f(\Delta)$.   □

Corollary 7.10 (Pseudo-Chevalley–Weil). *Let $f : X \to Y$ be a finite étale surjective morphism of finite type separated schemes over k. Then X is pseudo-Mordellic over k if and only if Y is pseudo-Mordellic over k.*

Proof. Since $f : X \to Y$ has finite fibres, the fibres of f are Mordellic over k. Therefore, if Y is pseudo-Mordellic over k, it easily follows that X is pseudo-Mordellic over k. Conversely, if X is pseudo-Mordellic over k, then it follows from Theorem 7.9 that Y is pseudo-Mordellic over k.   □

Corollary 7.11 (Pseudo-Chevalley–Weil, II). *Let $f : X \to Y$ be a finite étale surjective morphism of finite type separated schemes over k. Then X is pseudo-arithmetically hyperbolic over k if and only if Y is pseudo-arithmetically hyperbolic over k.*

Proof. Similar to the proof of Corollary 7.10.   □

Remark 7.12 (Birational Invariance). The birational invariance of the notion of pseudo-Mordellicity is essentially built into the definition. Indeed, the infinitude of the set of near-integral points is preserved under proper birational modifications. More precisely, let X and Y be proper integral varieties over k which are birational. Then X is pseudo-Mordellic over k if and only if Y is pseudo-Mordellic over k.

It is not clear to us whether the notion of pseudo-arithmetic hyperbolicity over k is a birational invariant for proper varieties over k, unless $k = \overline{\mathbb{Q}}$. Similarly, it is not so clear to us whether pseudo-arithmetically hyperbolic proper varieties are pseudo-groupless. On the other hand, this is not so hard to prove for pseudo-Mordellic varieties.

Theorem 7.13. *If X is a pseudo-Mordellic proper variety over k, then X is pseudo-groupless over k.*

Proof. The fact that an arithmetically hyperbolic variety is groupless is proven in [49, §3] using the potential density of rational points on an abelian variety over a field K of characteristic zero (Remark 3.7). The statement of the theorem is proven in [56] using similar arguments.   □

Remark 7.14. Let X be a proper surface over k. If X is pseudo-Mordellic over k, then X is of general type. To prove this, note that X is pseudo-groupless (Theorem 7.13), so that the claim follows from the fact that pseudo-groupless proper surfaces are of general type; see Remark 6.12.

Recall that a closed subvariety X of an abelian variety A is groupless modulo its special locus $\mathrm{Sp}(X)$, where $\mathrm{Sp}(X)$ is the union of translates of non-zero abelian subvarieties of A contained in X. (We are freely using here Kawamata–Ueno's theorem that $\mathrm{Sp}(X)$ is a closed subset of X.) This was proven in Lemma 4.8. In [34] Faltings proved the arithmetic analogue of this statement.

Theorem 7.15 (Faltings). *Let A be an abelian variety over k, and let $X \subset A$ be a closed subvariety. Then X is Mordellic modulo $\mathrm{Sp}(X)$.*

Lang and Vojta conjectured that a projective variety over $\overline{\mathbb{Q}}$ is pseudo-Mordellic if and only if it is of general type. We propose extending this to arbitrary algebraically closed fields of characteristic zero. As we also expect the notions of pseudo-arithmetic hyperbolicity and pseudo-Mordellicity to coincide, we include this in our version of the Lang–Vojta conjecture.

Conjecture 7.16 (Strong Lang–Vojta, II). *Let X be an integral projective variety over k. Then the following statements are equivalent.*

(1) The variety X is pseudo-Mordellic over k.
(2) The variety X is pseudo-arithmetically hyperbolic over k.
(3) The variety X is pseudo-groupless over k.
(4) The projective variety X is of general type over k.

This is a good time to collect examples of arithmetically hyperbolic varieties.

Example 7.17. It follows from Faltings's theorem [34] that a normal projective connected pseudo-groupless surface X over k with $\mathrm{h}^1(X, \mathcal{O}_X) > 2$ is pseudo-Mordellic. Let us prove this claim. To do so, let $\Delta \subset X$ be a proper closed subset such that X is groupless modulo Δ. Moreover, let A be the Albanese variety of X, let $p : X \to A$ be the canonical map (after choosing some basepoint in $X(k)$), and let Y be the image of X in A. Note that $\dim Y \geq 1$. If $\dim Y = 1$, then the condition on the dimension of A implies that Y is not an elliptic curve. In this case, since $\dim X = 2$ and $\dim Y = 1$, the claim follows from Faltings's (earlier) finiteness theorem for hyperbolic curves. However, if $\dim Y = 2$, we have to appeal to Faltings's Big Theorem. Indeed, in this case, the morphism $X \to Y$ is generically finite. Let $X \to X' \to Y$ be the Stein factorization of the morphism $X \to Y$, where $X' \to Y$ is a finite morphism with X' normal. Since X and X' are birational, it suffices to show that X' is pseudo-Mordellic (by the birational invariance of pseudo-Mordellicity and pseudo-grouplessness). Thus, we may and do assume that $X = X'$, so that $X \to A$ is finite. If the rational points on X are dense, then they are also dense in Y, so that Y is an abelian subvariety of A, contradicting our assumption that $\mathrm{h}^1(X, \mathcal{O}_X) = \dim A > 2$. Thus, the rational points on X are not dense. In particular, every irreducible component of the closure of a set of rational points on X is a curve of genus 1 (as X does not admit any curves of genus zero). Since X is pseudo-groupless, these components are contained in Δ.

Example 7.18. Let X be a smooth projective connected curve over k, let $n \geq 1$ be an integer, and let Δ be a proper closed subset of Sym^n_X. It follows from Faltings's theorem that Sym^n_X is groupless modulo Δ over k if and only if Sym^n_X is arithmetically hyperbolic modulo Δ over k.

Example 7.19 (Moriwaki). Let X be a smooth projective variety over k such that Ω^1_X is ample and globally generated. Then X is Mordellic by a theorem of Moriwaki [73]; see [7] for the analogous finiteness result in the logarithmic case.

Example 7.20. For every $\mathbb{Z}$-finitely generated normal integral domain A of characteristic zero, the set of A-isomorphism classes of smooth sextic surfaces in $\mathbb{P}^3_A$ is finite; see [54]. This finiteness statement can be reformulated as saying that the moduli stack of smooth sextic surfaces is Mordellic.

Example 7.21. Let X be a smooth proper hyperkaehler variety over k with Picard number at least three. Then X is not arithmetically hyperbolic; see [49].

7.3   Intermezzo: Arithmetic Hyperbolicity and Mordellicity

Let k be an algebraically closed field of characteristic zero. In this section, we show that the (a priori) difference between arithmetic hyperbolicity (modulo some subset) and Mordellicity is quite subtle, as this difference disappears in many well-studied cases.

The following notion of purity for models over $\mathbb{Z}$-finitely generated rings was first considered in [15] precisely to study the a priori difference between arithmetic hyperbolicity and Mordellicity.

Definition 7.22 (Pure Model). Let X be a variety over k and let $A \subset k$ be a subring. A model $\mathcal{X}$ for X over A is *pure over A* (or: *satisfies the extension property over A*) if, for every smooth finite type separated integral scheme T over A, every dense open subscheme $U \subset T$ with $T \setminus U$ of codimension at least two in T, and every A-morphism $f : U \to \mathcal{X}$, there is a (unique) morphism

$\overline{f} : T \to \mathcal{X}$ extending the morphism f. (The uniqueness of the extension $\overline{f}$ follows from our convention that a model for X over A is separated.)

Remark 7.23. Let X be a variety over k, and let $A \subset k$ be a subring. Let $\mathcal{X}$ be a pure model for X over A, and let $B \subset k$ be a subring containing A such that $\operatorname{Spec} B \to \operatorname{Spec} A$ is smooth (hence finite type). Then $\mathcal{X}_B$ is pure over B.

Definition 7.24. A variety X over k has an *arithmetically pure model* if there is a $\mathbb{Z}$-finitely generated subring $A \subset k$ and a pure model $\mathcal{X}$ for X over A.

Remark 7.25. Let X be a proper variety over k which has an arithmetically pure model. Then X has no rational curves. To prove this, assume that $\mathbb{P}^1_k \to X$ is a non-constant (hence finite) morphism, i.e., the proper variety X has a rational curve over k. Then, if we let 0 denote the point $(0 : 0 : 1)$ in $\mathbb{P}^2_k$, the composed morphism $\mathbb{P}^2_k \setminus \{0\} \to \mathbb{P}^1_k \to X$ does not extend to a morphism from $\mathbb{P}^2_k$ to X. Now, choose a $\mathbb{Z}$-finitely generated subring $A \subset k$ and a model $\mathcal{X}$ over A such that the morphism $\mathbb{P}^1_k \to X$ descends to a morphism $\mathbb{P}^1_A \to \mathcal{X}$ of A-schemes. Define $U = \mathbb{P}^2_A \setminus \{0\}$ and $T = \mathbb{P}^2_A$, where we let $\{0\}$ denote the image of the section of $\mathbb{P}^2_A \to \operatorname{Spec} A$ induced by 0 in $\mathbb{P}^2_k$. Since the morphism $U_k \to \mathcal{X}_k$ does not extend to a morphism $T_k \to X_k$, we see that the morphism $U \to \mathcal{X}$ does not extend to a morphism $T \to \mathcal{X}$, so that $\mathcal{X}$ is not pure. This shows that a proper variety over k with a rational curve has no arithmetically pure model.

Remark 7.26. Let X be a proper variety over k. A pure model for X over a $\mathbb{Z}$-finitely generated subring A of k might have rational curves in every special fibre (of positive characteristic). Examples of such varieties can be constructed as complete subvarieties of the moduli space of principally polarized abelian varieties.

Remark 7.27. Let X be a smooth projective variety over k. If $\Omega^1_{X/k}$ is ample, then X has an arithmetically pure model. Indeed, choose a $\mathbb{Z}$-finitely generated subring $A \subset k$ with A smooth over $\mathbb{Z}$ and a smooth projective model $\mathcal{X}$ for X over A such that $\Omega_{\mathcal{X}/A}$ is ample. Then, the geometric fibres of $\mathcal{X} \to \operatorname{Spec} A$ do not contain any rational curves, so that [37, Proposition 6.2] implies that $\mathcal{X}$ is a pure model for X over A.

Remark 7.28. Let $k \subset L$ be an extension of algebraically closed fields of characteristic zero, and let X be a variety over k. Then X has an arithmetically pure model if and only if X_L has an arithmetically pure model.

Theorem 7.29. *Let X be a variety over k which has an arithmetically pure model. Let $\Delta \subset X$ be a closed subset. Then X is Mordellic modulo Δ over k if and only if X is arithmetically hyperbolic modulo Δ over k.*

Proof. We follow the proof of [15, Theorem 8.10]. Suppose that X is arithmetically hyperbolic modulo Δ over k. Let $A \subset k$ be a $\mathbb{Z}$-finitely generated subring and let $\mathcal{X}$ be a pure model for X over A. It suffices to show that, for every $\mathbb{Z}$-finitely generated subring $B \subset k$ containing A, the set $\mathcal{X}(B)^{(1)} \setminus \Delta$ is finite. To do so, we may and do assume that $\operatorname{Spec} B \to \operatorname{Spec} A$ is smooth in which case it follows from the definition of a pure model that $\mathcal{X}(B)^{(1)} = \mathcal{X}(B)$. We conclude that

$$\mathcal{X}(B)^{(1)} \setminus \Delta = \mathcal{X}(B) \setminus \Delta$$

is finite. This shows that X is Mordellic modulo Δ over k. □

Lemma 7.30 (Affine Varieties). *Let X be an affine variety over k. Then X has an arithmetically pure model.*

Proof. Affine varieties have an arithmetically pure model by Hartog's Lemma. □

Lemma 7.31. *Let X be a variety over k which admits a finite morphism to some semi-abelian variety over k. Then X has an arithmetically pure model.*

Proof. Let G be a semi-abelian variety and let $X \to G$ be a finite morphism. It follows from Hartog's Lemma that X has an arithmetically pure model if and only if G has an arithmetically pure model. Choose a $\mathbb{Z}$-finitely generated subring and a model $\mathcal{G}$ for G over A such that $\mathcal{G} \to \operatorname{Spec} A$ is a semi-abelian scheme. Then, this model $\mathcal{G}$ has the desired extension property by [72, Lemma A.2], so that G (hence X) has an arithmetically pure model. □

Remark 7.32. Let X be a projective integral groupless surface over k which admits a non-constant map to some abelian variety. Then X has an arithmetically pure model by [15, Lemma 8.11].

Corollary 7.33. *Let X be an integral variety over k, and let $\Delta \subset X$ be a closed subset. Assume that one of the following statements holds.*

(1) The variety X is affine over k.
(2) There is a finite morphism $X \to G$ with G a semi-abelian variety over k.
(3) We have that X is a groupless surface which admits a non-constant morphism $X \to A$ with A an abelian variety over k.

Then X is arithmetically hyperbolic modulo Δ over k if and only if X is Mordellic modulo Δ over k.

Proof. Assume (1). Then the statement follows from Lemma 7.30 and Theorem 7.29. Similarly, if (2) holds, then the statement follows from Lemma 7.31 and Theorem 7.29. Finally, assuming (3), the statement follows from Remark 7.32 and Theorem 7.29. □

Remark 7.34. Let $g \geq 1$ and $N \geq 3$ be integers. Now, if X is the fine moduli space of g-dimensional principally polarized abelian schemes over k with level

$\overline{\mathbb{Q}}$ structure, then X has an arithmetically pure model. As is explained in [68], this is a consequence of Grothendieck's theorem on homomorphisms of abelian schemes [40]. The existence of such a model is used by Martin-Deschamps to deduce the Mordellicity of X_k over k from the Mordellicity of X over $\overline{\mathbb{Q}}$ (cf. Theorem 3.14).

8 Pseudo-Brody Hyperbolicity

The notion of pseudo-hyperbolicity appeared first in the work of Kiernan and Kobayashi [58] and afterwards in Lang [62]; see also [59]. We recall some of the definitions.

Definition 8.1. Let X be a variety over $\mathbb{C}$ and let Δ be a closed subset of X. We say that X is *Brody hyperbolic modulo* Δ if every holomorphic non-constant map $\mathbb{C} \to X^{\mathrm{an}}$ factors over Δ.

Definition 8.2. A variety X over $\mathbb{C}$ is *pseudo-Brody hyperbolic* if there is a proper closed subset $\Delta \subsetneq X$ such that X is Brody hyperbolic modulo Δ.

Green–Griffiths and Lang conjectured that a projective integral variety of general type is pseudo-Brody hyperbolic. The conjecture that a projective integral variety is of general type if and only if it is pseudo-Brody hyperbolic is commonly referred to as the Green–Griffiths–Lang conjecture.

Note that the notion of pseudo-Brody hyperbolicity is a birational invariant. More precisely, if X and Y are proper integral varieties over $\mathbb{C}$ which are birational, then X is pseudo-Brody hyperbolic if and only if Y is pseudo-Brody hyperbolic. Furthermore, just like the notions of pseudo-Mordellicity and pseudo-grouplessness, the notion of pseudo-Brody hyperbolicity descends along finite étale morphisms. That is, if $X \to Y$ is finite étale, then X is pseudo-Brody hyperbolic if and only if Y is pseudo-Brody hyperbolic. Also, it is not hard to show that, if a variety X is Brody hyperbolic modulo Δ, then X is groupless modulo Δ.

Note that a variety X is Brody hyperbolic (as defined in Section 2) if and only if X is Brody hyperbolic modulo the empty set. Bloch–Ochiai–Kawamata's theorem classifies Brody hyperbolic closed subvarieties of abelian varieties. In fact, their result is a consequence of the following more general statement (also proven in [57]).

Theorem 8.3 (Bloch–Ochiai–Kawamata). *Let X be a closed subvariety of an abelian variety A. Let $\mathrm{Sp}(X)$ be the special locus of X. Then $\mathrm{Sp}(X)$ is a closed subset of X and X is Brody hyperbolic modulo $\mathrm{Sp}(X)$.*

We now introduce the pseudo-analogue of Kobayashi hyperbolicity for algebraic varieties. Of course, these definitions make sense for complex-analytic spaces.

Definition 8.4. Let X be a variety over $\mathbb{C}$ and let Δ be a closed subset of X. We say that X is *Kobayashi hyperbolic modulo* Δ if, for every x and y in $X^{\mathrm{an}} \setminus \Delta^{\mathrm{an}}$ with $x \neq y$, the Kobayashi pseudo-distance $d_{X^{\mathrm{an}}}(p, q)$ is positive.

Definition 8.5. A variety X over $\mathbb{C}$ is *pseudo-Kobayashi hyperbolic* if there is a proper closed subset $\Delta \subsetneq X$ such that X is Kobayashi hyperbolic modulo Δ.

It is clear from the definitions and the fact that the Kobayashi pseudo-metric vanishes everywhere on $\mathbb{C}$, that a variety X which is Kobayashi hyperbolic modulo a closed subset $\Delta \subset X$ is Brody hyperbolic modulo Δ. Nonetheless, the notion of pseudo-Kobayashi hyperbolicity remains quite mysterious at the moment. Indeed, we do not know whether a pseudo-Brody hyperbolic projective variety X over $\mathbb{C}$ is pseudo-Kobayashi hyperbolic.

One can show that the notion of pseudo-Kobayashi hyperbolicity is a birational invariant. That is, if X and Y are proper integral varieties over $\mathbb{C}$ which are birational, then X is pseudo-Kobayashi hyperbolic if and only if Y is pseudo-Kobayashi hyperbolic; see [59]. Moreover, just like the notions of pseudo-Mordellicity and pseudo-grouplessness, pseudo-Kobayashi hyperbolicity descends along finite étale morphisms.

Yamanoi proved the pseudo-Kobayashi analogue of Bloch–Ochiai–Kawamata's theorem for closed subvarieties of abelian varieties; see [88, Theorem 1.2].

Theorem 8.6 (Yamanoi). *Let X be a closed subvariety of an abelian variety A. Let $\mathrm{Sp}(X)$ be the special locus of X. Then $\mathrm{Sp}(X)$ is a closed subset of X and X is Kobayashi hyperbolic modulo $\mathrm{Sp}(X)$.*

The Lang–Vojta conjecture and the Green–Griffiths conjecture predict that the above notions of hyperbolicity are equivalent. To state this conjecture, we will need one more definition. (Recall that k denotes an algebraically closed field of characteristic zero.)

Definition 8.7. A variety X over k is *strongly-pseudo-Brody hyperbolic* (resp. *strongly-pseudo-Kobayashi hyperbolic*) if, for every subfield $k_0 \subset k$, every model $\mathcal{X}$ for X over k_0, and every embedding $k_0 \to \mathbb{C}$, the variety $\mathcal{X}_{0,\mathbb{C}}$ is pseudo-Brody hyperbolic (resp. pseudo-Kobayashi hyperbolic).

Conjecture 8.8 (Strong Lang–Vojta, III). *Let X be an integral projective variety over k. Then the following statements are equivalent.*

(1) The variety X is strongly-pseudo-Brody hyperbolic over k.

(2) The variety X is strongly-pseudo-Kobayashi hyperbolic over k.

(3) The projective variety X is pseudo-Mordellic over k.

(4) The projective variety X is pseudo-arithmetically hyperbolic over k.

(5) The projective variety X is pseudo-groupless over k.

(6) The projective variety X is of general type over k.

As stated this conjecture does not predict that every conjugate of a pseudo-Brody hyperbolic variety is again pseudo-Brody hyperbolic. We state this as a separate conjecture, as we did in Conjecture 3.10 for Brody hyperbolic varieties.

Conjecture 8.9 (Conjugates of Pseudo-Brody Hyperbolic Varieties). *If X is an integral variety over k and σ is a field automorphism of k, then the following statements hold.*

(1) The variety X is pseudo-Brody hyperbolic if and only if X^σ is pseudo-Brody hyperbolic.

(2) The variety X is pseudo-Kobayashi hyperbolic if and only if X^σ is pseudo-Kobayashi hyperbolic.

We conclude this section with a brief discussion of a theorem of Kwack on the algebraicity of holomorphic maps to a hyperbolic variety, and a possible extension of his result to the pseudo-setting.

Remark 8.10 (Borel Hyperbolicity). Let X be a variety over $\mathbb{C}$ and let $\Delta \subset X$ be a closed subset. We extend the notion of Borel hyperbolicity introduced in [51] to the pseudo-setting and say that X is *Borel hyperbolic modulo Δ* if, for every reduced variety S over $\mathbb{C}$, every holomorphic map $f : S^{\mathrm{an}} \to X^{\mathrm{an}}$ with $f(S^{\mathrm{an}}) \not\subset \Delta^{\mathrm{an}}$ is the analytification of a morphism $\varphi : S \to X$. The proof of [51, Lemma 3.2] shows that, if X is Borel hyperbolic modulo Δ, then it is Brody hyperbolic modulo Δ. In [61] Kwack showed that, if X is a proper Kobayashi hyperbolic variety, then X is Borel hyperbolic (modulo the empty set). It seems reasonable to suspect that Kwack's theorem also holds in the pseudo-setting. Thus, we may ask: if X is Kobayashi hyperbolic modulo Δ, does it follow that X is Borel hyperbolic modulo Δ?

The reader interested in investigating further complex-analytic notions of hyperbolicity is also encouraged to have a look at the notion of taut-hyperbolicity modulo a subset introduced by Kiernan–Kobayashi [58]; see also [59, Chapter 5].

9 Algebraic Hyperbolicity

In the following three sections we investigate (a priori) different function field analogues of Mordellicity. Conjecturally, they are all equivalent notions. At this point it is also clear that hyperbolicity *modulo* a subset is more natural to study (especially from a birational perspective) which is why we will give the definitions in this more general context.

The notion we introduce in this section extends Demailly's notion of algebraic hyperbolicity [29, 50] to the pseudo-setting.

Definition 9.1 (Algebraic Hyperbolicity Modulo a Subset). Let X be a projective scheme over k and let Δ be a closed subset of X. We say that X is *algebraically hyperbolic over k modulo Δ* if, for every ample line bundle L on X, there is a real number $\alpha_{X,\Delta,L}$ depending only on X, Δ, and L such that, for every smooth projective connected curve C over k and every morphism $f : C \to X$ with $f(C) \not\subset \Delta$, the inequality

$$\deg_C f^*L \leq \alpha_{X,\Delta,L} \cdot \mathrm{genus}(C)$$

holds.

Definition 9.2. A projective scheme X is *pseudo-algebraically hyperbolic* (*over k*) if there is a proper closed subset Δ such that X is algebraically hyperbolic modulo Δ.

We will say that a projective scheme X is *algebraically hyperbolic* over k if it is algebraically modulo the empty subset. This terminology is consistent with that of [50].

The motivation for introducing and studying algebraically hyperbolic projective schemes are the results of Demailly stated below. They say that algebraic hyperbolicity lies between Brody hyperbolicity and grouplessness. In particular, the Lang–Vojta conjectures as stated in the previous sections imply that groupless projective varieties should be algebraically hyperbolic, and that algebraically hyperbolic projective varieties should be Brody hyperbolic. This observation is due to Demailly and allows one to split the conjecture that groupless projective varieties are Brody hyperbolic into two a priori different parts.

Before stating Demailly's theorems, we note that it is not hard to see that pseudo-algebraic hyperbolicity descends along finite étale maps, and that pseudo-algebraic hyperbolicity for projective schemes is a birational invariant; see [56, §4] for details. These two properties should be compared with their counterparts for pseudo-grouplessness, pseudo-Mordellicity, pseudo-Brody hyperbolicity, and pseudo-Kobayashi hyperbolicity.

Demailly's theorem for projective schemes reads as follows.

Theorem 9.3 (Demailly). *Let X be a projective scheme over $\mathbb{C}$. If X is Brody hyperbolic, then X is algebraically hyperbolic over $\mathbb{C}$.*

A proof of this is given in [29, Theorem 2.1] when X is smooth. The smoothness of X is, however, not used in its proof. We stress that it is *not* known whether a pseudo-Brody hyperbolic projective scheme is pseudo-algebraically hyperbolic. On the other hand, Demailly proved that algebraically hyperbolic projective schemes are groupless, and his proof can be adapted to show the following more general statement.

Theorem 9.4 (Demailly + ϵ). *Let X be a projective scheme over k and let $\Delta \subset X$ be a closed subset. If X is algebraically hyperbolic modulo Δ, then X is groupless modulo Δ.*

Proof. See [50] when $\Delta = \emptyset$. The more general statement is proven in [56]. The argument involves the multiplication maps on an abelian variety.   □

Combining Demailly's theorems with Bloch–Ochiai–Kawamata's theorem, we obtain that a closed subvariety of an abelian variety over k is algebraically hyperbolic over k if and only if it is groupless. The pseudo-version of this theorem is due to Yamanoi (see Section 13 for a precise statement).

10   Boundedness

To say that a projective variety X is algebraically hyperbolic (Definition 9.1) is to say that the degree of any curve C is bounded uniformly and linearly in the genus of that curve. The reader interested in understanding how far we are from proving that groupless projective schemes are algebraically hyperbolic is naturally led to studying variants of algebraic hyperbolicity in which one asks (in Definition 9.1 above) for "weaker" bounds on the degree of a map. This led the authors of [50] to introducing the notion of boundedness. To state their definition, we first recall some basic properties of moduli spaces of morphisms between projective schemes.

Let S be a scheme, and let $X \to S$ and $Y \to S$ be projective flat morphisms of schemes. By Grothendieck's theory of Hilbert schemes and Quot schemes [75], the functor

$$\mathrm{Sch}/S^{op} \to \mathrm{Sets}, \quad T \to S \mapsto \mathrm{Hom}_T(Y_T, X_T)$$

is representable by an S-scheme which we denote by $\underline{\mathrm{Hom}}_S(X, Y)$. Moreover, for $h \in \mathbb{Q}[t]$ a polynomial, the subfunctor parametrizing morphisms whose graph has Hilbert polynomial h is representable by a quasi-projective subscheme $\underline{\mathrm{Hom}}_S^h(Y, X)$ of $\underline{\mathrm{Hom}}_S(Y, X)$. Similarly, the subfunctor of $\underline{\mathrm{Hom}}_S(X, X)$ parametrizing automorphisms of X over S is representable by a locally finite type group scheme $\mathrm{Aut}_{X/S}$ over S. It is imperative to note that this group scheme need not be quasi-compact. In fact, for a K3 surface X over $\mathbb{C}$, the scheme $\mathrm{Aut}_{X/\mathbb{C}}$ is zero-dimensional. Nonetheless, there are K3 surfaces with infinitely many automorphisms. Thus, the automorphism group scheme of a projective scheme over k is not necessarily of finite type (even when it is zero-dimensional).

If $S = \mathrm{Spec}\, k$, $d \geq 1$ is an integer, and $X = Y = \mathbb{P}^1_k$, let $\underline{\mathrm{Hom}}_k^d(\mathbb{P}^1_k, \mathbb{P}^1_k)$ be the subscheme of $\underline{\mathrm{Hom}}_k(\mathbb{P}^1_k, \mathbb{P}^1_k)$ parametrizing morphisms of degree d. In particular, we have that $\underline{\mathrm{Hom}}_k^1(\mathbb{P}^1_k, \mathbb{P}^1_k) = \mathrm{Aut}_{\mathbb{P}^1_k/k} = \mathrm{PGL}_{2,k}$. For every $d \geq 1$, the quasi-projective scheme $\underline{\mathrm{Hom}}_k^d(\mathbb{P}_k^1, \mathbb{P}_k^1)$ is non-empty (and even positive-dimensional). If we identity the subscheme of $\underline{\mathrm{Hom}}_k(\mathbb{P}_k^1, \mathbb{P}_k^1)$ parametrizing constant morphisms with $\mathbb{P}_k^1$, then

$$\underline{\mathrm{Hom}}_k(\mathbb{P}^1, \mathbb{P}_k^1) = \mathbb{P}_k^1 \sqcup \mathrm{PGL}_{2,k} \sqcup \bigsqcup_{d=2}^{\infty} \underline{\mathrm{Hom}}_k^d(\mathbb{P}_k^1, \mathbb{P}_k^1).$$

It follows that the scheme $\underline{\mathrm{Hom}}_k(\mathbb{P}^1, \mathbb{P}_k^1)$ has infinitely many connected components. It is in particular not of finite type.

It turns out that studying projective varieties X over k for which *every* Hom-scheme $\mathrm{Hom}_k(Y, X)$ is of finite type is closely related to studying algebraically hyperbolic varieties. The aim of this section is to explain the connection in a systematic manner as is done in [15, 50, 56]. We start with the following definitions.

Definition 10.1 (Boundedness Modulo a Subset). Let $n \geq 1$ be an integer, let X be a projective scheme over k, and let Δ be a closed subset of X. We say that X is *n-bounded over k modulo Δ* if, for every normal projective variety Y of dimension at most n, the scheme $\underline{\mathrm{Hom}}_k(Y, X) \setminus \underline{\mathrm{Hom}}_k(Y, \Delta)$ is of finite type over k. We say that X is *bounded over k modulo Δ* if, for every $n \geq 1$, the scheme X is n-bounded modulo Δ.

Definition 10.2. Let $n \geq 1$ be an integer. A projective scheme X over k is *pseudo-n-bounded over k* if there is a proper closed subset Δ such that X is n-bounded modulo Δ.

Definition 10.3. A projective scheme X over k is *pseudo-bounded over k* if it is pseudo-n-bounded over k for every $n \geq 1$.

Remark 10.4. At the beginning of this section we discussed the structure of the scheme $\underline{\mathrm{Hom}}_k(\mathbb{P}_k^1, \mathbb{P}_k^1)$. From that discussion it follows that $\mathbb{P}_k^1$ is not 1-bounded over k. In particular, if X is a 1-bounded projective variety over k, then it has no rational curves. It is also not hard to show that $\mathbb{P}_k^1$ is not pseudo-1-bounded by showing that, for every x in $\mathbb{P}^1(k)$, there is a y in $\mathbb{P}^1(k)$ such that the set of morphisms $f : \mathbb{P}_k^1 \to \mathbb{P}_k^1$ with $f(y) = x$ is infinite. We refer the interested reader to Section 11 for a related discussion.

We say that X is *bounded* if it is bounded modulo the empty subset. We employ similar terminology for n-bounded. This terminology is consistent with that of [15, 50]. Let us start with looking at some implications and relations between these a priori different notions of boundedness.

Boundedness is a condition on moduli spaces of maps from higher-dimensional varieties. Although it might seem a priori stronger than 1-boundedness, Lang–Vojta's conjecture predicts their equivalence. In fact, we have the following result from [50] which shows the equivalence of three a priori different notions. In this theorem, the implications (2) $\implies$ (1) and (3) $\implies$ (1) are straightforward consequences of the definitions.

Theorem 10.5. *Let X be a projective scheme over k. Then the following are equivalent.*

(1) The projective scheme X is 1-bounded over k.
(2) The projective scheme X is bounded over k.
(3) For every ample line bundle $\mathcal{L}$ and every integer $g \geq 0$, there is an integer $\alpha(X, \mathcal{L}, g)$ such that, for every smooth projective connected curve C of genus g over k and every morphism $f : C \to X$, the inequality

$$\deg_C f^* \mathcal{L} \leq \alpha(X, \mathcal{L}, g)$$

holds.

Proof. The fact that a 1-bounded scheme is n-bounded for every $n \geq 1$ is proven by induction on n in [50, §9]. The idea is that, if $f_i : Y \to X$ is a sequence of morphisms from an n-dimensional smooth projective variety Y with pairwise distinct Hilbert polynomial, then one can find a smooth hyperplane section $H \subset Y$ such that the restrictions $f_i|_H$ of these morphisms f_i to H still have pairwise distinct Hilbert polynomial.

The fact that a bounded scheme satisfies the "uniform" boundedness property in (3) follows from reformulating this statement in terms of the quasi-compactness of the universal Hom-stack of morphisms of curves of genus g to X; see the proof of [50, Theorem 1.14] for details.   □

Studying boundedness is "easier" than studying boundedness modulo a subset Δ. Indeed, part of the analogue of this theorem for pseudo-boundedness (unfortunately) requires an assumption on the base field k.

Theorem 10.6. *Let X be a projective scheme over k, and let Δ be a closed subset of X. Assume that k is **uncountable**. Then X is 1-bounded modulo Δ if and only if X is bounded modulo Δ.*

Proof. This is proven in [15], and the argument is similar to the proof of Theorem 10.5. We briefly indicate how the uncountability of k is used.

Assume that X is 1-bounded modulo Δ. We show by induction on n that X is n-bounded modulo Δ over k. If $n = 1$, then this holds by assumption. Thus, let $n > 1$ be an integer and assume that X is $(n-1)$-bounded modulo Δ. Let Y be an n-dimensional projective reduced scheme and let $f_m : Y \to X$ be a sequence of morphisms with pairwise distinct Hilbert polynomial such that, for every $m = 1, 2, \ldots$, we have $f_m(Y) \not\subset \Delta$. Since k is **uncountable**, there is an ample divisor D in Y which is not contained in $f_m^{-1}(\Delta)$ for all $m \in \{1, 2, \ldots\}$. Now, the restrictions $f_m|_D : D \to X$ have pairwise distinct Hilbert polynomial and, for infinitely many m, we have that $f_m(D) \not\subset \Delta$. This contradicts the induction hypothesis. We conclude that X is bounded modulo Δ over k, as required.   □

The "pseudo" analogue of the equivalence between (2) and (3) in Theorem 10.5 holds without any additional assumption on k; see [15].

Theorem 10.7. *Let X be a projective scheme over k. Then X is bounded modulo Δ over k if and only if, for every ample line bundle $\mathcal{L}$ and every integer $g \geq 0$, there is an integer $\alpha(X, \mathcal{L}, g)$ such that, for every smooth projective connected curve C of genus g over k and every morphism $f : C \to X$ with $f(C) \not\subset \Delta$, the inequality*

$$\deg_C f^*\mathcal{L} \leq \alpha(X, \mathcal{L}, g)$$

holds.

It is not hard to see that being pseudo-n-bounded descends along finite étale maps. Also, if X and Y are projective schemes over k which are birational, then X is pseudo-1-bounded if and only if Y is pseudo-1-bounded; see [56, §4]. However, in general, it is not clear that pseudo-n-boundedness is a birational invariant (unless $n = 1$ or k is uncountable).

It is shown in [15, 50] that pseudo-algebraically hyperbolic varieties are pseudo-bounded. More precisely, one can prove the following statement.

Theorem 10.8. *If X is algebraically hyperbolic modulo Δ over k, then X is bounded modulo Δ.*

Proof. This is proven in three steps in [15, §9]. First, one chooses an uncountable algebraically closed field L containing k and shows that X_L is algebraically hyperbolic modulo Δ_L. Then, one makes the "obvious" observation that X_L is 1-bounded modulo Δ_L. Finally, as L is uncountable and X_L is 1-bounded modulo Δ_L, it follows from Theorem 10.6 that X_L is bounded modulo Δ_L.   □

Demailly proved that algebraically hyperbolic projective varieties are groupless (Theorem 9.4). His proof can be adapted to show the following more general statement.

Proposition 10.9 (Demailly + ϵ). *If X is 1-bounded modulo Δ over k, then X is groupless modulo Δ.*

11   Geometric Hyperbolicity

In the definition of Mordellicity over $\overline{\mathbb{Q}}$ one considers the "finiteness of arithmetic curves" on some model. On the other hand, the notions of algebraic hyperbolicity and boundedness require one to test "boundedness of curves". In this section we introduce a new notion in which one considers the "finiteness of pointed curves".

Definition 11.1 (Geometric Hyperbolicity Modulo a Subset). Let X be a variety over k and let Δ be a closed subset of X. We say that X is *geometrically hyperbolic over k modulo Δ* if, for every x in $X(k) \setminus \Delta$, every smooth connected curve C over k and every c in $C(k)$, we have that the set $\operatorname{Hom}_k((C,c),(X,x))$ of morphisms $f : C \to X$ with $f(c) = x$ is finite.

Definition 11.2. A variety X over k is *pseudo-geometrically hyperbolic over k* if there is a proper closed subset Δ such that X is geometrically hyperbolic modulo Δ.

We say that a variety X over k is *geometrically hyperbolic over k* if it is geometrically hyperbolic modulo the empty subset. At this point we should note that a projective scheme X over k is geometrically hyperbolic over k if and only if it is "$(1,1)$-bounded". The latter notion is defined in [50, §4], and the equivalence of these two notions is [50, Lemma 4.6] (see also Proposition 11.4 below). The terminology "$(1,1)$-bounded modulo Δ" is used in [15], and also coincides with being geometrically hyperbolic modulo Δ for projective schemes by the results in [15, §9].

Remark 11.3 (Geometric Hyperbolicity Versus Arithmetic Hyperbolicity). Let us say that a scheme T is an *arithmetic curve* if there is a number field K and a finite set of finite places S of K such that $T = \operatorname{Spec} \mathcal{O}_{K,S}$. Let X be a variety over $\overline{\mathbb{Q}}$. It is not hard to show that the following two statements are equivalent.

(1) The variety X is arithmetically hyperbolic (or Mordellic) over $\overline{\mathbb{Q}}$.
(2) For every arithmetic curve $\mathcal{C}$, every closed point c in $\mathcal{C}$, every model $\mathcal{X}$ for X over $\mathcal{C}$, and every closed point x of $\mathcal{X}$, the subset

$$\operatorname{Hom}_{\mathcal{C}}((\mathcal{C},c),(\mathcal{X},x)) \subset \mathcal{X}(\mathcal{C})$$

of morphisms $f : \mathcal{C} \to \mathcal{X}$ with $f(c) = x$ is finite.

Indeed, if (1) holds, then $\operatorname{Hom}_{\mathcal{C}}(\mathcal{C},\mathcal{X})$ is finite by definition, so that clearly the set

$$\operatorname{Hom}_{\mathcal{C}}((\mathcal{C},c),(\mathcal{X},x))$$

is finite. Conversely, assume that (2) holds. Now, let $\mathcal{C}$ be an arithmetic curve and let $\mathcal{X}$ be a model for X over $\mathcal{C}$. To show that $\mathcal{X}(\mathcal{C})$ is finite, let c be a closed point of $\mathcal{C}$ and let κ be its residue field. Then κ is finite and c lies in $\mathcal{C}(\kappa)$. In particular, the image of c along any morphism $\mathcal{C} \to \mathcal{X}$ is a κ-point of $\mathcal{X}$. This shows that

$$\mathcal{X}(\mathcal{C}) \subset \bigcup_{x \in \mathcal{X}(\kappa)} \operatorname{Hom}_{\mathcal{C}}((\mathcal{C},c),(\mathcal{X},x)).$$

Since $\mathcal{X}(\kappa)$ is finite and every set $\mathrm{Hom}_{\mathcal{C}}((\mathcal{C}, c), (\mathcal{X}, x))$ is finite, we conclude that $\mathcal{X}(\mathcal{C})$ is finite, as required.

The second statement allows one to see the similarity between geometric hyperbolicity and arithmetic hyperbolicity. Indeed, the variety X is geometrically hyperbolic over $\overline{\mathbb{Q}}$ if, for every integral algebraic curve C over $\overline{\mathbb{Q}}$, every closed point c in C, and every closed point x of X, the set

$$\mathrm{Hom}_k((C, c), (X, x)) = \mathrm{Hom}_C((C, c), (X \times C, (x, c)))$$

is finite.

Just like pseudo-grouplessness and pseudo-Mordellicity, it is not hard to see that pseudo-geometric hyperbolicity descends along finite étale morphisms. Also, if X and Y are projective varieties which are birational, then X is pseudo-geometrically hyperbolic if and only if Y is pseudo-geometrically hyperbolic.

The following proposition says that a projective scheme is geometrically hyperbolic if and only if the moduli space of pointed maps is of finite type. In other words, asking for boundedness of all pointed maps is equivalent to asking for the finiteness of all sets of pointed maps.

Proposition 11.4. *Let X be a projective scheme over k and let Δ be a closed subset of X. Then the following are equivalent.*

(1) For every smooth projective connected curve C over k, every c in $C(k)$ and every x in $X(k) \backslash \Delta$, the scheme $\underline{\mathrm{Hom}}_k((C, c), (X, x))$ is of finite type over k.

(2) The variety X is geometrically hyperbolic modulo Δ.

Proof. This is proven in [15, §9]. The proof is a standard application of the bend-and-break principle. Indeed, the implication (2) $\Longrightarrow$ (1) being obvious, let us show that (1) $\Longrightarrow$ (2). Thus, let us assume that X is not geometrically hyperbolic modulo Δ, so that there is a sequence $f_1, f_2, \ldots$ of pairwise distinct elements of $\mathrm{Hom}_k((C, c), (X, x))$, where C is a smooth projective connected curve over k, $c \in C(k)$ and $x \in X(k) \setminus \Delta$. Since $\underline{\mathrm{Hom}}_k((C, c), (X, x))$ is of finite type, the degree of all the f_i is bounded by some real number (depending only on X, Δ, c, x and C). In particular, it follows that some connected component of $\underline{\mathrm{Hom}}_k((C, c), (X, x))$ has infinitely many elements. As each connected component of $\underline{\mathrm{Hom}}_k((C, c), (X, x))$ is a finite type scheme over k, it follows from bend-and-break [27, Proposition 3.5] that there is a rational curve in X containing x. This contradicts the fact that every rational curve in X is contained in Δ (by Proposition 11.7). $\qquad\square$

This proposition has the following consequence.

Corollary 11.5. *Let X be a projective scheme over k and let Δ be a proper closed subset of X. If X is 1-bounded modulo Δ, then X is geometrically hyperbolic modulo Δ.*

Proof. If X is 1-bounded, then it is clear that, for every smooth projective connected curve C, every c in $C(k)$ and every x in $X(k) \setminus \Delta$, the scheme $\underline{\mathrm{Hom}}_k((C, c), (X, x))$ is of finite type over k. Indeed, the latter scheme is closed in the scheme $\underline{\mathrm{Hom}}_k(C, X)$, and contained in the quasi-projective subscheme $\underline{\mathrm{Hom}}_k(C, X) \setminus \underline{\mathrm{Hom}}_k(C, \Delta)$. Therefore, the result follows from Proposition 11.4.   □

Remark 11.6. Urata showed that a Brody hyperbolic projective variety over $\mathbb{C}$ is geometrically hyperbolic over $\mathbb{C}$; see [59, Theorem 5.3.10] (or the original [84]). Note that Corollary 11.5 generalizes Urata's theorem (in the sense that the assumption in Corollary 11.5 is a priori weaker than being Brody hyperbolic, and we also allow for an "exceptional set" Δ). Indeed, as a Brody hyperbolic projective variety is 1-bounded (even algebraically hyperbolic), Urata's theorem follows directly from Corollary 11.5.

Demailly's argument to show that algebraically hyperbolic projective varieties are groupless (Theorem 9.4) can be adapted to show that geometrically hyperbolic projective varieties are groupless; see [56] for a detailed proof.

Proposition 11.7. *Let X be a projective variety over k and let Δ be a closed subset of X. If X is geometrically hyperbolic modulo Δ over k, then X is groupless modulo Δ over k.*

12   The Conjectures Summarized

After a lengthy preparation, we are finally ready to state the complete version of Lang–Vojta's conjecture.

Conjecture 12.1 (Strong Lang–Vojta, IV). *Let X be an integral projective variety over k. Then the following statements are equivalent.*

(1) *The variety X is strongly-pseudo-Brody hyperbolic over k.*
(2) *The variety X is strongly-pseudo-Kobayashi hyperbolic.*
(3) *The projective variety X is pseudo-Mordellic over k.*
(4) *The projective variety X is pseudo-arithmetically hyperbolic over k.*
(5) *The projective variety X is pseudo-groupless over k.*
(6) *The projective variety X is pseudo-algebraically hyperbolic over k.*
(7) *The projective variety X is pseudo-bounded over k.*
(8) *The projective variety X is pseudo-1-bounded over k.*
(9) *The projective variety X is pseudo-geometrically hyperbolic over k.*
(10) *The projective variety X is of general type over k.*

Conjecture 12.1 is the final version of the Lang–Vojta conjecture for pseudo-hyperbolic varieties, and also encompasses Green–Griffiths's conjecture for projective varieties of general type. We note that one aspect of the Lang–Vojta conjecture and the Green–Griffiths conjecture that is ignored in this conjecture is whether the conjugate of a Brody hyperbolic variety is Brody hyperbolic (see Conjectures 3.10 and 8.9).

The following implications are known. First, (6) $\implies$ (7), (7) $\implies$ (8), (8) $\implies$ (9), and (9) $\implies$ (5). Also, (3) $\implies$ (4), (3) $\implies$ (5). Finally, (2) $\implies$ (1) and (1) $\implies$ (5). The following diagram summarizes these known implications. The content of the Strong Lang–Vojta conjecture is that all the notions appearing in this diagram are equivalent.

pseudo- $\implies$ pseudo-bounded $\implies$ pseudo-1-bounded $\implies$ pseudo-
algebraically   geometrically
hyperbolic   hyperbolic

$$\Downarrow$$

pseudo-Mordellic $\implies$ pseudo- $\qquad\qquad\implies$ pseudo-groupless
arithmetically
hyperbolic

$$\Uparrow$$

strongly-pseudo- $\implies$ stongly-pseudo-
Kobayashi   Brody hyperbolic
hyperbolic

We stress that the Strong Lang–Vojta conjecture is concerned with classifying projective varieties of general type via their complex-analytic or arithmetic properties. Recall that Campana's special varieties can be considered as being opposite to varieties of general type. As Campana's conjectures are concerned with characterizing special varieties via their complex-analytic or arithmetic properties, his conjectures should be considered as providing another part of the conjectural picture. We refer the reader to [16] for a discussion of Campana's conjectures.

The following conjecture is only concerned with hyperbolic varieties and is, therefore, a priori weaker than the Strong Lang-Vojta conjecture. It is not clear to us whether the Strong Lang–Vojta conjecture can be deduced from the following weaker version, as there are pseudo-hyperbolic projective varieties which are not birational to a hyperbolic projective variety.

Conjecture 12.2 (Weak Lang–Vojta, IV). *Let X be an integral projective variety over k. Then the following statements are equivalent.*

(1) The variety X is strongly-Brody hyperbolic over k.
(2) The variety X is strongly-Kobayashi hyperbolic over k.
(3) The projective variety X is Mordellic over k.
(4) The projective variety X is arithmetically hyperbolic over k.
(5) The projective variety X is groupless over k.
(6) The projective variety X is algebraically hyperbolic over k.

(7) The projective variety X is bounded over k.
(8) The projective variety X is 1-bounded over k.
(9) The projective variety X is geometrically hyperbolic over k.
(10) Every integral subvariety of X is of general type.

Remark 12.3 (Strong Implies Weak). Let us illustrate why the strong Lang–Vojta conjecture implies the Weak Lang–Vojta conjecture. To do so, let X be a projective variety. Assume that X is groupless. Then X is pseudo-groupless. Thus, by the Strong Lang–Vojta conjecture, we have that X is Mordellic modulo some proper closed subset $\Delta \subset X$. Now, since X is groupless, it follows that Δ is groupless. Repeating the above argument shows that Δ is Mordellic, so that X is Mordellic.

We know *more* about the Weak Lang–Vojta conjecture than we do about the Strong Lang–Vojta conjecture. Indeed, it is known that $(1) \iff (2)$ by Brody's Lemma. Also, it is not hard to show that $(2) \implies (5)$. Moreover, we know that $(3) \implies (4)$ and $(4) \implies (5)$. Of course, we also have that $(6) \implies (7)$, $(7) \implies (8)$, and $(8) \iff (9)$. In addition, we also have that $(1) \implies (6)$ and that $(10) \implies (5)$. Figure 1 summarizes these known implications.

Figure 2 below illustrates a projective variety which satisfies the Weak Lang–Vojta conjecture. The picture shows that this variety has infinitely many points valued in a number field (in orange), admits an entire curve (in blue), admits algebraic maps of increasing degree from some fixed curve (in red), and admits a non-constant map from an abelian variety (in green). It is therefore a non-Mordellic, non-Brody hyperbolic, non-bounded, and non-groupless projective variety.

12.1   The Conjecture on Exceptional Loci

We now define the exceptional loci for every notion that we have seen so far. As usual, we let k be an algebraically closed field of characteristic zero.

Definition 12.4. Let X be a variety over k.

- We define Δ_X^{gr} to be the intersection of all proper closed subset Δ such that X is groupless modulo Δ. Note that Δ_X^{gr} is a closed subset of X and that X is groupless modulo Δ_X^{gr}. We refer to Δ_X^{gr} as the *groupless-exceptional locus* of X.
- We define Δ_X^{ar-hyp} to be the intersection of all proper closed subsets Δ such that X is arithmetically hyperbolic modulo Δ. Note that X is arithmetically hyperbolic modulo Δ_X^{ar-hyp}. We refer to Δ_X^{ar-hyp} as the *arithmetic-exceptional locus* of X.

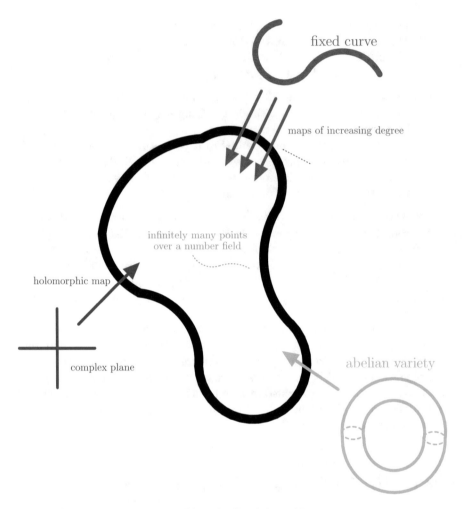

Fig. 1 A projective variety satisfying the Weak Lang-Vojta conjecture

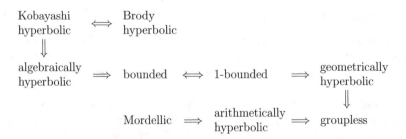

Fig. 2 Known implications between notions of hyperbolicity

- We define Δ_X^{Mor} to be the intersection of all proper closed subsets Δ such that X is Mordellic modulo Δ. Note that X is Mordellic modulo Δ_X^{Mor}. We refer to Δ_X^{Mor} as the *Mordellic-exceptional locus* of X.

Assuming X is a proper variety over k for a moment, it seems worthwhile stressing that Δ_X^{gr} equals the (Zariski) closure of Lang's algebraic exceptional set $\mathrm{Exc}_{alg}(X)$ as defined in [62, p. 160].

Definition 12.5. Let X be a variety over $\mathbb{C}$.

- We let Δ_X^{Br} be the intersection of all closed subsets Δ such that X is Brody hyperbolic modulo Δ. Note that Δ_X^{Br} is a closed subset of X and that X is Brody hyperbolic modulo Δ_X^{Br}. We refer to Δ_X^{Br} as the *Brody-exceptional locus* of X.
- We let Δ_X^{Kob} be the intersection of all closed subsets Δ such that X is Kobayashi hyperbolic modulo Δ. Note that Δ_X^{Kob} is a closed subset of X and that X is Kobayashi hyperbolic modulo Δ_X^{Kob}. We refer to Δ_X^{Kob} as the *Kobayashi-exceptional locus* of X.

We note that Δ_X^{Br} coincides with Lang's analytic exceptional set $\mathrm{Exc}(X)$ (defined in [62, p. 160]). Indeed, $\mathrm{Exc}(X)$ is defined to be the Zariski closure of the union of all images of non-constant entire curves $\mathbb{C} \to X^{\mathrm{an}}$.

Definition 12.6. Let X be a projective scheme over k.

- We define $\Delta_X^{alg-hyp}$ to be the intersection of all proper closed subsets Δ such that X is algebraically hyperbolic modulo Δ. Note that $\Delta_X^{alg-hyp}$ is a proper closed subset of X and that X is algebraically hyperbolic modulo $\Delta_X^{alg-hyp}$. We refer to $\Delta_X^{alg-hyp}$ as the *algebraic exceptional locus* of X.
- For $n \geq 1$, we define $\Delta_X^{n-bounded}$ to be the intersection of all proper closed subsets Δ such that X is n-bounded modulo Δ. Note that $\Delta_X^{n-bounded}$ is a proper closed subset of X and that X is n-bounded modulo $\Delta_X^{n-bounded}$. We refer to $\Delta_X^{n-bounded}$ as the *n-bounded-exceptional locus* of X.
- We define $\Delta_X^{bounded}$ to be the intersection of all proper closed subsets Δ such that X is bounded modulo Δ. Note that $\Delta_X^{bounded}$ is a proper closed subset of X and that X is bounded modulo $\Delta_X^{bounded}$. We refer to $\Delta_X^{bounded}$ as the *bounded-exceptional locus* of X.
- We define $\Delta_X^{geom-hyp}$ to be the intersection of all proper closed subsets Δ such that X is geometrically hyperbolic modulo Δ. Note that $\Delta_X^{geom-hyp}$ is a proper closed subset of X and that X is geometrically hyperbolic modulo $\Delta_X^{geom-hyp}$. We refer to $\Delta_X^{geom-hyp}$ as the *geometric-exceptional locus* of X.

The strongest version of Lang–Vojta's conjecture stated in these notes claims the equality of all exceptional loci. Note that these loci are all, by definition, closed subsets. This is to be contrasted with Lang's definition of his "algebraic exceptional set" (see [62, p. 160]).

Conjecture 12.7 (Strongest Lang–Vojta Conjecture). *Let k be an algebraically closed field of characteristic zero. Let X be an integral projective variety over k. Then the following three statements hold.*

(1) We have that

$$\Delta_X^{gr} = \Delta_X^{Mor} = \Delta_X^{geom-hyp} = \Delta_X^{1-bounded} = \Delta_X^{bounded} = \Delta_X^{alg-hyp}.$$

(2) The projective variety X is of general type if and only if $\Delta_X^{gr} \neq X$.
(3) If $k = \mathbb{C}$, then $\Delta_X^{gr} = \Delta_X^{Br} = \Delta_X^{Kob}$.

Remark 12.8 (Which Inclusions Do We Know?). Let X be a projective scheme over k. We have that

$$\Delta_X^{gr} \subset \Delta_X^{ar-hyp} \subset \Delta_X^{Mor},$$

and

$$\Delta_X^{gr} \subset \Delta_X^{geom-hyp} \subset \Delta_X^{1-bounded} \subset \Delta_X^{bounded} \subset \Delta_X^{alg-hyp}.$$

If k is uncountable, then

$$\Delta_X^{1-bounded} = \Delta_X^{bounded}.$$

If $k = \mathbb{C}$, then

$$\Delta_X^{gr} \subset \Delta_X^{Br} \subset \Delta_X^{Kob}.$$

Remark 12.9 (Reformulating Brody's Lemma). It is not known whether $\Delta_X^{Kob} \subset \Delta_X^{Br}$. Brody's lemma can be stated as saying that, if Δ_X^{Br} is empty, then Δ_X^{Kob} is empty.

Remark 12.10 (Reformulating Demailly's Theorem). It is not known whether $\Delta_X^{alg-hyp} \subset \Delta_X^{Kob}$. Demailly's theorem (Theorem 9.4) can be stated as saying that, if Δ^{Kob} is empty, then $\Delta^{alg-hyp}$ is empty.

13   Closed Subvarieties of Abelian Varieties

We have gradually worked our way towards the following theorem which says that the Strongest Lang–Vojta conjecture holds for closed subvarieties of abelian varieties. Recall that, for X a closed subvariety of an abelian variety A, the subset $\mathrm{Sp}(X)$ is defined to be the union of translates of positive-dimensional abelian subvarieties of A contained in A. It is a fundamental fact that $\mathrm{Sp}(X)$ is a closed subset of X. It turns out that $\mathrm{Sp}(X)$ is the "exceptional locus" of X in any sense of the word "exceptional locus".

Theorem 13.1 (Bloch–Ochiai–Kawamata,   Faltings,   Yamanoi, Kawamata–Ueno). *Let A be an abelian variety over k, and let $X \subset A$ be a closed subvariety. Then the following statements hold.*

(1) We have that $\mathrm{Sp}(X) \neq X$ of X equals if and only if X is of general type.
(2) We have that

$$\mathrm{Sp}(X) = \Delta_X^{gr} = \Delta_X^{Mor} = \Delta_X^{ar-hyp} = \Delta_X^{geom-hyp} = \Delta_X^{1-bounded}$$
$$= \Delta_X^{bounded} = \Delta_X^{alg-hyp}.$$

(3) If $k = \mathbb{C}$, then $\Delta_X^{gr} = \Delta_X^{Br} = \Delta_X^{Kob}$.

Proof. The fact that $\mathrm{Sp}(X) \neq X$ if and only if X is of general type is due to Kawamata–Ueno (see also Theorem 5.1). Moreover, an elementary argument (see Example 6.11) shows that X is groupless modulo $\mathrm{Sp}(X)$, so that $\Delta_X^{gr} \subset \mathrm{Sp}(X)$. On the other hand, it is clear from the definition that $\mathrm{Sp}(X) \subset \Delta_X^{gr}$. This shows that $\mathrm{Sp}(X) = \Delta_X^{gr}$.

By Faltings's theorem (Theorem 7.15), we have that X is Mordellic modulo $\mathrm{Sp}(X)$. This shows that $\Delta_X^{Mor} = \Delta_X^{ar-hyp} = \Delta_X^{gr} = \mathrm{Sp}(X)$. (One can also show that $\Delta_X^{ar-hyp} = \Delta_X^{Mor}$ without appealing to Faltings's theorem. Indeed, as X is a closed subvariety of an abelian variety, it follows from Corollary 7.33 that X is arithmetically hyperbolic modulo Δ if and only if X is Mordellic modulo Δ.)

It follows from Bloch–Ochiai–Kawamata's theorem that $\Delta_X^{Br} = \mathrm{Sp}(X)$. Yamanoi improved this result and showed that $\Delta_X^{Kob} = \mathrm{Sp}(X)$; see Theorem 8.6 (or the original [88, Theorem 1.2]). In his earlier work [87, Corollary 1.(3)], Yamanoi proved that $\Delta_X^{alg-hyp} = \mathrm{Sp}(X)$. Since

$$\Delta_X^{geom-hyp} \subset \Delta_X^{1-bounded} \subset \Delta_X^{bounded} \subset \Delta_X^{alg-hyp},$$

this concludes the proof.   □

14   Evidence for Lang–Vojta's Conjecture

In the previous sections, we defined every notion appearing in Lang–Vojta's conjecture, and we stated the "Strongest", "Stronger", and "Weakest" versions of Lang–Vojta's conjectures. We also indicated the known implications between these notions, and that the Strongest Lang–Vojta conjecture is known to hold for closed subvarieties of abelian varieties by work of Bloch–Ochiai–Kawamata, Faltings, Kawamata–Ueno, and Yamanoi.

In the following four sections, we will present some evidence for Lang–Vojta's conjectures. The results in the following sections are all in accordance with the Lang–Vojta conjectures.

15 Dominant Rational Self-maps of Pseudo-Hyperbolic Varieties

Let us start with a classical finiteness result of Matsumura [45, §11].

Theorem 15.1 (Matsumura). *If X is a proper integral variety of general type over k, then the set of dominant rational self-maps $X \dashrightarrow X$ is finite.*

Note that Matsumura's theorem is a vast generalization of the statement that a smooth curve of genus at least two has only finitely many automorphisms. Motivated by Lang–Vojta's conjecture, the arithmetic analogue of Matsumura's theorem is proven in [56] (building on the results in [49]) and can be stated as follows.

Theorem 15.2. *If X is a proper pseudo-Mordellic integral variety over k, then the set of rational dominant self-maps $X \dashrightarrow X$ is finite.*

Idea of Proof. We briefly indicate three ingredients of the proof of Theorem 15.2.

(1) First, one can use Amerik's theorem on dynamical systems [5] to show that every dominant rational self-map is a birational self-map of finite order whenever X is a pseudo-Mordellic projective variety.

(2) One can show that, if X is a projective integral variety over k such that $\mathrm{Aut}_k(X)$ is infinite, then $\mathrm{Aut}_k(X)$ has an element of infinite order. (It is crucial here that k is of characteristic zero.) This result is proven in [49].

(3) If X is a projective non-uniruled integral variety over k such that $\mathrm{Bir}_k(X)$ is infinite, then $\mathrm{Bir}_k(X)$ has a point of infinite order. To prove this, one can use Prokhorov–Shramov's notion of quasi-minimal models (see [79]) to reduce to the analogous finiteness result for automorphisms stated in (2). The details are in [56].

Combining (1) and (3), one obtains the desired result for pseudo-Mordellic projective varieties (Theorem 15.2). □

There is a similar finiteness statement for pseudo-algebraically hyperbolic varieties. This finiteness result is proven in [50] for algebraically hyperbolic varieties, and in [56] for pseudo-algebraically hyperbolic varieties.

Theorem 15.3. *If X is a projective pseudo-algebraically hyperbolic integral variety over k, then the set of dominant rational self-maps $X \dashrightarrow X$ is finite.*

In fact, more generally, we have the following a priori stronger result.

Theorem 15.4. *If X is a projective pseudo-1-bounded integral variety over k, then the set of dominant rational self-maps $X \dashrightarrow X$ is finite.*

Proof. For 1-bounded varieties this is proven in [50]. The more general statement for pseudo-1-bounded varieties is proven in [56] by combining Amerik's theorem [5] and Prokhorov–Shramov's theory of quasi-minimal models [79] with Weil's Regularization Theorem and properties of dynamical degrees of rational dominant self-maps.   □

As the reader may have noticed, for pseudo-Mordellic, pseudo-algebraically hyperbolic, and pseudo-1-bounded projective varieties we have satisfying results.

What do we know in the complex-analytic setting? We have the following result of Noguchi [59, Theorem 5.4.4] for Brody hyperbolic varieties.

Theorem 15.5 (Noguchi). *If X is a Brody hyperbolic projective integral variety over $\mathbb{C}$, then $\mathrm{Bir}_{\mathbb{C}}(X)$ is finite.*

First Proof of Theorem 15.5. Since a Brody hyperbolic projective integral variety over $\mathbb{C}$ is bounded by, for instance, Demailly's theorem (Theorem 9.4), this follows from Theorem 15.4.   □

Second Proof of Theorem 15.5. Let $Y \to X$ be a resolution of singularities of X. Note that, every birational morphism $X \dashrightarrow X$ induces a dominant rational map $Y \dashrightarrow X$. Since X has no rational curves (as X is Brody hyperbolic) and Y is smooth, by [50, Lemma 3.5], the rational map $Y \dashrightarrow X$ extends uniquely to a surjective morphism $Y \to X$.

Therefore, we have that

$$\mathrm{Bir}_{\mathbb{C}}(X) \subset \mathrm{Sur}_{\mathbb{C}}(Y, X)$$

Noguchi proved that the latter set is finite (see Theorem 16.1 below). He does so by showing that it is the set of $\mathbb{C}$-points on a finite type zero-dimensional scheme over $\mathbb{C}$. We discuss this result of Noguchi in more detail in the next section.   □

It is important to note that, in light of Green–Griffiths' and Lang–Vojta's conjectures, one expects an analogous finiteness result for pseudo-Brody hyperbolic varieties (as pseudo-Brody hyperbolic varieties should be of general type). This is, however, not known, and we state it as a separate conjecture.

Conjecture 15.6 (Pseudo-Noguchi, I). *If X is a pseudo-Brody hyperbolic projective integral variety over $\mathbb{C}$, then $\mathrm{Bir}_{\mathbb{C}}(X)$ is finite.*

Remark 15.7 (What Do We Not Know Yet?). First, it is not known whether the automorphism group of a groupless projective variety is finite. Also, it is not known whether a pseudo-Kobayashi hyperbolic projective variety has a finite automorphism group. Moreover, it is not known whether a geometric hyperbolic projective variety has only finitely many automorphisms. As these problems are unresolved, the finiteness of the set of birational self-maps is also still open.

16   Finiteness of Moduli Spaces of Surjective Morphisms

Our starting point in this section is the following finiteness theorem of Noguchi for dominant rational maps from a fixed variety to a hyperbolic variety (*formerly* a conjecture of Lang); see [59, §6.6] for a discussion of the history of this result.

Theorem 16.1 (Noguchi). *If X is a Brody hyperbolic proper variety over $\mathbb{C}$ and Y is a projective integral variety over $\mathbb{C}$, then the set of dominant rational maps $f : Y \dashrightarrow X$ is finite.*

In light of Lang–Vojta's conjecture, any "hyperbolic" variety should satisfy a similar finiteness property. In particular, one should expect similar (hence more general) results for bounded varieties, and such results are obtained in [50] over arbitrary algebraically closed fields k of characteristic zero.

Theorem 16.2. *If X is a 1-bounded projective variety over k and Y is a projective integral variety over k, then the set of dominant rational maps $f : Y \dashrightarrow X$ is finite.*

In particular, the same finiteness statement holds for bounded varieties and algebraically hyperbolic varieties. Indeed, such varieties are (obviously) 1-bounded.

Corollary 16.3. *If X is a bounded projective variety over k (e.g., algebraically hyperbolic variety over k) and Y is a projective integral variety over k, then the set of dominant rational maps $f : Y \dashrightarrow X$ is finite.*

We now make a "pseudo"-turn. In fact, the finiteness result of Noguchi should actually hold under the weaker assumption that X is only pseudo-Brody hyperbolic. To explain this, recall that Kobayashi–Ochiai proved a finiteness theorem for dominant rational maps from a given variety Y to a fixed variety of general type X which generalizes Matsumura's finiteness theorem for the group $\mathrm{Bir}_k(X)$ (Theorem 15.1).

Theorem 16.4 (Kobayashi–Ochiai). *Let X be a projective variety over k of general type. Then, for every projective integral variety Y, the set of dominant rational maps $f : Y \dashrightarrow X$ is finite.*

In light of Lang–Vojta's conjectures and Kobayashi–Ochiai's theorem, any "pseudo-hyperbolic" variety should satisfy a similar finiteness property. For example, Lang–Vojta's conjecture predicts a similar finiteness statement for pseudo-Brody hyperbolic projective varieties. We state this as a conjecture. Note that this conjecture is the "pseudo"-version of Noguchi's theorem (Theorem 16.1), and clearly implies Conjecture 15.6.

Conjecture 16.5 (Pseudo-Noguchi, II). *If X is a pseudo-Brody hyperbolic proper variety over $\mathbb{C}$ and Y is a projective integral variety over $\mathbb{C}$, then the set of dominant rational maps $f : Y \dashrightarrow X$ is finite.*

Now, as any "pseudo-hyperbolic" variety is pseudo-groupless, it is natural to first try and see what one can say about pseudo-groupless varieties. For simplicity, we will focus on surjective morphisms (as opposed to dominant rational maps) in the rest of this section.

There is a standard approach to establishing the finiteness of the set of surjective morphisms from one projective scheme to another. To explain this, let us recall some notation from Section 10. Namely, if X and Y are projective schemes over k, we let $\underline{\mathrm{Hom}}_k(Y, X)$ be the scheme parametrizing morphisms $X \to Y$. Note that $\underline{\mathrm{Hom}}_k(Y, X)$ is a countable disjoint union of quasi-projective schemes over k. Moreover, we let $\underline{\mathrm{Sur}}_k(Y, X)$ be the scheme parametrizing surjective morphisms $Y \to X$, and note that $\underline{\mathrm{Sur}}_k(Y, X)$ is a closed subscheme of $\underline{\mathrm{Hom}}_k(Y, X)$.

The standard approach to establishing the finiteness of the set $\mathrm{Sur}_k(Y, X)$ is to interpret it as the set of k-points on the scheme $\underline{\mathrm{Sur}}_k(Y, X)$. This makes it tangible to techniques from deformation theory. Indeed, to show that $\mathrm{Sur}_k(Y, X)$ is finite, it suffices to establish the following two statements:

(1) The tangent space to each point of $\underline{\mathrm{Sur}}_k(Y, X)$ is trivial;
(2) The scheme $\underline{\mathrm{Sur}}_k(Y, X)$ has only finitely many connected components.

It is common to refer to the first statement as a rigidity statement, as it boils down to showing that the objects parametrized by $\underline{\mathrm{Sur}}_k(Y, X)$ are infinitesimally rigid. Also, it is standard to refer to the second statement as being a boundedness property. For example, if Y and X are curves and X is of genus at least two, the finiteness of $\mathrm{Sur}_k(Y, X)$ is proven precisely in this manner; see [70, §II.8]. We refer the reader to [60] for a further discussion of the rigidity/boundedness approach to proving finiteness results for other moduli spaces.

We now focus on the rigidity of surjective morphisms $Y \to X$. The following rigidity theorem for pseudo-groupless varieties will prove to be extremely useful. This result is a consequence of a much more general statement about the deformation space of a surjective morphism due to Hwang–Kebekus–Peternell [44].

Theorem 16.6 (Hwang–Kebekus–Peternell + ϵ). *If Y is a projective normal variety over k and X is a pseudo-groupless projective variety over k, then the scheme $\underline{\mathrm{Sur}}_k(Y, X)$ is a countable disjoint union of zero-dimensional smooth projective schemes over k.*

Proof. As is shown in [56], this is a consequence of Hwang–Kebekus–Peternell's result on the infinitesimal deformations of a surjective morphism $Y \to X$. Indeed, since X is non-uniruled (Remark 6.7), for every such surjective morphism $f : Y \to X$, there is a finite morphism $Z \to X$ and a morphism $Y \to Z$ such that f is the composed map $Y \to Z \to X$. Moreover, the identity component $\mathrm{Aut}^0_{Z/k}$ of the automorphism group scheme surjects onto the connected component of f in $\underline{\mathrm{Hom}}_k(Y, X)$. Since X is pseudo-groupless, the same holds for Z. It is then not hard to verify that $\mathrm{Aut}^0_{Z/k}$ is trivial, so that the connected component of f in $\underline{\mathrm{Hom}}_k(Y, X)$ is trivial.   □

Remark 16.7. There are projective varieties X which are **not** pseudo-groupless over k, but for which the conclusion of the theorem above still holds. For example, a K3 surface or the blow-up of a simple abelian surface A in its origin. This means that the rigidity of surjective morphisms follows from properties strictly *weaker* than pseudo-hyperbolicity. We refer to [56] for a more general statement concerning rigidity of surjective morphisms.

When introducing the notions appearing in Lang–Vojta's conjecture, we made sure to emphasize that every one of these is pseudo-groupless. Thus, roughly speaking, any property we prove for pseudo-groupless varieties holds for all pseudo-hyperbolic varieties. This gives us the following rigidity statement.

Corollary 16.8 (Rigidity for Pseudo-Hyperbolic Varieties). *Let X be a projective integral variety over k and let Y be a projective normal variety over k. Assume that one of the following statements holds.*

(1)   The variety X is pseudo-groupless over k.
(2)   The variety X is pseudo-Mordellic over k.
(3)   The projective variety X is pseudo-algebraically hyperbolic over k.
(4)   The projective variety X is pseudo-1-bounded over k.
(5)   The projective variety X is pseudo-bounded over k.
(6)   The variety X is pseudo-geometrically hyperbolic over k.
(7)   The field k equals $\mathbb{C}$ and X is pseudo-Brody hyperbolic.

Then the scheme $\underline{\mathrm{Sur}}_k(Y, X)$ is a countable disjoint union of zero-dimensional smooth projective schemes over k.

Proof. Assume that either (1)–(6) or (7) holds. Then X is pseudo-groupless (as explained throughout these notes), so that the result follows from Theorem 16.6.   □

Proving the finiteness of $\mathrm{Sur}_k(Y, X)$ or, equivalently, the boundedness of $\underline{\mathrm{Sur}}_k(Y, X)$, for X pseudo-groupless or pseudo-Mordellic seems to be out of reach currently. However, for pseudo-algebraically hyperbolic varieties the desired finiteness property is proven in [56] and reads as follows.

Theorem 16.9. *If X is a pseudo-algebraically hyperbolic projective variety over k and Y is a projective integral variety over k, then the set of surjective morphisms $f : Y \to X$ is finite.*

A similar result can be obtained for pseudo-bounded varieties. The precise result can be stated as follows.

Theorem 16.10. *If X is a pseudo-bounded projective variety over k and Y is a projective integral variety over k, then the set of surjective morphisms $f : Y \to X$ is finite.*

To prove the analogous finiteness property for pseudo-1-bounded varieties, we require (as in the previous section) an additional uncountability assumption on the base field.

Theorem 16.11. *Assume k is uncountable. If X is a pseudo-1-bounded projective variety over k and Y is a projective integral variety over k, then the set of surjective morphisms $f : Y \to X$ is finite.*

We conclude with the following finiteness result for pseudo-algebraically hyperbolic varieties. It is proven in [56] using (essentially) the results in this section and the fact that pseudo-algebraically hyperbolic varieties are pseudo-geometrically hyperbolic.

Theorem 16.12. *If X is algebraically hyperbolic modulo Δ over k, then for every connected reduced projective variety Y over k, every non-empty closed reduced subset $B \subset Y$, and every reduced closed subset $A \subset X$ not contained in Δ, the set of morphisms $f : Y \to X$ with $f(B) = A$ is finite.*

Note that Theorem 16.12 can be applied with B a point or $B = Y$. This shows that the statement generalizes the finiteness result of this section.

17 Hyperbolicity Along Field Extensions

In this section we study how different notions of pseudo-hyperbolicity appearing in Lang–Vojta's conjectures (except for those that only make sense over $\mathbb{C}$ a priori) behave under field extensions. In other words, we study how the exceptional locus for each notion of hyperbolicity introduced in Section 12 behaves under field extensions.

Let us start with X a variety of general type over a field k, and let $k \subset L$ be a field extension. It is natural to wonder whether X_L is also of general type over L. A simple argument comparing the spaces of global sections of $\omega_{X/k}$ and $\omega_{X_L/L}$ shows that this is indeed the case. This observation is our starting point in this section. Indeed, the mere fact that varieties of general type remain varieties of general type after a field extension can be paired with the Strong Lang–Vojta conjecture to see that similar statements should hold for pseudo-groupless varieties, pseudo-Mordellic varieties, and so on.

The first three results we state in this section say that this "base-change" property can be proven in some cases. For proofs we refer to [15, 50, 56].

Theorem 17.1. *Let $k \subset L$ be an extension of algebraically closed fields of characteristic zero. Let X be a projective scheme over k and let Δ be a closed subset of X. Then the following statements hold.*

(1) If X is of general type over k, then X_L is of general type over L.
(2) If X is groupless modulo Δ, then X_L is groupless modulo Δ_L.
(3) If X is algebraically hyperbolic modulo Δ, then X_L is algebraically hyperbolic modulo Δ_L.
(4) If X is bounded modulo Δ, then X_L is bounded modulo Δ_L.

In this theorem we are missing (among others) the notions of 1-boundedness and geometric hyperbolicity. In this direction we have the following result; see [15, 46].

Theorem 17.2. *Let $k \subset L$ be an extension of **uncountable** algebraically closed fields of characteristic zero. Let X be a projective scheme over k and let Δ be a closed subset of X. Then the following statements hold.*

(1) If X is 1-bounded modulo Δ, then X_L is bounded modulo Δ_L.
(2) If X is geometrically hyperbolic modulo Δ, then X_L is geometrically hyperbolic modulo Δ_L.

If $\Delta = \emptyset$, then we do not need to impose uncountability.

Theorem 17.3. *Let $k \subset L$ be an extension of algebraically closed fields of characteristic zero. Let X be a projective scheme over k and let Δ be a closed subset of X. Then the following statements hold.*

(1) If X is 1-bounded, then X_L is bounded.
(2) If X is geometrically hyperbolic, then X_L is geometrically hyperbolic.

The reader will have noticed the absence of the notion of Mordellicity and arithmetic hyperbolicity above. The question of whether an arithmetically hyperbolic variety over $\overline{\mathbb{Q}}$ remains arithmetically hyperbolic over a larger field is not an easy one in general, as should be clear from the following remark.

Remark 17.4 (Persistence of Arithmetic Hyperbolicity). Let $f_1, \ldots, f_r \in \mathbb{Z}[x_1, \ldots, x_n]$ be polynomials, and let $X := Z(f_1, \ldots, f_r) = \operatorname{Spec}(\overline{\mathbb{Q}}[x_1, \ldots, x_n]/(f_1, \ldots, f_r)) \subset \mathbb{A}_{\overline{\mathbb{Q}}}^n$ be the associated affine variety over $\overline{\mathbb{Q}}$. To say that X is arithmetically hyperbolic over $\overline{\mathbb{Q}}$ is to say that, for every *number field* K and every finite set of finite places S of K, the set of $a = (a_1, \ldots, a_n) \in \mathcal{O}_{K,S}^n$ with $f_1(a) = \ldots = f_r(a) = 0$ is finite. On the other hand, to say that $X_{\mathbb{C}}$ is arithmetically hyperbolic over $\mathbb{C}$ is to say that, for every $\mathbb{Z}$-*finitely generated* subring $A \subset \mathbb{C}$, the set of $a = (a_1, \ldots, a_n) \in A^n$ with $f_1(a) = \ldots = f_r(a) = 0$ is finite.

Despite the apparent difference between being arithmetically hyperbolic over $\overline{\mathbb{Q}}$ and being arithmetically hyperbolic over $\mathbb{C}$, it seems reasonable to suspect their equivalence. For X a projective variety, the following conjecture is a consequence of the Weak Lang–Vojta conjecture for X. However, as it also seems reasonable in the non-projective case, we state it in this more generality.

Conjecture 17.5 (Persistence Conjecture). *Let $k \subset L$ be an extension of algebraically closed fields of characteristic zero. Let X be a variety over k and let Δ be a closed subset of X. If X is arithmetically hyperbolic modulo Δ over k, then X_L is arithmetically hyperbolic modulo Δ_L over L.*

Note that we will focus throughout on arithmetic hyperbolicity (as opposed to Mordellicity) as its persistence along field extensions is easier to study. The reader may recall that the difference between Mordellicity and arithmetic hyperbolicity disappears for many varieties (e.g., affine varieties); see Section 7.3 for a discussion.

This conjecture is investigated in [15, 46, 49, 52]. As a basic example, the reader may note that Faltings proved that a smooth projective connected curve of genus at least two over $\overline{\mathbb{Q}}$ is arithmetically hyperbolic over $\overline{\mathbb{Q}}$ in [32]. He then later explained in [33] that Grauert–Manin's function field version of the Mordell conjecture can be used to prove that a smooth projective connected curve of genus at least two over k is arithmetically hyperbolic over k.

In the rest of this section, we will present some results on the Persistence Conjecture. We start with the following result.

Theorem 17.6. *Let $k \subset L$ be an extension of algebraically closed fields of characteristic zero. Let X be an arithmetically hyperbolic variety over k such that X_L is geometrically hyperbolic over L. Then X_L is arithmetically hyperbolic over L.*

Note that Theorem 17.6 implies that the Persistence Conjecture holds for varieties over k which are geometrically hyperbolic over any field extension of L.

Theorem 17.6 is inspired by Martin-Deschamps's proof of the arithmetic Shafarevich conjecture over finitely generated fields (see also Remark 7.34). Indeed, in Szpiro's seminar [82], Martin-Deschamps gave a proof of the arithmetic Shafarevich conjecture by using a specialization argument on the moduli stack of principally polarized abelian schemes; see [68]. This specialization argument resides on Faltings's theorem that the moduli space of principally polarized abelian varieties of fixed dimension over $\mathbb{C}$ is geometrically hyperbolic over $\mathbb{C}$. We note that Theorem 17.6 is essentially implicit in her line of reasoning.

We will present applications of Theorem 17.6 to the Persistence Conjecture based on the results obtained in [49, 52]. However, before we give these applications, we mention the following result which implies that the Persistence Conjecture holds for normal projective surfaces with non-zero irregularity $h^1(X, \mathcal{O}_X)$.

Theorem 17.7. *Let X be a projective surface over k which admits a non-constant morphism to some abelian variety over k. Then X is arithmetically hyperbolic over k if and only if X_L is arithmetically hyperbolic over L.*

The proofs of Theorems 17.6 and 17.7 differ tremendously in spirit. In fact, we cannot prove Theorem 17.7 by appealing to the geometric hyperbolicity of X (as it is currently not known whether an arithmetically hyperbolic projective surface which admits a non-constant map to an abelian variety is geometrically hyperbolic). Instead, Theorem 17.7 is proven by appealing to the "mild boundedness" of abelian varieties; see [15]. More explicitly: in the proof of Theorem 17.7, we use that, for every smooth connected curve C over k, there exists an integer $n > 0$ and points $c_1, \ldots, c_n$ in $C(k)$ such that, for every abelian variety A over k and every $a_1, \ldots, a_n$ in $A(k)$, the set

$$\mathrm{Hom}_k((C, c_1, \ldots, c_n), (A, a_1, \ldots, a_n))$$

is finite. This finiteness property for abelian varieties can be combined with the arithmetic hyperbolicity of the surface X in Theorem 17.7 to show that the surface X is mildly bounded. The property of being mildly bounded is clearly *much* weaker than being geometrically hyperbolic, but it turns out to be enough to show the Persistence Conjecture; see [49, §4.1]. Note that it is a bit surprising that abelian varieties (as they are very far from being hyperbolic) satisfy some "mild" version of geometric hyperbolicity. We refer the reader to [49, §4] for the definition of what this notion entails, and to [15] for the fact that abelian varieties are mildly bounded.

We now focus as promised on the applications of Theorem 17.6. Our first application says that the Persistence Conjecture holds for all algebraically hyperbolic projective varieties.

Theorem 17.8. *Let X be a projective algebraically hyperbolic variety over k. Then X is arithmetically hyperbolic over k if and only if, for every algebraically closed field L containing k, the variety X_L is arithmetically hyperbolic over L.*

Proof. Since X is algebraically hyperbolic over k, it follows from (3) in Theorem 17.1 that X_L is algebraically hyperbolic over L. Since algebraically hyperbolic projective varieties are 1-bounded and thus geometrically hyperbolic (Corollary 11.5), the result follows from Theorem 17.6.   □

Our second application involves integral points on the moduli space of smooth hypersurfaces. We present the results obtained in [52] in the following section.

17.1   The Shafarevich Conjecture for Smooth Hypersurfaces

We explain in this section how Theorem 17.6 can be used to show the following finiteness theorem. This explanation will naturally lead us to studying integral points on moduli spaces.

Theorem 17.9. *Let $d \geq 3$ and $n \geq 2$ be integers. Assume that, for every number field K and every finite set of finite places S of K, the set of $\mathcal{O}_{K,S}$-isomorphism classes of smooth hypersurfaces of degree d in $\mathbb{P}^{n+1}_{\mathcal{O}_{K,S}}$ is finite. Then, for every $\mathbb{Z}$-finitely generated normal integral domain A of characteristic zero, the set of A-isomorphism classes of smooth hypersurfaces of degree d in $\mathbb{P}^{n+1}_{A}$ is finite.*

To prove Theorem 17.9, we (i) reformulate its statement in terms of the arithmetic hyperbolicity of an appropriate moduli space of smooth hypersurfaces, (ii) establish the geometric hyperbolicity of this moduli space, and (iii) apply Theorem 17.6. Indeed, the assumption in Theorem 17.9 can be formulated as saying that the (appropriate) moduli space of hypersurfaces is arithmetically hyperbolic over $\overline{\mathbb{Q}}$ and the conclusion of our theorem is then that this moduli space is also arithmetically hyperbolic over larger fields. To make these statements more precise, let $\mathrm{Hilb}_{d,n}$ be the Hilbert scheme of smooth hypersurfaces of degree d in $\mathbb{P}^{n+1}$. Note that $\mathrm{Hilb}_{d,n}$ is a smooth affine scheme over $\mathbb{Z}$. There is a natural action of the automorphism group scheme PGL_{n+2} of $\mathbb{P}^{n+1}_{\mathbb{Z}}$ on $\mathrm{Hilb}_{d,n}$. Indeed, given a smooth hypersurface H in $\mathbb{P}^{n+1}$ and an automorphism σ of $\mathbb{P}^{n+1}$, the resulting hypersurface $\sigma(H)$ is again smooth.

The quotient of a smooth affine scheme over $\mathbb{Z}$ by a reductive group (such as PGL_{n+2}) is an affine scheme of finite type over $\mathbb{Z}$ by Mumford's GIT. However, for the study of hyperbolicity and integral points, this quotient scheme is not very helpful, as the action of PGL_{n+2} on $\mathrm{Hilb}_{d,n}$ is not free. The natural solution it to consider the stacky quotient, as in [13, 14, 47]. However, one may avoid the use of stacks by adding level structure as in [48]. Indeed, by [48], there exists a smooth affine variety H' over $\mathbb{Q}$ with a free action by $\mathrm{PGL}_{n+2,\mathbb{Q}}$, and a finite étale $\mathrm{PGL}_{n+2,\mathbb{Q}}$-equivariant morphism $H' \to \mathrm{Hilb}_{d,n,\mathbb{Q}}$. Let $U_{d;n} := \mathrm{PGL}_{n+2,\mathbb{Q}} \backslash H'$ be the smooth affine quotient scheme over $\mathbb{Q}$. To prove Theorem 17.9, we establish the following result.

Theorem 17.10. *Let $d \geq 3$ and $n \geq 2$ be integers. Assume that $U_{d;n,\overline{\mathbb{Q}}}$ is arithmetically hyperbolic over $\overline{\mathbb{Q}}$. Then, for every algebraically closed field k*

of characteristic zero, the affine variety $U_{d;n,k}$ is arithmetically hyperbolic over k.

Proof. Let us write $U := U_{d;n,\overline{\mathbb{Q}}}$. The proof relies on a bit of Hodge theory. Indeed, we use Deligne's finiteness theorem for monodromy representations [28], the infinitesimal Torelli property for smooth hypersurfaces [35], and the Theorem of the Fixed Part in Hodge theory [80] to show that U_k is geometrically hyperbolic over k. Then, as U_k is geometrically hyperbolic over k, the result follows from Theorem 17.6. $\square$

We now explain how to deduce Theorem 17.9 from Theorem 17.10.

Proof of Theorem 17.9. Write $U := U_{d;n,\overline{\mathbb{Q}}}$. First, the assumption in Theorem 17.9 can be used to show that U is arithmetically hyperbolic over $\overline{\mathbb{Q}}$. Then, since U is arithmetically hyperbolic over $\overline{\mathbb{Q}}$, it follows from Theorem 17.10 that U_k is arithmetically hyperbolic for every algebraically closed field k of characteristic zero. Finally, to conclude the proof, let us recall that arithmetic hyperbolicity descends along finite étale morphisms of varieties (Remark 3.5). In [53], the analogous descent statement is proven for finite étale morphisms of algebraic stacks, after extending the notion of arithmetic hyperbolicity from schemes to stacks. Thus, by applying this "stacky" Chevalley–Weil theorem to the finite étale morphism $U_{d;n,k} \to [\mathrm{PGL}_{n+2,k}\backslash\mathrm{Hilb}_{d,n,k}]$ of stacks, where $[\mathrm{PGL}_{n+2,k}\backslash\mathrm{Hilb}_{d,n,k}]$ denotes the quotient stack, we obtain that the stack $[\mathrm{PGL}_{n+2,k}\backslash\mathrm{Hilb}_{d,n,k}]$ is arithmetically hyperbolic over k. Finally, the moduli-interpretation of the points of this quotient stack can be used to see that, for every $\mathbb{Z}$-finitely generated normal integral domain A of characteristic zero, the set of A-isomorphism classes of smooth hypersurfaces of degree d in $\mathbb{P}^{n+1}_{\mathcal{O}_{K,S}}$ is finite. This concludes the proof. $\square$

Remark 17.11 (Period Domains). Theorem 17.10 actually follows from a more general statement about varieties with a quasi-finite period map (e.g., Shimura varieties). Namely, in [52] it is shown that a complex algebraic variety with a quasi-finite period map is geometrically hyperbolic. For other results about period domains we refer the reader to the article of Bakker–Tsimerman in this book [12].

18   Lang's Question on Openness of Hyperbolicity

It is obvious that being hyperbolic is not stable under specialization. In fact, being pseudo-groupless is not stable under specialization, as a smooth proper curve of genus two can specialize to a tree of $\mathbb{P}^1$'s. Nonetheless, it seems reasonable to suspect that being hyperbolic (resp. pseudo-hyperbolic) is in fact stable under generization. The aim of this section is to investigate this property for all notions of hyperbolicity discussed in these notes. In fact, on [62, p. 176] Lang says "I do not clearly understand the extent to which hyperbolicity is open for the Zariski topology". This brings us to the following question of Lang and our starting point of this section.

Question 1 (Lang). Let S be a noetherian scheme over $\mathbb{Q}$ and let $X \to S$ be a projective morphism. Is the set of s in S such that $X_{\overline{k(s)}}$ is groupless a Zariski open subscheme of S?

Here we let $k(s)$ denote the residue field of the point s, and we let $k(s) \to \overline{k(s)}$ be an algebraic closure of $k(s)$. Note that one can ask similar questions for the set of s in S such that $X_{\overline{k(s)}}$ is algebraically hyperbolic or arithmetically hyperbolic, respectively.

Before we discuss what one may expect regarding Lang's question, let us recall what it means for a subset of a scheme to be a Zariski-countable open.

If $(X, \mathcal{T})$ is a noetherian topological space, then there exists another topology $\mathcal{T}^{\mathrm{cnt}}$, or $\mathcal{T}$-countable, on X whose closed sets are the countable union of $\mathcal{T}$-closed sets. If S is a noetherian scheme, a subset $Z \subset S$ is a Zariski-countable closed if it is a countable union of closed subschemes $Z_1, Z_2, \ldots \subset S$.

Remark 18.1 (What to Expect? I). We will explain below that the locus of s in S such that X_s is groupless is a Zariski-countable open of S, i.e., its complement is a countable union of closed subschemes. In fact, we will show similar statements for algebraic hyperbolicity, boundedness, geometric hyperbolicity, and the property of having only subvarieties of general type. Although this provides some indication that the answer to Lang's question might be positive, it is not so clear whether one should expect a positive answer to Lang's question. However, it seems plausible that, assuming the Strongest Lang–Vojta conjecture (Conjecture 12.1), one can use certain Correlation Theorems (see Ascher–Turchet [8]) to show that the answer to Lang's question is positive.

One can also ask about the pseudofied version of Lang's question.

Question 2 (Pseudo-Lang). Let S be a noetherian scheme over $\mathbb{Q}$ and let $X \to S$ be a projective morphism. Is the set of s in S such that $X_{\overline{k(s)}}$ is pseudo-groupless a Zariski open subscheme of S?

Again, one can ask similar questions for the set of s in S such that $X_{\overline{k(s)}}$ is pseudo-algebraically hyperbolic or pseudo-arithmetically hyperbolic, respectively.

Remark 18.2 (What to Expect? II). We will argue below that one should expect (in light of the Strong Lang–Vojta conjecture) that the answer to the Pseudo-Lang question is positive. This is because of a theorem of Siu–Kawamata–Nakayama on invariance of plurigenera.

What do we know about the above questions (Questions 1 and 2)? The strongest results we dispose of are due to Nakayama; see [74, Chapter VI.4].

In fact, the following theorem can be deduced from Nakayama's [74, Theorem VI.4.3]. (Nakayama's theorem is a generalization of theorems of earlier theorems of Siu and Kawamata on invariance of plurigenera.)

Theorem 18.3 (Siu, Kawamata, Nakayama). *Let S be a noetherian scheme over $\mathbb{Q}$ and let $X \to S$ be a projective morphism of schemes. Then, the set of s in S such that X_s is of general type is open in S.*

Thus, by Theorem 18.3, assuming the Strong Lang–Vojta conjecture (Conjecture 12.1), the answer to the Pseudo-Lang question should be positive. Also, assuming the Strong Lang–Vojta conjecture, the set of s in S such that $X_{\overline{k(s)}}$ is pseudo-algebraically hyperbolic should be open. Similar statements should hold for pseudo-Mordellicity and pseudo-boundedness. Although neither of these statements are known, some partial results are obtained in [15, §9].

In fact, as a consequence of Nakayama's theorem and the fact that the stack of proper schemes of general type is a countable union of finitely presented algebraic stacks, one can prove the following result.

Theorem 18.4 (Countable-Openness of Every Subvariety Being of General Type). *Let S be a noetherian scheme over $\mathbb{Q}$ and let $X \to S$ be a projective morphism. Then, the set of s in S such that every integral closed subvariety of X_s is of general type is Zariski-countable open in S.*

The countable-openness of the locus of every subvariety being of general type does not give a satisfying answer to Lang's question. However, it does suggest that every notion appearing in the Lang–Vojta conjecture should be Zariski-countable open. This expectation can be shown to hold for some notions of hyperbolicity. For example, given a projective morphism of schemes $X \to S$ with S a complex algebraic variety, one can show that the locus of s in S such that X_s is algebraically hyperbolic is an open subset of S in the countable-Zariski topology; see [15, 29]. This result is essentially due to Demailly.

Theorem 18.5. *Let S be a noetherian scheme over $\mathbb{Q}$ and let $X \to S$ be a projective morphism. Then, the set of s in S such that X_s is algebraically hyperbolic is Zariski-countable open in S.*

It is worth noting that this is not the exact result proven by Demailly, as it brings us to a subtle difference between the Zariski-countable topology on a variety X over $\mathbb{C}$ and the induced topology on $X(\mathbb{C})$. Indeed, Demailly proved that, if $k = \mathbb{C}$ and $S^{\text{not-ah}}$ is the set of s in S such that X_s is not algebraically hyperbolic, then $S^{\text{not-ah}} \cap S(\mathbb{C})$ is closed in the countable topology on $S(\mathbb{C})$. This, strictly speaking, does not imply that $S^{\text{not-ah}}$ is closed in the countable topology on S. For example, if S is an integral curve over $\mathbb{C}$ and η is the generic point of S, then $\{\eta\}$ is not a Zariski-countable open of S, whereas $\{\eta\} \cap S(\mathbb{C}) = \emptyset$ is a Zariski-countable open of $S(\mathbb{C})$.

In [15] similar results are obtained for boundedness and geometric hyperbolicity. The precise statements read as follows.

Theorem 18.6 (Countable-Openness of Boundedness). *Let S be a noetherian scheme over $\mathbb{Q}$ and let $X \to S$ be a projective morphism. Then, the set of s in S such that X_s is bounded is Zariski-countable open in S.*

Theorem 18.7 (Countable-Openness of Geometric Hyperbolicity). *Let S be a noetherian scheme over $\mathbb{Q}$ and let $X \to S$ be a projective morphism. Then, the set of s in S such that X_s is geometrically hyperbolic is Zariski-countable open in S.*

Remark 18.8 (What Goes into the Proofs of Theorems 18.5, 18.6, and 18.7?). The main idea behind all these proofs is quite simple. Let us consider Theorem 18.5. First, one shows that the set of s in S such that X_s is *not* algebraically hyperbolic is the image of countably many constructible subsets of S. This is essentially a consequence of the fact that the Hom-scheme between two projective schemes is a countable union of quasi-projective schemes. Then, it suffices to note that the set of s in S with X_s algebraically hyperbolic is stable under generization. This relies on compactness properties of the moduli stack of stable curves.

Concerning Lang's question on the locus of groupless varieties, we note that in [55] it is shown that the set of s in S such that X_s is groupless is open in the Zariski-countable topology on S.

Theorem 18.9 (Countable-Openness of Grouplessness). *Let S be a noetherian scheme over $\mathbb{Q}$ and let $X \to S$ be a projective morphism. Then, the set of s in S such that $X_{\overline{k(s)}}$ is groupless is Zariski-countable open in S.*

We finish these notes with a discussion of the proof of Theorem 18.9. It will naturally lead us to introducing a non-archimedean counterpart to Lang–Vojta's conjecture.

18.1   Non-archimedean Hyperbolicity and Theorem 18.9

Let S be a noetherian scheme over $\mathbb{Q}$ and let $X \to S$ be a projective morphism. Define S^{n-gr} to be the set of s in S such that $X_{\overline{k(s)}}$ is not groupless. Our goal is to prove Theorem 18.9, i.e., to show that S^{n-gr} is Zariski-countable closed, following the arguments of [55]. As is explained in Remark 18.8, we prove this in two steps.

First, one shows that S^{n-gr} is a countable union of constructible subsets. This step relies on some standard moduli-theoretic techniques. Basically, to say that X is not groupless over k is equivalent to saying that, there is an integer g such that the Hom-stack $\underline{\mathrm{Hom}}_{\mathcal{A}_g}(\mathcal{U}_g, X \times \mathcal{A}_g) \to \mathcal{A}_g$ has a non-empty fibre over some k-point of $\mathcal{A}_g$, where $\mathcal{A}_g$ is the stack of principally polarized g-dimensional abelian schemes over k, and $\mathcal{U}_g \to \mathcal{A}_g$ is the universal family. We will not discuss this argument and refer the reader to [55] for details on this part of the proof.

Once the first step is completed, to conclude the proof, it suffices to show that the notion of being groupless is stable under generization. To explain how to do this, we introduce a new notion of hyperbolicity for rigid analytic varieties (and also adic spaces) over a non-archimedean field K of characteristic zero; see [55, §2]. This notion is inspired by the earlier work of Cherry [20] (see also [6, 21, 22, 66, 67]).

If K is a complete algebraically closed non-archimedean valued field of characteristic zero and X is a finite type scheme over K, we let X^{an} be the associated rigid analytic variety over K. We say that a variety over K is K-*analytically Brody hyperbolic* if, for every finite type connected group scheme G over K, every morphism $G^{an} \to X^{an}$ is constant. It follows from this definition that a K-analytically Brody hyperbolic variety is groupless. It seems reasonable to suspect that the converse of this statement holds for projective varieties.

Conjecture 18.10 (Non-archimedean Lang–Vojta). *Let K be an algebraically closed complete non-archimedean valued field of characteristic zero, and let X be an integral projective variety over K. If X is groupless over K, then X is K-analytically Brody hyperbolic.*

In [20] Cherry proves this conjecture for closed subvarieties of abelian varieties. That is, Cherry proved the non-archimedean analogue of the Bloch–Ochiai–Kawamata theorem (Theorem 2.5) for closed subvarieties of abelian varieties.

In [55] it is shown that the above conjecture holds for projective varieties over a non-archimedean field K, assuming that K is of equicharacteristic zero and X is a "constant" variety over K (i.e., can be defined over the residue field of K). This actually follows from the following more general result.

Theorem 18.11. *Let K be an algebraically closed complete non-archimedean valued field of equicharacteristic zero with valuation ring $\mathcal{O}_K$, and let $\mathcal{X} \to \operatorname{Spec} \mathcal{O}_K$ be a proper flat morphism of schemes. If the special fibre $\mathcal{X}_0$ of $\mathcal{X} \to \operatorname{Spec} \mathcal{O}_K$ is groupless, then the generic fibre $\mathcal{X}_K$ is K-analytically Brody hyperbolic.*

Proof. This is the main result of [55] and is proven in three steps. Write $X := \mathcal{X}_K$.

First, one shows that every morphism $\mathbb{G}_{m,K}^{an} \to X^{an}$ is constant by considering the "reduction" map $X^{an} \to \mathcal{X}_0$ and a careful analysis of the residue fields of points in the image of composed map $\mathbb{G}_{m,K}^{an} \to X^{an} \to \mathcal{X}_0$; see [55, §5] for details. This implies that X has no rational curves.

Now, one wants to show that every morphism $A^{\mathrm{an}} \to X^{\mathrm{an}}$ with A some abelian variety over K is constant. Instead of appealing to GAGA and trying to use algebraic arguments, we appeal to the uniformization theorem of Bosch–Lütkebohmert for abelian varieties. This allows us to reduce to the case that A has good reduction over $\mathcal{O}_K$. In this reduction step we use that every morphism $\mathbb{G}^{\mathrm{an}}_{m,K} \to X^{\mathrm{an}}$ is constant (which is what we established in the first part of this proof); we refer the reader to [55, Theorem 2.18] for details.

Thus, we have reduced to showing that, for A an abelian variety over K *with good reduction over* $\mathcal{O}_K$, every morphism $A^{\mathrm{an}} \to X^{\mathrm{an}}$ is constant. To do so, as A has good reduction over $\mathcal{O}_K$, we may let $\mathcal{A}$ be a smooth proper model for A over $\mathcal{O}_K$. Note that the non-constant morphism $A^{\mathrm{an}} \to X^{\mathrm{an}}$ over K algebraizes by GAGA, i.e., it is the analytification of a non-constant morphism $A \to X$. By the valuative criterion of properness, there is a dense open $U \subset \mathcal{A}$ whose complement is of codimension at least two and a morphism $U \to \mathcal{X}$ extending the morphism $A \to X$ on the generic fibre. Now, since $\mathcal{X}_0$ is groupless, it has no rational curves. In particular, as $\mathcal{A} \to \mathrm{Spec}\,\mathcal{O}_K$ is smooth, the morphism $U \to \mathcal{X}$ extends to a morphism $\mathcal{A} \to \mathcal{X}$ by [37, Proposition 6.2]. However, since $\mathcal{X}_0$ is groupless, this morphism is constant on the special fibre. The latter implies (as $\mathcal{A} \to \mathrm{Spec}\,\mathcal{O}_K$ is proper) that the morphism on the generic fibre is constant; see [55, §3.2] for details. We have shown that, for every abelian variety A over K, every morphism $A \to X$ is constant and that every morphism $\mathbb{G}^{\mathrm{an}}_m \to X^{\mathrm{an}}$ is constant.

Finally, by adapting the proof of Lemma 4.4 one can show that the above implies that, for every finite type connected group scheme G over K, every morphism $G^{\mathrm{an}} \to X^{\mathrm{an}}$ is constant, so that X is K-analytically Brody hyperbolic (see [55, Lemma 2.14] for details). $\qquad\square$

To conclude the proof of Theorem 18.9, we point out that a straightforward application of Theorem 18.11 shows that being groupless is stable under generalization, as required. $\qquad\square$

An important problem in the study of non-archimedean hyperbolicity at this moment is finding a "correct" analogue of the Kobayashi pseudo-metric (if there is any at all). Cherry defined an analogue of the Kobayashi metric but it does not have the right properties, as he showed in [21] (see also [55, §3.5]). A "correct" analogue of the Kobayashi metric in the non-archimedean context would most likely have formidable consequences. Indeed, it seems reasonable to suspect that a K-analytic Brody hyperbolic projective variety is in fact "Kobayashi hyperbolic" over K and that "Kobayashi hyperbolic" projective varieties over K are bounded over K by some version of the Arzelà–Ascoli theorem.

Acknowledgments We thank Marc-Hubert Nicole for his patience and guidance when writing these notes.

We would also like to thank Carlo Gasbarri, Nathan Grieve, Aaron Levin, Steven Lu, Marc-Hubert Nicole, Erwan Rousseau, and Min Ru for organizing the research workshop *Diophantine Approximation and Value Distribution Theory* and giving us the opportunity to give the mini-course on which these notes are based.

These notes would not exist without the input of Kenneth Ascher, Raymond van Bommel, Ljudmila Kamenova, Robert Kucharczyk, Daniel Litt, Daniel Loughran, Siddharth Mathur, Jackson Morrow, Alberto Vezzani, and Junyi Xie.

We are grateful to the organizers Philipp Habegger, Ronan Terpereau, and Susanna Zimmermann of the 7th Swiss-French workshop in Algebraic Geometry in 2018 in Charmey for giving me the opportunity to speak on hyperbolicity of moduli spaces. Part of these notes are also based on the mini-course I gave in Charmey.

We are especially grateful to Raymond van Bommel for providing the diagrams and figure in Section 12.

We gratefully acknowledge support from SFB/Transregio 45.

References

1. D. Abramovich. Uniformity of stably integral points on elliptic curves. *Invent. Math.*, 127(2):307–317, 1997.

2. Dan Abramovich. A high fibered power of a family of varieties of general type dominates a variety of general type. *Invent. Math.*, 128(3):481–494, 1997.

3. Dan Abramovich and Kenji Matsuki. Uniformity of stably integral points on principally polarized abelian varieties of dimension ≤ 2. *Israel J. Math.*, 121:351–380, 2001.

4. Dan Abramovich and José Felipe Voloch. Lang's conjectures, fibered powers, and uniformity. *New York J. Math.*, 2:20–34, electronic, 1996.

5. Ekaterina Amerik. Existence of non-preperiodic algebraic points for a rational self-map of infinite order. *Math. Res. Lett.*, 18(2):251–256, 2011.

6. Ta Thi Hoai An, W. Cherry, and Julie Tzu-Yueh Wang. Algebraic degeneracy of non-Archimedean analytic maps. *Indag. Math. (N.S.)*, 19(3):481–492, 2008.

7. K. Ascher, K. DeVleming, and A. Turchet. Hyperbolicity and uniformity of varieties of log general type. *arXiv:1807.05946*.

8. K. Ascher and A. Turchet. Hyperbolicity of Varieties of Log General Type, chapter in this book.

9. Kenneth Ascher and Amos Turchet. A fibered power theorem for pairs of log general type. *Algebra Number Theory*, 10(7):1581–1600, 2016.

10. P. Autissier. Géométries, points entiers et courbes entières. *Ann. Sci. Éc. Norm. Supér. (4)*, 42(2):221–239, 2009.

11. Pascal Autissier. Sur la non-densité des points entiers. *Duke Math. J.*, 158(1):13–27, 2011.

12. B. Bakker and J. Tsimerman. Lectures on the Ax-Schanuel conjecture, chapter in this book.

13. O. Benoist. Séparation et propriété de Deligne-Mumford des champs de modules d'intersections complètes lisses. *J. Lond. Math. Soc. (2)*, 87(1):138–156, 2013.

14. O. Benoist. Quelques espaces de modules d'intersections complètes lisses qui sont quasi-projectifs. *J. Eur. Math. Soc.*, 16(8):1749–1774, 2014.

15. Raymond van Bommel, Ariyan Javanpeykar, and Ljudmila Kamenova. Boundedness in families with applications to arithmetic hyperbolicity. *arXiv:1907.11225*.

16. F. Campana. Arithmetic Aspects of Orbifold Pairs, chapter in this book.

17. F. Campana. Orbifoldes géométriques spéciales et classification biméromorphe des variétés kählériennes compactes. *J. Inst. Math. Jussieu*, 10(4):809–934, 2011.

18. Frédéric Campana. Orbifolds, special varieties and classification theory. *Ann. Inst. Fourier (Grenoble)*, 54(3):499–630, 2004.
19. L. Caporaso, J. Harris, and B. Mazur. Uniformity of rational points. *J. Amer. Math. Soc.*, 10(1):1–35, 1997.
20. W. Cherry. Non-Archimedean analytic curves in abelian varieties. *Math. Ann.*, 300(3):393–404, 1994.
21. William Cherry. A non-Archimedean analogue of the Kobayashi semi-distance and its non-degeneracy on abelian varieties. *Illinois J. Math.*, 40(1):123–140, 1996.
22. William Cherry and Min Ru. Rigid analytic Picard theorems. *Amer. J. Math.*, 126(4):873–889, 2004.
23. G. Cornell and J. H. Silverman, editors. *Arithmetic geometry.* Springer-Verlag, New York, 1986. Papers from the conference held at the University of Connecticut, Storrs, Connecticut, July 30–August 10, 1984.
24. Pietro Corvaja. *Integral points on algebraic varieties.* HBA Lecture Notes in Mathematics. Springer, Singapore; Hindustan Book Agency, New Delhi, 2016. An introduction to Diophantine geometry, Electronic version of [MR3497029], IMSc Lecture Notes in Mathematics.
25. Pietro Corvaja and Umberto Zannier. *Applications of Diophantine approximation to integral points and transcendence*, volume 212 of *Cambridge Tracts in Mathematics.* Cambridge University Press, Cambridge, 2018.
26. Pranabesh Das and Amos Turchet. Invitation to integral and rational points on curves and surfaces. In *Rational points, rational curves, and entire holomorphic curves on projective varieties*, volume 654 of *Contemp. Math.*, pages 53–73. Amer. Math. Soc., Providence, RI, 2015.
27. O. Debarre. *Higher-dimensional algebraic geometry.* Universitext. Springer-Verlag, New York, 2001.
28. P. Deligne. Un théorème de finitude pour la monodromie. In *Discrete groups in geometry and analysis (New Haven, Conn., 1984)*, volume 67 of *Progr. Math.*, pages 1–19. Birkhäuser Boston, Boston, MA, 1987.
29. Jean-Pierre Demailly. Algebraic criteria for Kobayashi hyperbolic projective varieties and jet differentials. In *Algebraic geometry—Santa Cruz 1995*, volume 62 of *Proc. Sympos. Pure Math.*, pages 285–360. Amer. Math. Soc., Providence, RI, 1997.
30. Mireille Deschamps. Courbes de genre géométrique borné sur une surface de type général [d'après F. A. Bogomolov]. In *Séminaire Bourbaki, 30e année (1977/78)*, volume 710 of *Lecture Notes in Math.*, pages Exp. No. 519, pp. 233–247. Springer, Berlin, 1979.
31. Simone Diverio and Erwan Rousseau. *Hyperbolicity of projective hypersurfaces*, volume 5 of *IMPA Monographs.* Springer, [Cham], 2016. [Second edition].
32. G. Faltings. Endlichkeitssätze für abelsche Varietäten über Zahlkörpern. *Invent. Math.*, 73(3):349–366, 1983.
33. G. Faltings. Complements to Mordell. In *Rational points (Bonn, 1983/1984)*, Aspects Math., E6, pages 203–227. Vieweg, Braunschweig, 1984.
34. G. Faltings. The general case of S. Lang's conjecture. In *Barsotti Symposium in Algebraic Geometry (Abano Terme, 1991)*, volume 15 of *Perspect. Math.*, pages 175–182. Academic Press, San Diego, CA, 1994.
35. H. Flenner. The infinitesimal Torelli problem for zero sets of sections of vector bundles. *Math. Z.*, 193(2):307–322, 1986.
36. Gerhard Frey and Moshe Jarden. Approximation theory and the rank of abelian varieties over large algebraic fields. *Proc. London Math. Soc. (3)*, 28:112–128, 1974.
37. O. Gabber, Q. Liu, and D. Lorenzini. Hypersurfaces in projective schemes and a moving lemma. *Duke Math. J.*, 164(7):1187–1270, 2015.

38. Carlo Gasbarri. On some differences between number fields and function fields. *Riv. Math. Univ. Parma (N.S.)*, 7(1):1–18, 2016.
39. A. Grothendieck. *Revêtements étales et groupe fondamental (SGA I) Fasc. II: Exposés 6, 8 à 11*, volume 1960/61 of *Séminaire de Géométrie Algébrique*. Institut des Hautes Études Scientifiques, Paris, 1963.
40. A. Grothendieck. Un théorème sur les homomorphismes de schémas abéliens. *Invent. Math.*, 2:59–78, 1966.
41. B. Hassett and Y. Tschinkel. Abelian fibrations and rational points on symmetric products. *Internat. J. Math.*, 11(9):1163–1176, 2000.
42. Brendan Hassett. Correlation for surfaces of general type. *Duke Math. J.*, 85(1):95–107, 1996.
43. Fei Hu, Sheng Meng, and De-Qi Zhang. Ampleness of canonical divisors of hyperbolic normal projective varieties. *Math. Z.*, 278(3-4):1179–1193, 2014.
44. Jun-Muk Hwang, Stefan Kebekus, and Thomas Peternell. Holomorphic maps onto varieties of non-negative Kodaira dimension. *J. Algebraic Geom.*, 15(3):551–561, 2006.
45. Shigeru Iitaka. *Algebraic geometry*, volume 76 of *Graduate Texts in Mathematics*. Springer-Verlag, New York-Berlin, 1982. An introduction to birational geometry of algebraic varieties, North-Holland Mathematical Library, 24.
46. A. Javanpeykar and A. Levin. Urata's theorem in the logarithmic case and applications to integral points. *arXiv:2002.11709*.
47. A. Javanpeykar and D. Loughran. Complete intersections: moduli, Torelli, and good reduction. *Math. Ann.*, 368(3-4):1191–1225, 2017.
48. A. Javanpeykar and D. Loughran. The moduli of smooth hypersurfaces with level structure. *Manuscripta Math.*, 154(1-2):13–22, 2017.
49. Ariyan Javanpeykar. Arithmetic hyperbolicity: endomorphisms, automorphisms, hyperkähler varieties, geometricity. *arXiv:1809.06818*.
50. Ariyan Javanpeykar and Ljudmila Kamenova. Demailly's notion of algebraic hyperbolicity: geometricity, boundedness, moduli of maps. *Math. Z.*, 296(3–4), 1645–1672, 2020.
51. Ariyan Javanpeykar and Robert Kucharczyk. Algebraicity of analytic maps to a hyperbolic variety. *Math. Nachr.*, 293(8), 1490–1504, 2020.
52. Ariyan Javanpeykar and Daniel Litt. Integral points on algebraic subvarieties of period domains: from number fields to finitely generated fields. *arXiv:1907.13536*.
53. Ariyan Javanpeykar and Daniel Loughran. Arithmetic hyperbolicity and a stacky Chevalley-Weil theorem. *arXiv:1808.09876*.
54. Ariyan Javanpeykar and Daniel Loughran. Good reduction of Fano threefolds and sextic surfaces. *Ann. Sc. Norm. Super. Pisa Cl. Sci. (5)*, 18(2):509–535, 2018.
55. Ariyan Javanpeykar and Alberto Vezzani. Non-archimedean hyperbolicity and applications. *arXiv:1808.09880*.
56. Ariyan Javanpeykar and Junyi Xie. Finiteness properties of pseudo-hyperbolic varieties. *IMRN, to appear, arXiv:1909.12187*.
57. Yujiro Kawamata. On Bloch's conjecture. *Invent. Math.*, 57(1):97–100, 1980.
58. Peter Kiernan and Shoshichi Kobayashi. Satake compactification and extension of holomorphic mappings. *Invent. Math.*, 16:237–248, 1972.
59. Shoshichi Kobayashi. *Hyperbolic complex spaces*, volume 318 of *Grundlehren der Mathematischen Wissenschaften*. Springer-Verlag, Berlin, 1998.
60. Sándor J. Kovács. Subvarieties of moduli stacks of canonically polarized varieties: generalizations of Shafarevich's conjecture. In *Algebraic geometry—Seattle 2005. Part 2*, volume 80 of *Proc. Sympos. Pure Math.*, pages 685–709. Amer. Math. Soc., Providence, RI, 2009.
61. Myung H. Kwack. Generalization of the big Picard theorem. *Ann. of Math. (2)*, 90:9–22, 1969.

62. S. Lang. Hyperbolic and Diophantine analysis. *Bull. Amer. Math. Soc. (N.S.)*, 14(2):159–205, 1986.
63. B. Lawrence and A. Venkatesh. Diophantine problems and p-adic period mappings. *Invent. Math.*, 221(3), 893–999, 2020.
64. Robert Lazarsfeld. *Positivity in algebraic geometry. I. Classical setting: line bundles and linear series. Ergebnisse der Mathematik und ihrer Grenzgebiete. 3. Folge. A Series of Modern Surveys in Mathematics*, 48. Springer-Verlag, Berlin, 2004.
65. Robert Lazarsfeld. *Positivity in algebraic geometry. II. Positivity for vector bundles, and multiplier ideals. Ergebnisse der Mathematik und ihrer Grenzgebiete. 3. Folge. A Series of Modern Surveys in Mathematics*, 49. Springer-Verlag, Berlin, 2004.
66. Aaron Levin and Julie Tzu-Yueh Wang. On non-Archimedean curves omitting few components and their arithmetic analogues. *Canad. J. Math.*, 69(1):130–142, 2017.
67. Chien-Wei Lin and Julie Tzu-Yueh Wang. Generalizations of rigid analytic Picard theorems. *Proc. Amer. Math. Soc.*, 138(1):133–139, 2010.
68. M. Martin-Deschamps. Conjecture de Shafarevich pour les corps de fonctions sur **Q**. *Astérisque*, (127):256–259, 1985. Seminar on arithmetic bundles: the Mordell conjecture (Paris, 1983/84).
69. M. Martin-Deschamps. La construction de Kodaira-Parshin. *Astérisque*, (127):261–273, 1985. Seminar on arithmetic bundles: the Mordell conjecture (Paris, 1983/84).
70. B. Mazur. Arithmetic on curves. *Bull. Amer. Math. Soc.*, 14(2):207–259, 1986.
71. Michael McQuillan. Diophantine approximations and foliations. *Inst. Hautes Études Sci. Publ. Math.*, (87):121–174, 1998.
72. Shinichi Mochizuki. Topics in absolute anabelian geometry I: generalities. *J. Math. Sci. Univ. Tokyo*, 19(2):139–242, 2012.
73. Atsushi Moriwaki. Remarks on rational points of varieties whose cotangent bundles are generated by global sections. *Math. Res. Lett.*, 2(1):113–118, 1995.
74. Noboru Nakayama. *Zariski-decomposition and abundance*, volume 14 of *MSJ Memoirs*. Mathematical Society of Japan, Tokyo, 2004.
75. Nitin Nitsure. Construction of Hilbert and Quot schemes. In *Fundamental algebraic geometry*, volume 123 of *Math. Surveys Monogr.*, pages 105–137. Amer. Math. Soc., Providence, RI, 2005.
76. Junjiro Noguchi, Jörg Winkelmann, and Katsutoshi Yamanoi. Degeneracy of holomorphic curves into algebraic varieties. *J. Math. Pures Appl. (9)*, 88(3):293–306, 2007.
77. Junjiro Noguchi, Jörg Winkelmann, and Katsutoshi Yamanoi. The second main theorem for holomorphic curves into semi-abelian varieties. II. *Forum Math.*, 20(3):469–503, 2008.
78. Junjiro Noguchi, Jörg Winkelmann, and Katsutoshi Yamanoi. Degeneracy of holomorphic curves into algebraic varieties II. *Vietnam J. Math.*, 41(4):519–525, 2013.
79. Yuri Prokhorov and Constantin Shramov. Jordan property for groups of birational selfmaps. *Compos. Math.*, 150(12):2054–2072, 2014.
80. Wilfried Schmid. Variation of Hodge structure: the singularities of the period mapping. *Invent. Math.*, 22:211–319, 1973.
81. The Stacks Project Authors. *Stacks Project*. http://stacks.math.columbia.edu, 2015.
82. Lucien Szpiro, editor. *Séminaire sur les pinceaux arithmétiques: la conjecture de Mordell*. Société Mathématique de France, Paris, 1985. Papers from the seminar held at the École Normale Supérieure, Paris, 1983–84, Astérisque No. 127 (1985).

83. E. Ullmo. Points rationnels des variétés de Shimura. *Int. Math. Res. Not.*, (76):4109–4125, 2004.

84. Toshio Urata. Holomorphic mappings into taut complex analytic spaces. *Tôhoku Math. J. (2)*, 31(3):349–353, 1979.

85. P. Vojta. A higher-dimensional Mordell conjecture. In *Arithmetic geometry (Storrs, Conn., 1984)*, pages 341–353. Springer, New York, 1986.

86. Paul Vojta. A Lang exceptional set for integral points. In *Geometry and analysis on manifolds*, volume 308 of *Progr. Math.*, pages 177–207. Birkhäuser/Springer, Cham, 2015.

87. Katsutoshi Yamanoi. Holomorphic curves in algebraic varieties of maximal Albanese dimension. *Internat. J. Math.*, 26(6):1541006, 45, 2015.

88. Katsutoshi Yamanoi. Pseudo Kobayashi hyperbolicity of subvarieties of general type on abelian varieties. *J. Math. Soc. Japan*, 71(1):259–298, 2019.

Hyperbolicity of Varieties of Log General Type

Kenneth Ascher and Amos Turchet

MSC codes 1G35, 14G40, 14J20, 11J97, 32Q45

1 Introduction

Diophantine Geometry aims to describe the sets of rational and/or integral points on a variety. More precisely one would like *geometric* conditions on a variety X that determine the distribution of rational and/or integral points. Here geometric means conditions that can be checked on the algebraic closure of the field of definition.

Pairs, sometimes called log pairs, are objects of the form (X, D) where X is a projective variety and D is a reduced divisor. These objects naturally arise in arithmetic when studying integral points, and play a central role in geometry, especially in the minimal model program and the study of moduli spaces of higher dimensional algebraic varieties. They arise naturally in the study of integral points since, if one wants to study integral points on a quasi-projective variety V, this can be achieved by studying points on (X, D), where $(X \setminus D) \cong V$, the variety X is a smooth projective compactification of V, and the complement D is a normal crossings divisor. In this case, (X, D) is referred to as a *log pair*.

K. Ascher
Department of Mathematics, Princeton University, Princeton, NJ 08544, USA
e-mail: kascher@princeton.edu

A. Turchet (✉)
Centro di Ricerca Matematica Ennio De Giorgi, Scuola Normale Superiore, Palazzo Puteano, Piazza dei cavalieri, 3, 56100 Pisa, Italy
e-mail: amos.turchet@sns.it

© Springer Nature Switzerland AG 2020
M.-H. Nicole (ed.), *Arithmetic Geometry of Logarithmic Pairs and Hyperbolicity of Moduli Spaces*, CRM Short Courses,
https://doi.org/10.1007/978-3-030-49864-1_4

The goal of these notes is threefold:

(1) to present an introduction to the study of rational and integral points on curves and higher dimensional varieties and pairs;
(2) to introduce various notions of hyperbolicity for varieties and pairs, and discuss their conjectural relations;
(3) and to show how geometry influences the arithmetic of algebraic varieties and pairs using tools from birational geometry.

Roughly speaking, a k-rational point of an algebraic variety is a point whose coordinates belong to k. One of the celebrated results in Diophantine geometry of curves is the following.

Theorem 1.1. *If $\mathcal{C}$ is a geometrically integral smooth projective curve over a number field k, then the following are equivalent:*

(1) $g(\mathcal{C}) \geq 2$,
(2) *the set of L-rational points is finite for every finite extension L/k [Faltings' theorem; arithmetic hyperbolicity],*
(3) *every holomorphic map $\mathbb{C} \to \mathcal{C}_{\mathbb{C}}^{\mathrm{an}}$ is constant [Brody hyperbolicity], and*
(4) *the canonical bundle $\omega_{\mathcal{C}}$ is ample.*

In particular, one can view the above theorem as saying that various notions of hyperbolicity coincide for projective curves. One of the major open questions in this area is how the above generalizes to higher dimensions. The following conjecture we state is related to the Green–Griffiths–Lang conjecture.

Conjecture 1.2. *Let X be a projective geometrically integral variety over a number field k. Then, the following are equivalent:*

(1) *X is arithmetically hyperbolic,*
(2) *$X_{\mathbb{C}}$ is Brody hyperbolic, and*
(3) *every integral subvariety of X is of general type.*

We recall that a variety is of general type if there exists a desingularization with big canonical bundle. This conjecture is very much related to conjectures of Bombieri, Lang, and Vojta postulating that varieties of general type (resp. log general type) do not have a dense set of rational (resp. integral) points. While Conjecture 1.2 is essentially wide open, it is known that if the cotangent bundle Ω_X^1 is sufficiently positive, all three conditions are satisfied. In particular, the latter two are satisfied if Ω_X^1 is ample, and the first is satisfied if in addition Ω_X^1 is globally generated. We do note, however, that there are examples of varieties that are (e.g. Brody) hyperbolic but for which Ω_X^1 is not ample (see Example 6.11).

In any case, it is natural to ask what can be said about hyperbolicity for quasi-projective varieties. One can rephrase Conjecture 1.2 for quasi-projective varieties V, and replace (3) with the condition that all subvarieties are of *log general type*. We recall that a variety V is of log general type if there exists a desingularization $\widetilde{V}$, and a projective embedding $\widetilde{V} \subset Y$ with $Y \setminus \widetilde{V}$ a divisor of normal crossings, such that $\omega_Y(D)$ is big. It is then natural to ask if positivity of the *log cotangent bundle* implies hyperbolicity in this setting.

The first obstacle, is that the log cotangent sheaf is *never ample*. However, one can essentially ask that this sheaf is "as ample as possible" (see Definition 7.11 for the precise definition of almost ample). It turns out that, with this definition, quasi-projective varieties with almost ample log cotangent bundle are Brody hyperbolic (see [41, Section 3]). In recent joint work with Kristin DeVleming [10], we explore, among other things, the consequences for hyperbolicity that follow from such a positivity assumption. We prove that quasi-projective varieties with positive log cotangent bundle are arithmetically hyperbolic (see Theorem 8.1), and that all their subvarieties are of log general type (see Theorem 7.14).

Theorem 1.3 ([10]). *Let (X, D) be a log smooth pair with almost ample $\Omega_X^1(\log D)$. If $Y \subset X$ is a closed subvariety, then*

(1) *all pairs (Y, E), where $E = (Y \cap D)_{red}$, with $Y \not\subset D$ are of log general type.*
(2) *If in addition $\Omega_X^1(\log D)$ is globally generated, and $V \cong (X \setminus D)$ is a smooth quasi-projective variety over a number field k, then for any finite set of places S, the set of S-integral points $V(\mathcal{O}_{k,S})$ is finite.*

The main focus of these notes is to present, in a self-contained manner, the proofs of these statements. In particular, we review the notions of ampleness, almost ampleness, and global generation for vector bundles (see Section 7). The proof of the second statement heavily relies upon the theory of semi-abelian varieties and the quasi-Albanese variety, and so we develop the necessary machinery (see Section 8).

Along the way, we discuss the related conjecture of Lang (see Conjecture 5.1), which predicts that varieties of general type *do not* have a dense set of rational points. We discuss the (few) known cases in Section 5.2. A related, more general conjecture due to Vojta (see Conjecture 9.5) suggests that one can control the heights of points on varieties of general (resp. log general) type. We discuss this conjecture, and we introduce the theory of heights in Section 9.1.

We are then naturally led to discuss what happens in the *function field* case (see Section 10). In this setting, the analogue of Faltings' Theorem is known (see Theorem 10.10). Similarly, positivity assumptions on the cotangent bundle lead to hyperbolicity results. In this context, we discuss a theorem of Noguchi (see Theorem 10.16) and give insight into what is expected in the quasi-projective setting.

We first got interested in studying positivity of the log cotangent bundle to understand "uniformity" of integral points as it relates to the Lang–Vojta conjecture. Consequently, we end these notes with a short section discussing and summarizing some key results in this area.

1.1 Outline

The road map of these notes is the following:

§2 Rational points with a focus on projective curves.
§3 Integral points with a focus on quasi-projective curves.
§4 Tools from positivity and birational geometry.
§5 Lang's conjecture and some known cases.
§6 Hyperbolicity of projective varieties and positivity of vector bundles.
§7 Hyperbolicity for quasi-projective varieties.
§8 Semi-abelian varieties, the quasi-Albanese, and arithmetic hyperbolicity of quasi-projective varieties.
§9 Vojta's conjecture and the theory of heights.
§10 Diophantine geometry over function fields.
§11 Some known consequences of Lang's conjecture.

1.2 Notation

We take this opportunity to set some notation.

1.2.1 Geometry

Divisors will refer to Cartier divisors, and $\mathrm{Pic}(X)$ will denote the Picard group, i.e. the group of isomorphism classes of line bundles on X. We recall that a reduced divisor is of *normal crossings* if each point étale locally looks like the intersection of coordinate hyperplanes.

1.2.2 Arithmetic

Throughout k will denote a number field, i.e. a finite extension of $\mathbb{Q}$. We will denote by M_k the set of places of k, i.e. the set of equivalence classes of absolute values of k. We will denote by $\mathcal{O}_k = \{\alpha \in k : |\alpha|_v \leq 1 \text{ for every } v \in M_k\}$ the ring of integers of k, with $\mathcal{O}_k^*$ the group of units, and with $\mathcal{O}_{k,S}$ the ring of S-integers in k, i.e. $\mathcal{O}_{k,S} = \{\alpha \in k : |\alpha|_v \leq 1 \text{ for every } v \notin S\}$. We will denote an algebraic closure of k by $\overline{k}$.

2 Rational Points on Projective Curves

Some of the main objects of study in Diophantine Geometry are *integral* and *rational points* on varieties. An *algebraic variety* is the set of common solutions to a system of polynomial equations with coefficients in R, where R is usually a field or a ring. In these notes we will consider fields that are finite extensions of $\mathbb{Q}$, and rings that are finite extensions of $\mathbb{Z}$.

If X is a variety defined over a number field k, i.e. defined by equations with coefficients in k, then the set of k-*rational* points of X is the set of solutions with coordinates in k. In a similar way, one can consider the set of *integral* points of X as the set of solutions with coordinates that belong to the ring of integers of k. However it is sometimes more convenient to take an approach that does not depend on the particular choice of coordinates used to present X.

Definition 2.1 (Rational Points). Let X be a projective variety defined over k. If $P \in X(\overline{k})$ is an algebraic point, then the residue field $k(P)$ is a finite extension of k. We say that P is k-*rational* if $k(P) = k$.

Remark 2.2. The above notion is intrinsic in the sense that it depends only on the function field of X, which is independent of the embedding in projective space. In this case a rational point corresponds to a morphism $\mathrm{Spec}\, k \to X$ (Exercise).

For a non-singular curve $\mathcal{C}$ defined over a number field k, the genus governs the distribution of the k-rational points: if a curve is rational, i.e. it has genus zero, then the set of k-rational points $\mathcal{C}(k)$ is dense, at most after a quadratic extension of k. Similarly, if the curve has genus one, then at most after a finite extension of k, the set of k-rational points $\mathcal{C}(k)$ is dense (see [37] for a gentle introduction and proofs of these statements). In the genus one case one can prove a stronger statement, originally proven by Mordell, and extended by Weil to arbitrary abelian varieties, namely that the set of k-rational points forms a finitely generated abelian group. We can summarize this in the following proposition.

Proposition 2.3. *Let $\mathcal{C}$ be a non-singular projective curve of genus $g(\mathcal{C})$ defined over a number field k.*

- *If $g(\mathcal{C}) = 0$, then $\mathcal{C}(k)$ is dense, after at most a quadratic extension of k.*
- *If $g(\mathcal{C}) = 1$, then $\mathcal{C}(k)$ is a finitely generated group of positive rank, (possibly) after a finite extension of k.*

In 1922, Mordell conjectured that a projective curve $\mathcal{C}$ of genus $g(\mathcal{C}) > 1$ has finitely many k-points. This was proven by Faltings' [45].

Theorem 2.4 (Faltings' Theorem [45], Formerly Mordell's Conjecture). *Let $\mathcal{C}$ be a non-singular projective curve $\mathcal{C}$ defined over a number field k. If $g(\mathcal{C}) > 1$, then the set $\mathcal{C}(k)$ is finite.*

The original proof of Faltings reduced the problem to the Shafarevich conjecture for abelian varieties, via Parhsin's trick. The argument uses very refined and difficult tools like Arakelov Theory on moduli spaces, semistable abelian schemes, and p-divisible groups, and therefore such a proof is outside the scope of these notes. A different proof was given shortly after by Vojta in [97] using ideas from Diophantine approximation while still relying on Arakelov theory. Faltings in [46] gave another simplification, eliminating the use of the arithmetic Riemann–Roch Theorem for arithmetic threefolds in Vojta's proof, and was able to extend these methods to prove a conjecture of Lang. Another simplification of both Vojta and Faltings' proofs was given by Bombieri in [12] combining ideas from Mumford [78] together with the ones in the aforementioned papers.

The above results leave open many other Diophantine questions: when is the set $\mathcal{C}(k)$ empty? Is there an algorithm that produces a set of generators for $E(k)$, for an elliptic curve E defined over k? Is there an algorithm that computes the set $\mathcal{C}(k)$ when it is finite (Effective Mordell)? We will not address these questions in this notes, but we will mention the very effective Chabauty–Coleman–Kim method that in certain situations can give answer to the latter question (see [23, 25, 61, 62, 75]).

2.1 Geometry Influences Arithmetic

In order to generalize, at least conjecturally, the distribution of k-rational points on curves to higher dimensional varieties, it is convenient to analyze the interplay between the arithmetic and the geometric properties of curves, following the modern philosophy that the geometric invariants of an algebraic variety determine arithmetic properties of the solution set.

We start by recalling the definition of the canonical sheaf.

Definition 2.5. Let X be a non-singular variety over k of $\dim X = n$. We define the *canonical sheaf* of X to be $\omega_X = \bigwedge^n \Omega^1_{X/k}$, where $\Omega^1_{X/k}$ denotes the sheaf of relative differentials of X.

If $\mathcal{C}$ is a curve, then $\omega_X = \Omega^1_{\mathcal{C}}$ is an invertible sheaf whose sections are the global 1-forms on $\mathcal{C}$. In this case, we call any divisor in the linear equivalence class a *canonical divisor*, and denote the divisor by $K_{\mathcal{C}}$.

Example 2.6. Let $\mathcal{C} \cong \mathbb{P}^1$, with coordinates $[x : y]$. In the open affine U_x given by $x \neq 0$ we can consider the global coordinate $t = y/x$ and the global differential form dt. We can extend dt as a *rational* differential form $s \in \Omega^1_{\mathbb{P}^1}$, noting that it will possibly have poles. To compute its associated divisor we note that in the locus $U_x \cap U_y$, i.e. where $x \neq 0$ and $y \neq 0$, the section s is invertible. In the intersection, the basic formula $d(1/t) = -dt/t^2$ shows that the divisor associated with s is $-2P$, where $P = [0 : 1]$. In particular, given that any two points are linearly equivalent on $\mathbb{P}^1$, $K_{\mathbb{P}^1} \sim -P_1 - P_2$ for two points on $\mathbb{P}^1$, and $\deg K_{\mathbb{P}^1} = -2$.

More generally, any divisor D on a smooth curve $\mathcal{C}$ is a weighted sum of points, and its degree is the sum of the coefficients. In the case of the canonical divisor, if $\mathcal{C}$ is a curve of genus g, then $K_\mathcal{C}$ has $\deg(K_\mathcal{C}) = 2g - 2$ (see [53, Example IV.1.3.3]).

Given Theorem 2.4 and Proposition 2.3, one can see that the positivity of the canonical divisor $K_\mathcal{C}$ determines the distribution of k-rational points $\mathcal{C}(k)$. In particular, the set $\mathcal{C}(k)$ is finite if and only if $\deg(K_\mathcal{C}) > 0$.

There is a further geometric property of curves that mimics the characterization of rational points given above. If $\mathcal{C}$ is a non-singular curve defined over a number field k, holomorphic maps to the corresponding Riemann Surface $\mathcal{C}_\mathbb{C}$ (viewed either as the set of complex points $\mathcal{C}(\mathbb{C})$ together with its natural complex structure, or as the analytification of the algebraic variety $\mathcal{C}$) are governed by the genus $g(\mathcal{C})$. If the genus is zero or one, the universal cover of $\mathcal{C}_C$ is either the Riemann sphere or a torus, and therefore there exist non-constant holomorphic maps $\mathbb{C} \to \mathcal{C}$ with dense image. On the other hand, if $g(\mathcal{C}) \geq 2$, the universal cover of $\mathcal{C}_\mathbb{C}$ is the unit disc and, by Liouville's Theorem, every holomorphic map to $\mathcal{C}_\mathbb{C}$ has to be constant, since its lift to the universal cover is constant.

Varieties X where every holomorphic map $\mathbb{C} \to X$ is constant play a fundamental role in complex analysis/geometry so we recall here their definition.

Definition 2.7. Let X be a complex analytic space. We say that X is *Brody Hyperbolic* if every holomorphic map $X \to \mathbb{C}$ is constant. We say that X is *Kobayashi hyperbolic* if the Kobayashi pseudo-distance is a distance (see [63] for definition and properties).

Remark 2.8. When X is compact the two notions are equivalent by [15] and we will only say that X is hyperbolic. For more details about the various notions of hyperbolicity and their connection with arithmetic and geometric properties of varieties we refer to the chapter by Javanpeykar in this volume [57].

Given a non-singular projective curve $\mathcal{C}$ defined over a number field, the previous discussion can be summarized in the following table:

$g(\mathcal{C})$	$\deg(K_\mathcal{C})$	Complex hyperbolicity	Density of k-points
0	$-2 < 0$	Not hyperbolic	Potentially dense
1	0	Not hyperbolic	Potentially dense
≥ 2	$2g - 2 > 0$	Hyperbolic	Finite

Remark 2.9. In the table, $\deg(K_{\mathcal{C}}) > 0$ is equivalent to requiring that ω_C is ample, since for curves a divisor is ample if and only if its degree is positive (see Section 4.1.1).

3   Integral Points on Curves

The previous section deals with the problem of describing the set $\mathcal{C}(k)$, which we can think of the k-solutions of the polynomial equations that define $\mathcal{C}$. An analogous problem, fundamental in Diophantine Geometry, is the study of the integral solutions of these equations, or equivalently of the integral points of $\mathcal{C}$. However, the definition of integral point is more subtle.

Example 3.1. Consider $\mathbb{P}^1_{\mathbb{C}}$ as the set $\{[x_0 : x_1] : x_0, x_1 \in \mathbb{C}\}$; if $k \subset \mathbb{C}$ is a number field, then the set of rational points, as defined in Definition 2.1, is the subset $\mathbb{P}^1_{\mathbb{C}}(k) \subset \mathbb{P}^1_{\mathbb{C}}$ consisting of the points $[x_0 : x_1]$ such that both coordinates are in k.

Remark 3.2. As points in $\mathbb{P}^n$ are equivalence classes, we are implicitly assuming the choice of a representative with k-rational coordinates. For example the point $[\sqrt{2} : \sqrt{2}]$ is a k-rational point, being nothing but the point $[1 : 1]$.

Let us focus on the case $k = \mathbb{Q}$. We want to identify the *integral* points: it is natural to consider points $[x_0 : x_1]$ in which both coordinates are *integral*, i.e. $x_0, x_1 \in \mathcal{O}_k = \mathbb{Z}$. In this case, by definition of projective space, this is equivalent to considering points of the form $[a : b]$ in which $\gcd(a, b) = 1$.

Now consider the problem of characterizing integral points among $\mathbb{Q}$-rational points, i.e. given a point $P = [\frac{a}{b} : \frac{c}{d}] \in \mathbb{P}^1(\mathbb{Q})$, when is this point integral (assuming we already took care of common factors)? One answer is to require that $b = d = 1$; however, the point P can also be written as $P = [ad : bc]$, and since we assumed that we already cleared any common factor in the fractions, we have $\gcd(ad, bc) = 1$. So in particular, *every* rational point is integral!

Example 3.3. Let us consider the *affine* curve $\mathbb{A}^1_{\mathbb{C}} \subset \mathbb{P}^1_{\mathbb{C}}$, as the set $\{[a : 1] : a \in \mathbb{C}\}$. Then $\mathbb{A}^1(k) = \{[a : 1] : a \in k\}$. The integral points should correspond to $\{[a : 1] : a \in \mathcal{O}_k\}$. Now we can ask the same question as in Example 3.1, specializing again to $k = \mathbb{Q}$: namely how can we characterize integral points on $\mathbb{A}^1$ among its $\mathbb{Q}$-rational points? Given a rational point $P = [\frac{a}{b} : 1]$, we can require that $b = 1$. This is equivalent to asking that for every prime $\mathfrak{p} \in \mathbb{Z}$, the prime $\mathfrak{p}$ does not divide b. In the case in which $\mathfrak{p} \mid b$, we can rewrite $P = [a : b]$ and we can see that the reduction modulo $\mathfrak{p}$ of P, i.e. the point whose coordinates are the reduction modulo $\mathfrak{p}$ of the coordinates of P, is the point $[1 : 0]$ which does not belong to $\mathbb{A}^1$! This shows that one characterization of integral points is the set of k-rational points

Fig. 1 An illustration of
a model (from [4])

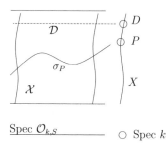

whose reduction modulo every prime $\mathfrak{p}$ is still a point of $\mathbb{A}^1$. More precisely, let $D = [0 : 1] \in \mathbb{P}^1 \setminus \mathbb{A}^1$ be the point at infinity: integral points $\mathbb{A}^1(\mathcal{O}_k)$ are precisely the k-rational points $\mathbb{A}^1(k)$ such that their reduction modulo every prime of $\mathcal{O}_k$ is disjoint from D.

The previous example gives an intuition for a coordinate-free definition of integral points. Note that, given an affine variety $X \subset \mathbb{A}^n$ defined over the ring of integers $\mathcal{O}_k$ of a number field k one could try to mimic Definition 2.1 as follows: define the ring $\mathcal{O}_k[X]$ to be the image of the ring $\mathcal{O}_k[T_1, \ldots, T_n]$ inside the coordinate ring $k[X]$ of X. Now given a rational point $P = (p_1, \ldots, p_n) \in X(k)$, it is integral when all the coordinates are in $\mathcal{O}_k$. Therefore the point P defines a specialization morphism $\varphi_p : \mathcal{O}_k[T_1, \ldots, T_n] \to \mathcal{O}_k$ which induces, by passing to the quotient, a morphism $\varphi_p : \mathcal{O}_k[X] \to \mathcal{O}_k$. The construction can be reversed to show that indeed every such morphism corresponds to an integral point. Note that this definition depends on the embedding $X \subset \mathbb{A}^n$.

We will instead pursue a generalization that is based on Example 3.3. Recall that the characterization we obtained of $\mathbb{A}^1(\mathcal{O}_k)$ made use of the reduction modulo primes of points. To give a more intrinsic and formal definition we introduce the notion of *models*.

Definition 3.4 (Models). Let X be a quasi-projective variety defined over a number field k. A *model* of X over the ring of integers $\mathcal{O}_k$ is a variety $\mathcal{X}$ with a dominant, flat, finite type map $\mathcal{X} \to \operatorname{Spec} \mathcal{O}_k$ such that the generic fiber $\mathcal{X}_\eta$ is isomorphic to X. See Figure 1 for an illustration taken from [4].

Over every prime $\mathfrak{p}$ of $\mathcal{O}_k$ we have a variety $\mathcal{X} \times_{\mathcal{O}_k} \operatorname{Spec} k_\mathfrak{p}$ defined over the residue field $k_\mathfrak{p}$ which is the "reduction modulo $\mathfrak{p}$ of X," while the generic fiber over (0) is isomorphic to the original X. This will make precise the notion of reduction modulo $\mathfrak{p}$ of a prime.

Given a rational point $P \in X(\kappa)$, since $X \cong \mathcal{X}_\eta$, this gives a point in the generic fiber of a model $\mathcal{X}$ of X. If the model $\mathcal{X}$ is proper, the point P, which corresponds to a map $\operatorname{Spec} k \to X$, will extend to a section $\sigma_P : \operatorname{Spec} \mathcal{O}_k \to \mathcal{X}$, therefore defining the reduction modulo a prime of P: it is just the point $P_\mathfrak{p} = \sigma(\mathcal{O}_k) \cap \mathcal{X}_\mathfrak{p}$. Therefore, given a *proper* model of X, there is a well-defined notion of the reduction modulo a prime of k-rational points.

The last ingredient we need to define integral points is the analogue of the point $D = [0 : 1] \in \mathbb{P}^1 \setminus \mathbb{A}^1$ in Example 3.3. In the example we used that the affine curve $\mathbb{A}^1$ came with a (natural) compactification, namely $\mathbb{P}^1$, and a divisor "at infinity," namely D. This motivates the idea that to study integral points, the geometric objects that we need to consider are pairs of a variety and a divisor, i.e. objects of the form (X, D) where X is a projective variety (corresponding to the compactification of the affine variety) and D is a divisor (corresponding to the divisor "at infinity").

Definition 3.5. A *pair* is a couple (X, D) where X is a (geometrically integral) projective variety defined over a field k and a normal crossing divisor D. A model $(\mathcal{X}, \mathcal{D}) \to \operatorname{Spec} \mathcal{O}_k$ of the pair, is a proper model $\mathcal{X} \to \operatorname{Spec} \mathcal{O}_k$ of X together with a model $\mathcal{D} \to \operatorname{Spec} \mathcal{O}_k$ of the divisor D such that $\mathcal{D}$ is a Cartier divisor of $\mathcal{X}$.

Remark 3.6. Given a non-singular affine variety Y defined over a field of characteristic 0, combining the theorems of Nagata and Hironaka, we can always find a non-singular projective compactification $X \supset Y$ such that $D = X \setminus Y$ is a simple normal crossing divisor. Therefore we can identify (non-canonically) every non-singular affine variety Y as the pair (X, D). This gives a way to characterize the set of integral points on Y as the set of rational points of X whose reduction modulo every prime does not specialize to (the reduction of) D. Using models, this gives a formal intrinsic defining of integral points.

Definition 3.7. Given a pair (X, D), the *set of D-integral points* of X, or equivalently the *set of integral points* of $Y = X \setminus D$, with respect to a model $(\mathcal{X}, \mathcal{D})$ of (X, D), is the set of sections $\operatorname{Spec} \mathcal{O}_k \to \mathcal{X} \setminus \mathcal{D}$. We will denote this set by $X(\mathcal{O}_{k,D})$ or $Y(\mathcal{O}_k)$ for $Y = X \setminus D$.

Remark 3.8.

- In the case in which Y is already projective, i.e. $X = Y$ and $D = \emptyset$, then the set of integral points coincide with the set of *all* sections $\operatorname{Spec} \mathcal{O}_k \to \mathcal{X}$. Since the model $\mathcal{X} \to \operatorname{Spec} \mathcal{O}_k$ is proper, this is equivalent to the set of sections of the generic fiber, i.e. maps $\operatorname{Spec} k \to X$, which is exactly the set $X(k)$. This shows that for projective varieties, the set of rational points coincide with the set of integral points.
- The definition of integral points depends on the choice of the model! Different choices of models might give different sets of integral points (see Example 3.9).
- When we consider an affine variety V given inside an affine space $\mathbb{A}^n$ as the vanishing of a polynomial f with coefficients in $\mathcal{O}_k$, this corresponds to a pair (X, D) where X is the projective closure of V (i.e. the set of solutions to the equations obtained by homogenizing f), and D the boundary divisor. In this case there is a natural model of (X, D) given by the model induced by the natural model of $\mathbb{P}^n$ over $\operatorname{Spec} \mathcal{O}_k$, i.e. $\mathbb{P}^n_{\mathcal{O}_k}$ and the closure of D inside it. Then one can show (exercise) that the set of

integral points with respect to this model coincide with solutions of $f = 0$ with integral coordinates (see discussion before Definition 3.4).

The following example shows that the definition of integral points as above truly depends on the choice of model.

Example 3.9 (Abramovich). Consider an elliptic curve given as $E : y^2 = x^3 + Ax + B$ with $A, B \in \mathbb{Z}$, and as usual the origin will be the point at infinity. Since E is given as the vanishing of a polynomial equation, the homogenization defines a closed subset of $\mathbb{P}^2_{\mathbb{Q}}$. Moreover, since the coefficients are all integers we get that the same equation defines a model $\mathcal{E}$ of E which is the closure of E inside $\mathbb{P}^2_{\mathbb{Z}}$, i.e. the standard model of $\mathbb{P}^2_{\mathbb{Q}}$ over $\operatorname{Spec} \mathbb{Z}$. Let $P \in E(\mathbb{Q})$ be a rational point which is not integral with respect to $D = \{0_E\}$. This means that there exists a prime $\mathfrak{p} \in \operatorname{Spec} \mathbb{Z}$ such that P reduces to the origin modulo $\mathfrak{p}$. In particular, the section $P : \operatorname{Spec} \mathbb{Z} \to \mathcal{E}$ intersects the zero section over the prime $\mathfrak{p}$. Call Q the point of intersection.

Consider now the blow up $\pi : \mathcal{E}' \to \mathcal{E}$ of $\mathcal{E}$ at Q: by definition of the blow up $\mathcal{E}'$ is also a model of E. To see this, observe that the composition of π with the model map $\mathcal{E} \to \operatorname{Spec} \mathbb{Z}$ is still flat and finite type; moreover the point we blow up was in a special fiber so it did not change the generic fiber, which is still isomorphic to E. In this new model the lift of the section $P : \operatorname{Spec} \mathbb{Z} \to \mathcal{E}'$ does not intersect the zero section over the prime $\mathfrak{p}$. We can repeat this process for every point where the section $P(\operatorname{Spec} \mathbb{Z})$ intersects the zero section, which will result in a different model for which the point P is now an integral point! This shows that the notion of integral points depends on the model chosen.

The following example motivates studying (S, D)-integral points, where S is a finite set of places.

Example 3.10. Sometimes it is useful to consider rational points that fail to be integral only for a specific set of primes in $\mathcal{O}_k$. For example the equation $2x + 2y = 1$ does not have any integral solutions while having infinitely many rational solutions. However, one sees that it has infinitely many solutions in the ring $\mathbb{Z}[\frac{1}{2}]$, which is finitely generated over $\mathbb{Z}$. A solutions in $\mathbb{Z}[\frac{1}{2}]$, e.g. $(\frac{1}{4}, \frac{1}{4})$, fails to be integral only with respect to the prime 2. More precisely consider the model of $\mathcal{C} : 2x + 2y - z = 0$ in $\mathbb{P}^2_{\mathbb{Z}}$ and of the divisor $\mathcal{D} = [1 : -1 : 0]$, then $P = [1 : 1 : 4]$ is a $\mathbb{Q}$-rational point of $\mathcal{C}$ but it is not integral, since the reduction of P modulo 2 is the point $P_2 = [1 : 1 : 0] = [1 : -1 : 0]$ over $\mathbb{F}_2$. On the other hand, for every prime $\mathfrak{p} \neq 2$, the reduction modulo $\mathfrak{p}$ of P is disjoint from $\mathcal{D}$.

Analogously, the rational point P gives rise, since the model is proper, to a morphism $P : \operatorname{Spec} \mathbb{Z} \to \mathcal{C}$ which is not disjoint from $\mathcal{D}$, but such that the intersection $P(\operatorname{Spec} \mathbb{Z}) \cap \mathcal{D}$ is supported over the prime 2.

This motivates the following definition:

Definition 3.11. Let S be a finite set of places of k, and let $(\mathcal{X}, \mathcal{D})$ be a model of a pair (X, D) defined over k. An (S, D)-integral point is an integral point $P : \operatorname{Spec} \mathcal{O}_k \to \mathcal{X}$ such that the support of $P^*\mathcal{D}$ is contained in S. We denote the set of (S, D)-integral points of (X, D) as $X(\mathcal{O}_{S,D})$. If $Y = X \setminus D$, then we denote the set of (S, D)-integral points of Y as $Y(\mathcal{O}_S)$.

Remark 3.12. One can also define the set of (S, D)-integral points as sections $\operatorname{Spec} \mathcal{O}_{k,S} \to \mathcal{X}$ that do not intersect $\mathcal{D}$, where $\mathcal{O}_{k,S}$ is the ring of S-integers.

Now that we have an intrinsic definition for integral points we can concentrate on the problem of describing the set of (S, D)-integral points on affine curves. As in the case of rational points on projective curves, the distribution of integral points will be governed by the geometry of the affine curve, i.e. of the corresponding pair. For one-dimensional pairs, the fundamental invariant is the Euler characteristic, or equivalently the degree of the log canonical divisor.

Definition 3.13. Given a non-singular projective curve $\mathcal{C}$ and a pair $(\mathcal{C}, D)$, the Euler Characteristic of $(\mathcal{C}, D)$ is the integer $\chi_D(\mathcal{C}) = 2g(\mathcal{C}) - 2 + \#D$, which corresponds to the degree of the *log canonical divisor* $K_\mathcal{C} + D$.

The Euler Characteristic encodes information of both the genus of the projective curve $\mathcal{C}$ and of the divisor D, and its sign determines the arithmetic of the affine curve $\mathcal{C} \setminus D$.

Theorem 3.14. *Given a pointed projective curve $(\mathcal{C}, D)$ defined over a number field k and a finite set of places S the following hold:*

- *If $2g(\mathcal{C}) - 2 + \#D \leq 0$, then the set of (S, D)-integral points is dense, possibly after a finite extension of k and/or S;*
- *If $2g(\mathcal{C}) - 2 + \#D > 0$, then the set of (S, D)-integral points is finite (Siegel's Theorem).*

We treat the case of non-positive Euler Characteristic in the following example.

Example 3.15. When $\mathcal{C}$ is smooth projective, in order for $\chi_D(\mathcal{C})$ to be non-positive, there are only four cases to consider: if $g(\mathcal{C}) = 0$, then $\#D \leq 2$, and if the $g(\mathcal{C}) = 1$, then $D = \emptyset$. For projective curves, i.e. when $D = \emptyset$, Proposition 2.3 shows that, up to a finite extension of k, the rational points are infinite. We showed that for projective varieties, integral and rational points coincide, which implies that in these cases the set of integral points is dense, up to a finite extension of the base field.

If we consider affine curves, i.e. such that $D \neq \emptyset$, then there are only two remaining cases that we have to discuss, namely $\mathbb{A}^1 = (\mathbb{P}^1, P)$ and $\mathbb{G}_m = (\mathbb{P}^1, P + Q)$.

We saw in Example 3.3 that integral points on $\mathbb{A}^1$ are infinite, and more generally we have that $\mathbb{A}^1(\mathcal{O}_{S,D}) \cong \mathcal{O}_{k,S}$. In the case of the multiplicative group $\mathbb{G}_m$ the integral points correspond to the group of S-units $\mathcal{O}_{k,S}^*$. To see this consider $\mathbb{G}_m$ as the complement of the origin in $\mathbb{A}^1$, i.e. $\mathbb{P}^1 \backslash \{[0:1], [1:0]\}$. Then a point $[a:b] \in \mathbb{G}_m(k)$ is (S, D)-integral if, for every $\mathfrak{p}$ in $\mathcal{O}_{k,S}$, we have that $\mathfrak{p}$ does not divide neither a or b, i.e. a and b are both S-units. Finally, Dirichlet's Unit Theorem implies that the group of S-units is finitely generated and has positive rank as soon as $\#S \geq 2$. In particular, for every number field k, there exists a finite extension for which the rank of $\mathcal{O}_{k,S}^*$ is positive, and therefore such that $\mathbb{G}_m(\mathcal{O}_{S,D})$ is infinite.

In the following example, we show that the set of (S, D)-integral points on the complement of three points in $\mathbb{P}^1$ is finite.

Example 3.16 ($\mathbb{P}^1$ and Three Points). Consider the case of $(\mathbb{P}^1, D)$ where $D = [0] + [1] + [\infty]$ over a number field k. In this case $\deg(K_{\mathbb{P}^1} + D) > 0$, therefore Siegel's Theorem tells us that the number of (S, D)-integral points is finite, for every finite set of places S of k. This can be deduced directly in this case using the S-unit equation Theorem as follows.

Integral points in the complement of $[\infty]$ are integral points in $\mathbb{A}^1$ with respect to the divisor $[0] + [1]$, i.e. $u \in k$ such that $u \in \mathcal{O}_S^*$ (which corresponds to integrality with respect to $[0]$) and such that $1 - u \in \mathcal{O}_S^*$ (which corresponds to integrality with respect to $[1]$). Then if we define $v = 1 - u \in \mathcal{O}_S^*$, the set of (S, D)-integral points correspond to solutions in S-units of $x + y = 1$. The S-unit Theorem (see e.g. [27, Theorem 1.2.4]) then implies that the set of solutions is always finite, for every set of places S.

Siegel's Theorem [90] (and for general number fields and set of places S in [65, 72]) on the finiteness of the set of (S, D)-integral points is the analogue of Faltings' Theorem 2.4. We give here a brief sketch of the proof. For the details we refer to [56, D.9].

Sketch of Proof of Siegel's Theorem. We will focus on the case $g(\mathcal{C}) \geq 1$; the case of genus zero can be treated via finiteness of solutions of S-unit equations, see [56, Theorem D.8.4]. We can always assume that $\mathcal{C}$ has at least one rational point, and use the point to embed $\mathcal{C} \to \mathrm{Jac}(\mathcal{C})$, in its Jacobian.

Suppose that (x_i) is an infinite sequence of integral points on $\mathcal{C} \backslash \mathcal{D}$. Then by completeness of $\mathcal{C}(k_v)$, with $v \in S$, up to passing to a subsequence, (x_i) converges to a limit $\alpha \in \mathcal{C}(\overline{k_v})$. In the embedding $\mathcal{C} \subset \mathrm{Jac}(\mathcal{C})$, we see that for every positive integer m the sequence (x_i) becomes eventually constant in $\mathrm{Jac}(\mathcal{C})/m\,\mathrm{Jac}(\mathcal{C})$, which is finite by the Weak Mordell–Weil Theorem. In particular we can write $x_i = my_i + z$, for some fixed $z \in \mathrm{Jac}(\mathcal{C})$.

Let $\varphi_m : \mathrm{Jac}(\mathcal{C}) \to \mathrm{Jac}(\mathcal{C})$ be the multiplication by m map and define $\psi(x) = \varphi_m(x) + z = m.x + z$. Then (y_i) are a sequence of integral points (since ψ is unramified, and applying Chevalley–Weil Theorem, see [27, Theorem

1.3.1]) on $\psi^*(\mathcal{C})$ that converges to some $\beta \in \mathrm{Jac}(\mathcal{C})$ (eventually up to passing to an extension).

By definition of the canonical height on $\mathrm{Jac}(\mathcal{C})$ (with respect to a fixed symmetric divisor) one has that $\hat{h}(\psi(y_i)) \gg m^2 \hat{h}(y_i)$. By increasing m one gets very good approximations to α which eventually contradicts Roth's Theorem [56, Theorem D.2.1]. □

The sketch of the proof illustrates a couple of very powerful ideas in Diophantine Geometry: the use of abelian varieties (here played by $\mathrm{Jac}(\mathcal{C})$ as ambient spaces with extra structure), the use of the so-called height machine, and techniques from Diophantine approximation. A different proof that avoids the use of the embedding in the Jacobian, thus allowing generalization to higher dimensions, has been given more recently by Corvaja–Zannier in [28], replacing Roth's Theorem by the use of Schmidt's Subspace Theorem (see [103, Chapter 3] for more details).

Finally, we can ask about hyperbolicity properties of affine curves, as in Definition 2.7; it is easy to see that both $\mathbb{A}^1$ and $\mathbb{G}_m$ are not hyperbolic (via the exponential map), while on the other hand the complement of any number of points in a curve of genus one is hyperbolic (again applying Liouville's Theorem). Therefore, if $(\mathcal{C}, D)$ is a pair of a non-singular projective curve $\mathcal{C}$ and a reduced divisor D, both defined over a number field k, and S is a finite set of places containing the Archimedean ones, we can summarize the result described in the previous sections in the following table:

$\chi_D(\mathcal{C}) = \deg(K_{\mathcal{C}} + D)$	Complex hyperbolicity	(S, D)-integral points
≤ 0	Not hyperbolic	Potentially dense
> 0	Hyperbolic	Finite

4   Positivity of the Canonical Sheaf

As we saw for curves, hyperbolicity was governed by the positivity of the canonical sheaf. In particular, we saw if $g(C) \leq 1$, then $\deg \omega_C \leq 0$ (and C is *not* hyperbolic), and if $g(C) \geq 2$, then $\deg \omega_C > 0$ (and C *is* hyperbolic). Conjecturally, positivity of the canonical sheaf governs hyperbolicity of algebraic varieties. Before introducing the conjectures, we give a few examples of canonical sheaves on proper algebraic varieties. Recall we saw earlier that for a curve C, the canonical sheaf $\deg \omega_C = 2g - 2$.

Example 4.1 ([53, Example II.8.20.1]). First consider the Euler sequence

$$0 \to \mathcal{O}_{\mathbb{P}^n} \to \mathcal{O}_{\mathbb{P}^n}(1)^{\oplus(n+1)} \to \mathcal{T}_{\mathbb{P}^n} \to 0,$$

where $\mathcal{T}_{\mathbb{P}^n}$ denotes the tangent sheaf. Taking highest exterior powers, we see that $\omega_{\mathbb{P}^n} = \mathcal{O}_{\mathbb{P}^n}(-n-1)$.

Example 4.2. If A is an abelian variety, then the tangent bundle of A is trivial. In particular, $\omega_X = \mathcal{O}_X$.

A standard way to calculate the canonical sheaf of algebraic varieties is via the *adjunction formula*, which relates the canonical sheaf of a variety to the canonical sheaf of a hypersurface inside the variety.

If X is smooth and projective, and Y is a smooth subvariety, then there is an inclusion map $i : Y \hookrightarrow X$. If we denote by $\mathcal{I}$ the ideal sheaf of $Y \subset X$, then the conormal exact sequence gives (where Ω_X denotes the cotangent sheaf on X)

$$0 \to \mathcal{I}/\mathcal{I}^2 \to i^*\Omega_X \to \Omega_Y \to 0.$$

In particular, taking determinants yields

$$\omega_Y = i^*\omega_X \otimes \det(\mathcal{I}/\mathcal{I}^2)^{\vee}.$$

If we let the subvariety Y to be a divisor $D \subset X$, then one obtains the following.

Proposition 4.3 (Adjunction Formula). *Let X be a smooth projective variety with D a smooth divisor on X. Then*

$$K_D = (K_X + D)|_D.$$

Example 4.4. We can use the adjunction formula to calculate that the canonical sheaf of X a smooth hypersurface of degree d in $\mathbb{P}^n$ is $\omega_X \cong \mathcal{O}_X(d-n-1)$. We note one can do similar calculations in the case of complete intersections.

Now that we have shown some examples of computing canonical sheaves; we introduce the notions from birational geometry we will use to understand positivity of the canonical sheaf and hyperbolicity. Our main reference is [69].

4.1 Notions from Birational Geometry

Let X be a projective variety and let L be a line bundle on X. For each $m \geq 0$ such that $h^0(X, L^{\otimes m}) \neq 0$, the linear system $|L^{\otimes m}|$ induces a rational map

$$\phi_m = \phi_{|L^{\otimes m}|} : X \dashrightarrow \mathbb{P}H^0(X, L^{\otimes m}).$$

We denote by $Y_m = \phi_m(X, L)$ the closure of its image.

Definition 4.5. Let X be normal. The *Iitaka dimension* of (X, L) is

$$\kappa(X, L) = \max_{m>0}\{\dim \phi_m(X, L)\},$$

as long as $\phi_m(X, L) \neq \emptyset$ for some m. Otherwise, we define $\kappa(X, L) = -\infty$.

In particular, either $\kappa(X, L) = -\infty$ or $0 \leq \kappa(X, L) \leq \dim X$.

Remark 4.6. If X is not normal, consider the normalization $\nu : X^\nu \to X$ and take $\kappa(X^\nu, \nu^* L)$.

Example 4.7 (Kodaira Dimension). If X is a smooth projective variety and K_X is the canonical divisor, then $\kappa(X, K_X) = \kappa(X)$ is the *Kodaira dimension* of X.

The Kodaira dimension is a birational invariant, and the Kodaira dimension of a singular variety X is defined to be $\kappa(X')$ where X' is any desingularization of X. However, care needs to be taken in this case. When X is not smooth, the dualizing sheaf ω_X can exist as a line bundle on X, but $\kappa(X, \omega_X) > \kappa(X)$. This is the case, e.g. if X is the cone over a smooth plane curve of large degree (see [69, Example 2.1.6]).

4.1.1   Positivity of Line Bundles

Definition 4.8. A line bundle L on a projective variety X is *ample* if for any coherent sheaf F on X, there exists an integer n_F such that $F \otimes L^{\otimes n}$ is generated by global sections for $n > n_F$. Equivalently, L is ample if a positive tensor power is very ample, i.e. there is an embedding $j : X \to \mathbb{P}^N$ such that $L^{\otimes n} = j^*(\mathcal{O}_{\mathbb{P}^N}(1))$.

The following result is a standard way for checking if a divisor is ample.

Theorem 4.9 (Nakai–Moishezon). *Let X be a projective scheme and let D be a divisor. The divisor D is ample if and only if $D^{\dim Y}.Y > 0$ for all subvarieties $Y \subset X$.*

Corollary 4.10. *If X is a surface, then a divisor D is ample if and only if $D^2 > 0$ and $D . C > 0$ for all curves $C \subset X$.*

Example 4.11. By Riemann–Roch, a divisor D on a curve C is ample if and only if $\deg D > 0$.

Example 4.12. We saw in Example 4.1 that $\omega_{\mathbb{P}^n} = \mathcal{O}_{\mathbb{P}^n}(-n-1)$. Therefore, we see that $\omega_{\mathbb{P}^n}$ is *never ample* for any n, as no power of $\omega_{\mathbb{P}^n}$ will have nonzero sections. It is not so hard to see that $-\omega_{\mathbb{P}^n}$ is ample for all n. This is referred to as *anti-ample*.

Example 4.13. In Example 4.4 we computed the canonical sheaf for hypersurfaces of degree d in $\mathbb{P}^n$ using the adjunction formula. From this we see that

(1) If $d \leq n$, then ω_X is anti-ample.
(2) if $d = n + 1$, then $\omega_X = \mathcal{O}_X$, and thus is not ample.
(3) If $d \geq n + 2$, then ω_X is very ample (exercise using Serre Vanishing).

Definition 4.14. A line bundle L on a projective variety X is *big* if $\kappa(X, L) = \dim X$. A Cartier divisor D on X is *big* if $\mathcal{O}_X(D)$ is big.

Remark 4.15. There are some standard alternative criteria for big divisors. One is that there exists a constant $C > 0$ such that $h^0(X, \mathcal{O}_X(mD)) \geq C \cdot m^n$ or all sufficiently large m (see [69, Lemma 2.2.3]). Another is that mD can be written as the sum of an ample plus effective divisor (Kodaira's Lemma, see [69, Corollary 2.2.7]).

Definition 4.16. We say that X is of *general type* if $\kappa(X) = \dim(X)$, i.e. ω_X is *big*.

Example 4.17. We see immediately that ample implies big so that varieties with ample canonical sheaves are of general type. In this case, some power $\omega_X^{\otimes m}$ for $m \gg 0$ embeds X into a projective space.

Example 4.18. For curves big is the same as ample, so general type is equivalent to $g(C) \geq 2$.

Some examples of varieties of general type are high degree hypersurfaces (in $\mathbb{P}^3$ we require $d \geq 5$) and products of varieties of general type (e.g. product of higher genus curves). It is worth noting that projective space $\mathbb{P}^n$ and abelian varieties are *not* of general type.

Remark 4.19. There exist big divisors that are not ample. One of the standard ways to obtain examples is to note that bigness is preserved under pullback via birational morphisms, but ampleness is not. Suppose X and Y are proper, and $f : X \to Y$ is a birational morphism. A divisor D on Y is big if and only if f^*D is big on X. This is easy to see using Definition 4.14, since X and Y have isomorphic dense open subsets.

We now give an example to show that ampleness is not preserved. Suppose H be a line in $\mathbb{P}^2$, and let $f : X \to \mathbb{P}^2$ be the blowup of $\mathbb{P}^2$ at a point with exceptional divisor E. Then f^*H is big by the above discussion, but f^*H is *not ample* since the projection formula gives that $(f^*H).E = 0$, and thus violates Theorem 4.9.

4.2   Log General Type

As we saw for proper curves, hyperbolicity was essentially governed by the positivity of the canonical sheaf. For affine curves, we saw that hyperbolicity was governed by positivity of the log canonical divisor. As a result, we discuss a mild generalization that will be needed later—the notion of *log general type* for quasi-projective varieties. Recall that we saw in Remark 3.6 that given a quasi-projective variety V, one can always relate it to a pair (X, D) of a smooth projective variety and normal crossings divisor D.

Definition 4.20. We say that V (or the pair (X, D)) is of *log general type* if $\omega_X(D)$ is big.

Of course any pair (X, D) with X of general type will be of log general. Perhaps more interesting examples are when X does not have its own positivity properties.

Example 4.21 (Curves). A pointed curve $(C, D = \sum p_i)$ is of log general type if:

- $g(C) = 0$ and $\#(D) \geq 3$,
- $g(C) = 1$ and $\#(D) \geq 1$, or
- $g(C) \geq 2$.

This is because $\deg \omega_C = 2g - 2$ and so $\deg \omega_C(D) = 2g - 2 + \#D$.

Example 4.22. If $X = \mathbb{P}^2$ and D is a normal crossings divisor, then the pair (X, D) is of log general type if the curve D has $\deg(D) \geq 4$. Again, this is because $\omega_X(D) \cong \mathcal{O}_{\mathbb{P}^2}(-3 + \deg(D))$. More generally, if $X = \mathbb{P}^n$, then one requires $\deg D \geq n + 2$.

As we will see in the next section, there are conjectural higher dimensional analogues of Faltings' Theorem which assert hyperbolicity properties of projective varieties X which are of general type. In the quasi-projective setting, there are also conjectural analogues that ask for log general type.

5   Lang's Conjecture

We are now in a position to state the conjectural higher dimensional generalization of Faltings' Theorem. The main idea is that varieties of general type should satisfy an analogous arithmetic behavior to curves of high genus.

The first conjecture that we mention is due to Bombieri (in the case of surfaces) and Lang: Bombieri addressed the problem of degeneracy of rational points in varieties of general type in a lecture at the University of Chicago in 1980, while Lang gave more general conjectures centered on the relationship between the distribution of rational points with hyperbolicity and Diophantine approximation (see [68] and [66]). The conjecture reads as follows:

Conjecture 5.1 (Lang's Conjecture, Bombieri–Lang for Surfaces). *Let X be a (projective) variety of general type over a number field k. Then $X(k)$ is not Zariski dense.*

We note that one cannot expect that $X(k)$ would be finite once $\dim X \geq 2$, as varieties of general type can, for example, contain rational curves, which in turn (potentially) contain infinitely many rational points.

5.1 Generalizations of Lang and Other Applications

From the previous discussion we have seen that Conjecture 5.1 conjecturally extends Faltings' Theorem 2.4. It is natural to ask whether a similar extension exists for Siegel's Theorem 3.14. Indeed such a generalization exists: the role of curves with positive Euler Characteristic is played now by pairs of log general type. Then, the conjectural behavior of integral points is summarized in the following conjecture due to Vojta, and, in the following reformulation, using ideas of Lang.

Conjecture 5.2 (Lang–Vojta). *Let X be a quasi-projective variety of log general type defined over a number field k and let $\mathcal{O}_{S,k}$ the ring of S-integers for a finite set of places of k containing the Archimedean ones. Then the set $X(\mathcal{O}_{S,k})$ is not Zariski dense.*

As was with the Lang conjecture for projective varieties, the finiteness result of Siegel becomes non-density. In higher dimensions, the positivity of the log canonical divisor is not sufficient to exclude the presence of infinitely many integral points. In particular, varieties of log general type of dimension at least two can contain (finitely many) curves that are *not* of log general type.

Example 5.3 ($\mathbb{P}^2$ and 4 Lines). We can consider $D = x_0 x_1 x_2 (x_0 + x_1 + x_2)$ as a divisor in $\mathbb{P}^2_{\mathbb{Q}}$, and S a finite set of places. Then (S, D)-integral points are a subset of points of the form $[x_0 : x_1 : x_2]$ where x_0, x_1, and x_2 are S-units. This is equivalent to requiring that the points are integral with respect to the three lines $x_0 = 0$, $x_1 = 0$, and $x_2 = 0$. In particular we can consider points of the form $(x_0 : x_1 : x_2) = (1 : x : y)$ with $x, y \in \mathcal{O}_S^*$. The integrality with respect to the fourth line implies that $1 + x + y$ is not 0 modulo every $\mathfrak{p} \notin S$. So if we define $z := 1 + x + y$, then z is a S-unit and we have that

$$z - x - y = 1$$

which is the classical S-unit equation to be solved in units. Then, as an application of Schmidt's subspace theorem [96, Theorem 2.2.1], one gets that there are only finitely many solutions outside the three trivial families:

$$
\begin{cases} z = 1 \\ x = u \\ y = -u \end{cases}
\qquad
\begin{cases} z = u \\ x = -1 \\ y = u \end{cases}
\qquad
\begin{cases} z = u \\ y = -1 \\ x = u \end{cases}
$$

These families correspond to curves in X with non-positive Euler characteristic (since they intersect the divisor D in only two points—passing through two of the singular points of D). In particular by Theorem 3.14, there will be infinitely many (S, D)-integral points contained in these curves, up to a finite extension of $\mathbb{Q}$.

Conjecture 5.2 is a consequence of a more general conjecture, proposed by Paul Vojta and related to his "landmark Ph.D. Thesis," which gave the basis for a systematic treatment of analogies between value distribution theory and Diophantine geometry over number fields. Based on this analogy Vojta formulated a set of far-reaching conjectures. For a detailed description we refer to [96] as well as chapters in the books [13, 56, 83]. We will discuss this in Section 10.

Finally we mention that more recently Campana has proposed a series of conjectures based on a functorial geometric description of varieties that aims at classify completely the arithmetic behavior based on geometric data. For this new and exciting developments we refer to Campana's chapter [19] in this book.

5.2   Known Cases of Lang's Conjecture

As noted above, Faltings' second proof of the Mordell conjecture followed from his resolution of the following conjecture of Lang.

Theorem 5.4 ([46, 47]). *Let A be an abelian variety over a number field K and let X be a geometrically irreducible closed subvariety of A which is not a translate of an abelian subvariety over $\overline{K}$. Then $X \cap A(K)$ is not Zariski dense in X.*

In particular, one has the following corollary.

Corollary 5.5 (Faltings). *Let A be an abelian variety defined over a number field K. If X is a subvariety of A which does not contain any translates of abelian subvarieties of A, then $X(K)$ is finite.*

Using this result, Moriwaki proved the following result, whose generalization is one of the main results in these notes.

Theorem 5.6 ([76]). *Let X be a projective variety defined over a number field k such that Ω_X^1 is ample and globally generated. Then $X(k)$ is finite.*

Sketch of Proof. Using Faltings' Theorem 5.4, and the Albanese variety, one can show that if X is a projective variety with Ω_X^1 globally generated, then every irreducible component of $\overline{X(k)}$ is geometrically irreducible and isomorphic to an abelian variety. We will see in Proposition 6.9, that if Ω_X^1 is ample, then all subvarieties of X are of general type, and so X does not contain any abelian varieties. Therefore by Corollary 5.5, the set $X(k)$ is finite. □

For curves $\Omega_C^1 \cong \omega_C$, but for higher dimensional varieties X, assuming positivity of the vector bundle Ω_X^1 is a stronger condition than assuming positivity of ω_X. In the following section, we will review positivity for vector bundles.

5.3 Known Cases of the Lang–Vojta Conjecture

In the context of degeneracy of S-integral points, as predicted by Conjecture 5.2, the analogue of Theorem 5.4 is the following result due to Vojta. For the definition of semi-abelian varieties see Definition 8.2.

Theorem 5.7 ([98, 99]). *Let $X \subset A$ be an irreducible subvariety of a semi-abelian variety A defined over a number field k. If X is not a translate of a semi-abelian subvariety, then for every ring of S-integers $\mathcal{O}_{k,S}$, the set of integral points $X(\mathcal{O}_{k,S})$ is not Zariski dense in X.*

Corollary 5.8. *In the above setting, if X does not contain any translate of a semi-abelian subvariety of A, then $X(\mathcal{O}_{D,S})$ is a finite set.*

In a parallel direction, the Lang–Vojta conjecture is known when the divisor D has several components: we discussed one example of this in Example 5.3. Such results arise from the higher dimensional extension of a method developed by Corvaja and Zannier in [28] to give a new proof of Siegel's Theorem. In [29], Corvaja and Zannier prove a general result that implies non-density of S-integral points on surface pairs (X, D) where D has at least two components that satisfy a technical condition on their intersection numbers. This result has been generalized by the same authors, Levin, Autissier et al., extending the method both to higher dimensions as well as refining the conditions on the divisor D; see e.g. [11, 30, 32, 35, 71]—we refer to [27, 34] for surveys of known results.

6 Hyperbolicity of Projective Varieties and Ample Cotangent Bundles

The goal of this section is to understand the assumptions in Theorem 5.6—namely ampleness and global generation for vector bundles. Recall that the definition of ampleness for line bundles was given in Definition 4.8.

Definition 6.1. Let E be a vector bundle on a projective variety X and let H be an ample line bundle. We say that

- E is *globally generated* if there exists a positive integer $a > 0$ and a surjective map
  $\mathcal{O}_X^a \to E \to 0$.
- E is *ample* if there exists a positive integer $a > 0$ such that the sheaf $\mathrm{Sym}^a(E) \otimes H^{-1}$ is globally generated, and
- E is *big* if there exists a positive integer $a > 0$ such that the sheaf $\mathrm{Sym}^a(E) \otimes H^{-1}$ is generically globally generated.

Remark 6.2. These definitions are independent of the choice of ample line bundle H (see [95, Lemma 2.14a].

We note that there are alternative ways to describe ampleness and bigness for vector bundles (see [52]).

Proposition 6.3. *A vector bundle E be on a projective variety X is ample if and only if $\mathcal{O}_{\mathbb{P}(E)}(1)$ is ample on $\mathbb{P}(E)$.*

Remark 6.4. One can try to make the above definition for big, namely that E is big if and only if $\mathcal{O}_{\mathbb{P}(E)}(1)$ is big, but this definition *does not* always coincide with the above definition (see Example 6.5). We will call E *weakly big* if $\mathcal{O}_{\mathbb{P}(E)}(1)$ is big to distinguish the two notions.

Example 6.5. The vector bundle $E = \mathcal{O}_{\mathbb{P}^1} \oplus \mathcal{O}_{\mathbb{P}^1}(1)$ is weakly big, but not big as in Definition 6.1. This is because any symmetric power will have a $\mathcal{O}_{\mathbb{P}^1}$ summand, which will become negative when tensoring with H^{-1}. In particular, it will never be generically globally generated. The fact that E is weakly big follows from the following calculation (or see Remark 4.19). The nth symmetric power is $\mathrm{Sym}^n(E) = \mathcal{O}_{\mathbb{P}^2} + \mathcal{O}_{\mathbb{P}^1}(1) + \cdots + \mathcal{O}_{\mathbb{P}^1}(n)$ so $h^0(\mathrm{Sym}^n(E)) = 1 + 2 + \cdots + (n+1) = cn^2 + \ldots$, and therefore grows like a degree 2 polynomial. If $X = \mathbb{P}(E)$ then $X = F_1$, the Hirzebruch surface. Consider the natural map $f : F_1 \to \mathbb{P}^1$ then $f_*(\mathcal{O}(n)) = \mathrm{Sym}^n(E)$, and so $h^0(F_1, \mathcal{O}(n)) = h^0(\mathbb{P}^1, \mathrm{Sym}^n(E))$.

We will need the following fact repeatedly:

Proposition 6.6 ([52, Proposition 2.2 & 4.1]). *Any quotient of an ample vector bundle is ample. The restriction of an ample vector bundle is ample.*

Sketch of Proof of 6.6. We show the quotient result, the other result is similar. If $A \to B$ is a surjective map of vector bundles, then $\mathrm{Sym}^n(A) \otimes F \to \mathrm{Sym}^n(B) \otimes F$ is surjective. So if the former is globally generated, so is the latter.   □

Example 6.7. Let $X = \mathbb{P}^1$. Recall that any vector bundle of rank r on $\mathbb{P}^1$ can be decomposed as the sum of r line bundles. Then $\mathcal{O}_X \oplus \mathcal{O}_X$ is globally generated (but not ample nor big). The vector bundle $\mathcal{O}_X(1) \oplus \mathcal{O}_X(1)$ is ample.

Example 6.8. Let $X = \mathbb{P}^n$. Then T_X is ample by the Euler sequence combined with Proposition 6.6, and $T_X(-1)$ is globally generated, but neither ample nor big. To see it is globally generated, note that tensoring the Euler sequence with $\mathcal{O}_X(-1)$, we obtain

$$\mathcal{O}_X(-1) \to \mathcal{O}_X^{\otimes(n+1)} \to T_X(-1) \to 0.$$

The fact that it is not ample follows since the restriction of $T_X(-1)$ to a line $l \subset \mathbb{P}^n$ is $\mathcal{O}_l(1) \oplus \mathcal{O}_l^{\otimes(n+1)}$. One can also show that $T_{\mathbb{P}^n}(-1)$ is not big (see [55, Remark 2.4]).

We are now ready to prove the main result in this section.

Proposition 6.9 ([70, Example 6.3.28]). *Let X be a smooth projective variety with ample cotangent bundle Ω_X^1. Then all irreducible subvarieties of X are of general type.*

Proof. Let $Y_0 \subset X$ be an irreducible subvariety of dimension d, and let $\mu : Y \to Y_0$ be a resolution of singularities. Then there is a generically surjective homomorphism $\mu^*\Omega_X^d \to \Omega_Y^d = \mathcal{O}_Y(K_Y)$. Since Ω_X^d is ample, the pullback $\mu^*\Omega_X^d$ is big (see Remark 4.19) and thus $\mathcal{O}_Y(K_Y)$ is also big.   □

Remark 6.10.

(1) It is worth noting that the converse is not true, that is there are hyperbolic varieties X such that the cotangent bundle Ω_X^1 is not ample, see Example 6.11.
(2) In general, it is not so easy to find varieties X with ample cotangent bundle Ω_X^1 (see [70, Section 6.3.B]).

We saw in Theorem 5.6, that if we also assume that Ω_X^1 is *globally generated*, we can obtain finiteness of integral points (unconditionally with respect to Lang's conjecture). Finally, although we will not prove it, we recall that Kobayashi proved (see [70, Theorem 6.3.26]) that if X is a smooth projective variety with ample cotangent bundle Ω_X^1, then X is Brody hyperbolic (see also [41, Proposition 3.1]). Again, there are examples of Brody hyperbolic varieties for which Ω_X^1 is not ample (see Example 6.11).

Example 6.11 ([70, Remark 6.3.27]). Let B be a curve of genus $g(B) \geq$ 2 and consider the variety $X = B \times B$. Then X is hyperbolic but Ω^1_X is *not* ample as its restriction to $Y = B \times \{pt\}$ has a trivial quotient. On the other hand, consider a holomorphic map $\mathbb{C} \to X = B \times B$. Since B is hyperbolic, the map cannot be contained in a fiber. Consider the composition $\mathbb{C} \to B \times B \to B$. This is a holomorphic map to a curve of genus ≥ 2 and therefore by Liouville's theorem must be constant.

Example 6.12 ([70, Construction 6.3.37]).

(1) Let X_1, X_2 be smooth projective surfaces (over $\mathbb{C}$) of general type with $c_1(X_i)^2 > 2c_2(X_i)$. Then a complete intersection of two general sufficiently positive divisors in $X_1 \times X_2$ is a surface X with Ω^1_X ample.
(2) Let $f : X \to B$ be a non-isotrivial family of smooth projective curves of genus $g \geq 3$ over a smooth curve of genus $g(B) \geq 2$. Then Ω^1_X is ample. These are called Kodaira surfaces.

Example 6.13 ([70, Example 6.3.38]). Let $Y_1, \ldots, Y_m$ be smooth projective varieties of dimension $d \geq 1$ with big cotangent bundle (e.g. if Y_i are surfaces of general type with $c_1(Y)^2 > c_2(Y)$) and let

$$X \subseteq Y_1 \times \cdots \times Y_m$$

be a general complete intersection of sufficiently high multiples of an ample divisor. Then if

$$\dim X \leq \frac{d(m+1)+1}{2(d+1)},$$

then X has ample cotangent bundle Ω^1_X.

Example 6.14 ([39]). If X is the complete intersection of $e \geq n$ sufficiently ample general divisors in a simple abelian variety of dimension $n + e$, then the cotangent bundle Ω^1_X is ample.

Debarre conjectured that if $X \subset \mathbb{P}^r$ is the complete intersection of $e \geq r/2$ hypersurfaces of sufficiently high degree, then the cotangent bundle Ω^1_X is ample [39]. This conjecture is now a theorem of Brotbek and Darondeau [16] and independently, Xie [102]. We state one of the related results.

Theorem 6.15 ([16, 102]). *In every smooth projective variety M for each $n \leq \dim(M)/2$, there exist smooth subvarieties of dimension n with ample cotangent bundles.*

Remark 6.16. Brotbek–Darondeau prove Debarre's conjecture without providing effective bounds. Xie provides effective bounds, and the work of Deng [42] improves these bounds. Work of Coskun–Riedl improves the bound in many cases [36].

In the next section, we shift our focus to quasi-projective varieties.

7 Hyperbolicity of Log Pairs

Using the ideas introduced in the above section, we now wish to understand positivity conditions on a pair (X, D) that will guarantee hyperbolicity. We first define what hyperbolicity means for quasi-projective varieties.

Definition 7.1. Let X be a projective geometrically integral variety over a field k, let D be a normal crossings divisor on X, and let $V = X \setminus D$.

- $V = (X \setminus D)$ is *arithmetically hyperbolic* if $V(\mathcal{O}_D)$ is finite.
- $V_{\mathbb{C}}$ is *Brody hyperbolic* if every holomorphic map $\mathbb{C} \to V_{\mathbb{C}}$ is constant.

Then the conjectures in the spirit of Green–Griffiths–Lang–Vojta assert that the above are equivalent, and are additionally equivalent to all subvarieties of V being of log general type. Recall that aside from the canonical sheaf, the main player to study hyperbolicity was the cotangent sheaf Ω_X^1. We now consider the generalization to pairs.

Definition 7.2. Let X be a smooth projective variety and let $D = \sum_{j=1}^r D_j$ be a reduced normal crossings divisor on X. The *log cotangent bundle* $\Omega_X^1(\log D)$ denotes the sheaf of differential forms on X with logarithmic poles along D.

For example, if $\dim X = n$ and $U \subset X$ is an open set such that $D|_U = z_1 z_2 \cdots z_k = 0$ (with $k < n$), then

$$H^0(U, \Omega_X^1(\log D)) = \mathrm{Span}\{\frac{dz_1}{z_1}, \ldots, \frac{dz_k}{z_k}, dz_{k+1}, \ldots, dz_n\}.$$

The natural idea would be to ask whether or not ampleness of the log cotangent bundle $\Omega_X^1(\log D)$ implies the desired hyperbolicity properties. It turns out that the log cotangent bundle $\Omega_X^1(\log D)$ is *never* ample. Indeed, there are non-ample quotients coming from D which violate the quotient property from Proposition 6.6.

Proposition 7.3 ([10, 17]). *Let X be a smooth variety of $\dim X > 1$ and $D \neq \emptyset$ a normal crossings divisor on X. Then the log cotangent sheaf $\Omega_X^1(\log D)$ is never ample.*

Proof. Suppose that the log cotangent bundle $\Omega_X^1(\log D)$ were ample. Consider the following exact sequence (see [44, Proposition 2.3]):

$$0 \to \Omega_X^1 \to \Omega_X^1(\log D) \to \oplus_{j=1}^r D_j \to 0,$$

where D_j are the components of D. Now consider the restriction of this sequence to a component $D_i \subseteq D$, and tensor with $\mathcal{O}_{D_i}$.
    In this way one obtains a surjection

$$A \to \mathcal{O}_{D_i} + Q \to 0,$$

where A is an ample sheaf (being the restriction of an ample sheaf), and Q is a torsion sheaf supported at $D_i \cap D_j$, whenever $D_i \cap D_j \neq \emptyset$. We note that if there is no common point between any two irreducible components of D, in particular if D is irreducible, then the second term of the sequence will just be $\mathcal{O}_{D_i}$. Then, since $\mathcal{O}_{D_i} \oplus Q$ (and in particular $\mathcal{O}_{D_i}$) is *not* ample, there cannot be a surjection from an ample sheaf (this would violate Proposition 6.6) and thus the log cotangent bundle $\Omega^1_X(\log D)$ cannot be ample.   □

Instead, we can ask what happens if the log cotangent bundle $\Omega^1_X(\log D)$ is, in a sense, as ample as possible. Before introducing our notion of *almost ample*, we recall the definition of the augmented base locus. Let $\mathrm{Bs}(D)$ denote the base locus of D.

Definition 7.4. The *stable base locus* of a line bundle L on a projective variety X is the Zariski closed subset defined as

$$\mathbf{B}(L) := \bigcap_{m \in \mathbb{N}} \mathrm{Bs}(mL),$$

and the *augmented base locus* (aka non-ample locus) of L is

$$\mathbf{B}_+(L) := \bigcap_{m \in \mathbb{N}} \mathbf{B}(mL - A),$$

where A is any ample line bundle on X.

Remark 7.5. If E is a vector bundle on X we define the augmented base locus as $\pi(\mathbf{B}_+(\mathcal{O}(1)_{\mathbb{P}(E)})$ where $\pi : \mathbb{P}(E) \to X$. Note that $\mathbf{B}_+(L)$ is empty if and only if L is ample, and that $\mathbf{B}_+(L) \neq X$ if and only if L is big (see [43, Example 1.7].

Example 7.6 (Nef and Big Divisors). Let L be a big and nef divisor on X. We define the *null locus* $\mathrm{Null}(L) \subseteq X$ to be the union of all positive dimensional subvarieties $V \subseteq X$ with $(L^{\dim V} \cdot V) = 0$. Then this is a proper algebraic subset of X (see [70, Lemma 10.3.6]), and a theorem of Nakamaye (see [70, Theorem 10.3.5]) says that

$$\mathbf{B}_+(L) = \mathrm{Null}(L).$$

Example 7.7 (Surfaces). If X is a smooth surface and D is a big divisor on X, then there exists a *Zariski decomposition* $D = P + N$, where P is the *nef* part, and N is the *negative* part. In this case, one can prove (see [43, Example 1.11]) that

$$\mathbf{B}_+(D) = \mathrm{Null}(P).$$

Theorem 7.8 ([14]). *Let $\mathcal{L}$ be a big line bundle on a normal projective variety X. The complement of the augmented base locus is the largest Zariski open set such that the morphism $\phi_m(X, L)$ is an isomorphism onto its image.*

Before giving our definition of almost ample, we recall the notion of augmented base loci for coherent sheaves.

Definition 7.9 ([14, Definition 2.4]). Let X be a normal projective variety, let $\mathcal{F}$ be a coherent sheaf, and A an ample line bundle. Let $r = p/q \in \mathbb{Q} > 0$. The *augmented base locus* of $\mathcal{F}$ is

$$\mathbf{B}_+(\mathcal{F}) := \bigcap_{r \in \mathbb{Q} > 0} \mathbf{B}(\mathcal{F} - rA),$$

where $\mathbf{B}(\mathcal{F} - rA) = \mathbf{B}(\mathrm{Sym}^{[q]} \mathcal{F} \otimes A^{-p})$.

Remark 7.10.

(1) The augmented base locus does not depend on the choice of an ample divisor A.
(2) By [14, Proposition 3.2], if $\mathcal{F}$ is a coherent sheaf and $\pi : \mathbb{P}(\mathcal{F}) = \mathbb{P}(\mathrm{Sym}\,\mathcal{F}) \to X$ is the canonical morphism, then $\pi(\mathbf{B}_+(\mathcal{O}_{\mathbb{P}(\mathcal{F})}(1))) = \mathbf{B}_+(\mathcal{F})$, i.e. the non-ample locus of $\mathcal{F}$.

Definition 7.11. Let (X, D) be a pair of a smooth projective variety and a normal crossings divisor D. We say that the log cotangent sheaf $\Omega_X^1(\log D)$ is *almost ample* if

(1) $\Omega_X^1(\log D)$ is big, and
(2) $\mathbf{B}_+(\Omega_X^1(\log D)) \subseteq \mathrm{Supp}(D)$.

Remark 7.12.

(1) We can define the above notion more generally for singular varieties (see [10]), e.g. varieties with lc and slc singularities coming from moduli theory. This is necessary to obtain the uniformity results in loc. cit.; however, it is unnecessary for the proof of Theorem 8.1.
(2) When X is smooth, our notion does not quite coincide with almost ample as in [17, Definition 2.1]. If the log cotangent sheaf is almost ample in the sense of [17], then it is almost ample in our sense. However, our definition is a priori weaker.
(3) Brotbek–Deng proved that for any smooth projective X there exists a choice of D so that the log cotangent bundle $\Omega_X^1(\log D)$ is almost ample (see Theorem 7.13).
(4) For a log smooth pair (X, D) with almost ample $\Omega_X^1(\log D)$, the complement $X \setminus D$ is Brody hyperbolic by the base locus condition (see e.g. [41, Proposition 3.3]).

We now state the above theorem of Brotbek–Deng.

Theorem 7.13 ([17, Theorem A]). *Let Y be a smooth projective variety of* $\dim Y = n$ *and let* $c > n$. *Let* $\mathcal{L}$ *be a very ample line bundle on* Y. *For any* $m \geq (4n)^{n+2}$ *and for general hypersurfaces* $H_1, \ldots, H_c \in |\mathcal{L}^m|$, *writing* $D = \sum_{i=1}^{c} H_i$, *the logarithmic cotangent bundle* $\Omega_Y^1(\log D)$ *is almost ample. In particular,* $Y \setminus D$ *is Brody hyperbolic.*

We now present the proof that almost ample log cotangent implies that all subvarieties are of log general type. We note that the proof of the statement in full generality is outside the scope of these notes, and so we present a simplified proof which works for log smooth pairs.

Theorem 7.14 ([10, Theorem 1.5]). *Let (X, D) be a log smooth pair. If (X, D) has almost ample log cotangent* $\Omega_X^1(\log D)$, *then all pairs (Y, E) where* $E := (Y \cap D)_{red}$ *with $Y \subset X$ irreducible and not contained in D are of log general type.*

Proof. Consider a log resolution $(\widetilde{Y}, \widetilde{E}) \to (Y, E)$, which gives a map $\phi : (\widetilde{Y}, \widetilde{E}) \to (X, D)$. Since Y is not contained in D, by the definition of almost ample, Y is not contained in the base locus of $\Omega_X^1(\log D)$. This gives a map $\phi^*(\Omega_X^1(\log D)) \to \Omega_{\widetilde{Y}}^1(\log \widetilde{E})$. The image of this map is a big subsheaf of $\Omega_{\widetilde{Y}}^1(\log \widetilde{E})$, being a quotient of a big sheaf, and thus its determinant is also big. By [20, Theorem 4.1] (see also [86, Theorem 1]) $K_{\widetilde{Y}} + \widetilde{E}$ is big, and so (Y, E) is of log general type. $\qquad\square$

Remark 7.15. In the above proof, we used that the quotient of a big sheaf is a big sheaf. We stress that this is not true for weakly big. The key idea is to be big there is a generically surjective map, and this map remains generically surjective when restricting to a subvariety not contained in the base locus (in this case a subvariety contained in the divisor D).

In [10], we prove this statement in further generality. Namely, we prove the result for pairs with singularities which arise from moduli theory (i.e. lc and slc singularities). This is necessary for the proofs of uniformity in loc. cit. We now show an alternative proof for Theorem 7.14 in the case $\dim X = 2$, which avoids the use of [20].

Alternative Proof of Theorem 7.14 If $\dim X = 2$. By assumption, $\Omega_X^1(\log D)$ is big and so its restriction to any subvariety $Y \not\subset D$ is still big. Since Y is a curve, big is equivalent to ample, and so the restriction is actually ample. Consider the normalization $\phi : Y^v \to Y$ and denote by E^v the divisor $E^v = \phi^{-1}(E) \cup \{$ exceptional set of $\phi\}$. Since $\Omega_X^1(\log D)|_Y$ is ample, its pullback $\phi^*(\Omega_X^1(\log D)|_Y)$ is big. There is a generically surjective map (see [51, Theorem 4.3])

$$\phi^*(\Omega_X^1(\log D)|_Y) \to \Omega_{Y^v}^1(E^v) = \mathcal{O}_{Y^v}(K_{Y^v} + E^v).$$

Therefore we see that $\mathcal{O}_{Y^v}(K_{Y^v} + E^v)$ is big and so (Y, E) is of log general type. $\qquad\square$

8 Semi-abelian Varieties and the Quasi-Albanese Map

In this section we introduce semi-abelian varieties, and prove a generalization of Moriwaki's theorem. In particular, we show that the Lang–Vojta conjecture holds for varieties which have almost ample and globally generated log cotangent bundle.

Theorem 8.1 ([10]). *Let V be a smooth quasi-projective variety with log smooth compactification (X, D) over a number field K. If the log cotangent sheaf $\Omega^1_X(\log D)$ is globally generated and almost ample, then for any finite set of places S the set of S-integral points $V(\mathcal{O}_{K,S})$ is finite.*

We begin with the definition of a semi-abelian variety. Our discussion follows [48].

Definition 8.2. A *semi-abelian variety* is an irreducible algebraic group A which, after a suitable base change, can be realized as an extension of an abelian variety by a linear torus, i.e. the middle term of an exact sequence

$$1 \to \mathbb{G}_m^r \to A \to A_0 \to 1,$$

where A_0 is an abelian variety.

Example 8.3. Immediate examples of semi-abelian varieties are tori and abelian varieties. Any product of a torus with an abelian variety is a semi-abelian variety called *split*.

By Vojta's generalization of Faltings' theorem (see Theorem 5.7 and Corollary 5.8), one way to obtain finiteness of the set of integral points is to consider varieties $X \setminus D$ that satisfy the following two conditions:

(1) $X \setminus D$ embeds in a semi-abelian variety as a proper subscheme;
(2) $X \setminus D$ does not contain any subvariety which is isomorphic to (a translate of) a semi-abelian variety.

Clearly the two conditions imply that the set of D-integral points on X is finite. This strategy has some similarity with the proof of Siegel's Theorem using the Roth's Theorem, where one make use of the embedding of the curve in its Jacobian.

To embed a pair in a semi-abelian variety we will use the theory of (quasi-)Albanese maps. Recall that every variety admits a universal morphism to an abelian variety, called the *Albanese map*. The same is true for quasi-projective varieties where the universal morphism instead maps to a semi-abelian variety.

Definition 8.4 ([48, Definition 2.15]). Let V be a smooth variety. The *quasi-Albanese map*

$$\alpha : V \to \mathcal{A}_V$$

is a morphism to a semi-abelian variety $\mathcal{A}_V$ such that

(1) For any other morphism $\beta : V \to B$ to a semi-abelian variety B, there is a morphism $f : \mathcal{A}_V \to B$ such that $\beta = f \circ \alpha$, and
(2) the morphism f is uniquely determined.

The semi-abelian variety $\mathcal{A}_V$ is called the *quasi-Albanese variety* of X and was constructed originally by Serre in [87, Théorème 7].

Remark 8.5.

- If $V = \mathcal{C}$ is a projective curve, then $\mathcal{A}_V$ is the abelian variety $\mathrm{Jac}(\mathcal{C})$.
- If $V = X \setminus D$ is rational, then $\mathcal{A}_V$ is a torus. (Exercise)
- There is no semi-abelian subvariety of $\mathcal{A}_V$ containing $\alpha(V)$.

8.1   Construction of $\mathcal{A}_V$

We briefly sketch the construction of $\mathcal{A}_V$ for a smooth quasi-projective variety V defined over the complex numbers. More generally if V is defined over a perfect field k one can define more abstractly the Albanese variety to be the dual of the Picard variety of X. In what follows we use the standard notation $q(V) = \dim H^0(X, \Omega_X^1(\log D))$ and $q(X) = \dim H^0(X, \Omega_X^1)$.

Let $\{\omega_1, \ldots, \omega_{q(X)}, \varphi_1, \ldots, \varphi_\delta\}$ be a basis of $H^0(X, \Omega_X^1(\log D))$. The quasi-Albanese variety of V is $\mathcal{A}_V \cong \mathbb{C}^{q(V)}/L$, where L is the lattice defined by the periods, i.e. the integrals of the basis elements of $H^0(X, \Omega_X^1(\log D))$ evaluated on a basis of the free part of $H_1(V, \mathbb{Z})$. Then $\mathcal{A}_V$ is a semi-abelian variety [48, Lemma 3.8]. If $0 \in V$ is a point of V, then the map $\alpha : V \to \mathcal{A}_V$ is defined as

$$P \mapsto \left(\int_0^P \omega_1, \ldots, \int_0^P \omega_{q(X)}, \int_0^P \varphi_1, \ldots \int_0^P \varphi_\delta \right).$$

The map α is well defined [48, Lemma 3.9] and one can check that α is an algebraic map [48, Lemma 3.10]. In particular $\dim \mathcal{A}_V = q(V)$. We will denote by $d(V) = \dim \alpha(V)$.

We see now that in order to use the quasi-Albanese variety we will need to impose some condition on the positivity of the sheaf $\Omega_X^1(\log D)$. The main idea is that the geometric conditions on the log cotangent sheaf will ensure that we can embed V inside its quasi-Albanese as a proper subvariety and then ensure that it does not contain any proper semi-abelian subvariety and conclude using Vojta's Theorem 5.7.

8.2 Proof of the Main Theorem

We begin with the following.

Proposition 8.6. *Let V be a smooth quasi-projective variety over a number field K. If $d(V) < q(V)$, then the closure of $V(\mathcal{O}_{S,K})$ in V is a proper closed subset.*

Proof. Assume $V(\mathcal{O}_S) \neq \emptyset$. Since $d(V) < q(V)$, the quasi-Albanese map α is not surjective. In particular $\alpha(V)$ is a proper subvariety of a semi-abelian variety $\mathcal{A}_V$. If $V(\mathcal{O}_S)$ is dense in V, then so is its image $\alpha(V)(\mathcal{O}_S)$ in $\alpha(V)$. By Vojta's Theorem (Theorem 5.7), the image $\alpha(V)$ is a semi-abelian subvariety of $\mathcal{A}_V$. This is a contradiction by Remark 8.5 (alternatively think about $\alpha(V)(\mathcal{O}_S)$ generating $\mathcal{A}_V(\mathcal{O}_S)$). □

Now we discuss the consequences of $\Omega_X^1(\log D)$ being almost ample and globally generated.

Lemma 8.7. *Let V be a smooth quasi-projective variety with log smooth compactification (X, D) over a field k of characteristic zero. If the sheaf $\Omega_X^1(\log D)$ is almost ample and globally generated, then $q(V) \geq 2 \dim V$.*

Proof. If $P = \mathrm{Proj}(\Omega_X^1(\log D))$ and $L = \mathcal{O}_P(1)$, then since $\Omega_X^1(\log D)$ is globally generated there is a morphism $\phi_{|L|} : P \to \mathbb{P}^N$ where $\phi_{|L|}^* \mathcal{O}_{\mathbb{P}^N}(1) = L$ and $N = \dim_k H^0(P, L) - 1$. Furthermore, by definition L is big. Then the map $\phi_{|L|}$ is generically finite which implies that

$$\dim P = \dim \phi_{|L|}(P) \leq N = \dim_k H^0(P, L) - 1.$$

Noting that $\dim P = 2 \dim V - 1$ we obtain that

$$q(V) = \dim H^0(\overline{X}, \Omega_X^1(\log D)) \geq 2 \dim V.$$

□

Theorem 8.8. *Let $V \cong (X \setminus D)$ be a log smooth variety over a number field k, let $\mathcal{A}_V$ be a semi-abelian variety, and let $\alpha : V \to \mathcal{A}_V$ be a morphism. If $\alpha^*(\Omega_{\mathcal{A}_V}^1) \to \Omega_V^1$ is a surjective map of sheaves, then every irreducible component of $V(\mathcal{O}_S)$ is geometrically irreducible and isomorphic to a semi-abelian variety.*

Proof. Let Y be an irreducible component of $\overline{V(\mathcal{O}_S)}$. Since $Y(\mathcal{O}_S)$ is dense in Y, we see that Y is geometrically irreducible. We are thus left to show that Y is isomorphic to a semi-abelian variety. For this we will use [48, Theorem 4.2] and so it suffices to show that Y is smooth and $\alpha|_Y$ is étale.

Let $B = \alpha(Y)$. Since $Y(k)$ is dense in Y, so is $B(k)$ in B. By Vojta's theorem (see Theorem 5.7), B is a translated of a semi-abelian subvariety of $\mathcal{A}_V$. Consider the following diagram:

$$
\begin{array}{ccc}
\alpha^*(\Omega^1_{\mathcal{A}_V})|_Y & \longrightarrow & \Omega^1_V|_Y \\
\downarrow & & \downarrow \\
(\alpha|_Y)^*(\Omega^1_B) & \longrightarrow & \Omega^1_Y
\end{array}
$$

We know that $h : (\alpha|_Y)^*(\Omega^1_B) \to \Omega^1_Y$ is surjective. On the other hand, $\mathrm{rank}(\Omega^1_B) \leq \mathrm{rank}(\Omega^1_Y)$ and the former is locally free. Therefore h is actually an isomorphism. Therefore Y is smooth over k and $\alpha|_Y$ is étale. Thus we conclude the result by [48, Theorem 4.2]. $\qquad\square$

Corollary 8.9. *Let $V \cong (X \setminus D)$ be a log smooth variety over a number field k. If the log cotangent sheaf $\Omega^1_X(\log D)$ is globally generated, then for every finite set of places S, every irreducible component of $\overline{V(\mathcal{O}_S)}$ is geometrically irreducible and isomorphic to a semi-abelian variety.*

Proof. Consider the quasi-Albanese map $\alpha : V \to \mathcal{A}_V$. Since $\Omega^1_X(\log D)$ is globally generated and $H^0(V, \Omega^1_V) \otimes \mathcal{O}_{\mathcal{A}_V} \cong \Omega^1_{\mathcal{A}_V}$ by [48, Lemma 3.12] the map $\alpha^*(\Omega^1_{\mathcal{A}_V}) \to \Omega^1_V$ is surjective. Therefore applying Lemma 8.7 gives the desired result. $\qquad\square$

Proof 1 of Theorem 8.1. For a smooth variety V with log smooth completion (X, D), assuming that $\Omega^1_X(\log D)$ is almost ample implies there are no semi-abelian varieties inside V (see Theorem 7.14). Therefore, the set $V(\mathcal{O}_S)$ is finite when $\Omega^1_X(\log D)$ is globally generated and almost ample by Corollary 8.9. $\qquad\square$

We now give a proof that does not use Theorem 7.14.

Proof 2 of Theorem 8.1. Assume that $\overline{V(\mathcal{O}_S)}$ has an irreducible component Y of dimension $\dim Y \geq 1$. Let $(\overline{Y}, E)$ denote the completion of Y. Note that $\overline{Y}$ is geometrically irreducible. Furthermore, $\Omega^1_X(\log D)|_{\overline{Y}}$ is almost ample and globally generated. Therefore $\Omega^1_{\overline{Y}}(\log E)$ is almost ample and globally generated as well. By Lemma 8.7, $q(Y) \geq 2 \dim Y$. Therefore, by Proposition 8.6, $Y(\mathcal{O}_S)$ is not dense in Y, which is a contradiction. $\qquad\square$

9 Vojta's Conjecture

The goal of this section is to introduce Vojta's conjecture and the relevant height machinery to present a result analogous to Theorem 8.1 in the function field setting. This result gives a height bound for integral points that is predicted by Vojta's main conjecture (Conjecture 9.5). We will see in this section that this "main" conjecture implies Conjecture 5.2.

9.1 Vojta's Conjecture and the Theory of Heights

We will now recall the basic definition needed to state the main conjecture whose specific reformulation will imply Conjecture 5.2. The main technical tool is the concept of *height*, that plays a fundamental role in almost all results in Diophantine Geometry. The idea is that a height function measures the "arithmetic complexity" of points. In the classical case of $\mathbb{P}^n$ the *logarithmic height* is defined as

$$H(x_0 : \cdots : x_n) = \max_i(|x_i|)$$

for a rational point $(x_0 : \cdots : x_n) \in \mathbb{P}^n(\mathbb{Q})$ with integer coordinates without common factors. Weil extended this notion to treat arbitrary height functions on algebraic varieties defined over number fields. In this language, the logarithmic height on $\mathbb{P}^n$ is the height associated with a hyperplane divisor over $\mathbb{Q}$.

Definition 9.1 (Weil's Height Machinery). Let X be a smooth projective algebraic variety defined over a number field k. There exists a (unique) map

$$h_{X,\_} : \operatorname{Pic}(X) \to \{\text{functions } X(\overline{k}) \to \mathbb{R}\}$$

well defined up to bounded functions, i.e. modulo $O(1)$, whose image $h_{X,D}$ for a class $D \in \operatorname{Pic}(X)$ is called a *Weil height* associated with D. The map $h_{X,\_}$ satisfies

(a) the map $D \mapsto h_{X,D}$ is an homomorphism modulo $O(1)$;
(b) if $X = \mathbb{P}^n$ and $H \in \operatorname{Pic}(\mathbb{P}^n)$ is the class of some hyperplane in $\mathbb{P}^n$, then $h_{X,H}$ is the usual *logarithmic height* in the projective space;
(c) Functoriality: for each $\overline{k}$-morphism $f : X \to Y$ of varieties and for each $D \in \operatorname{Pic}(Y)$ the following holds:

$$h_{X,f^*D} = h_{Y,D} + O(1).$$

By abuse of notation, for a divisor D, we will denote the height corresponding to the class $\mathcal{O}(D) \in \operatorname{Pic}(X)$ with $h_{X,D}$. The previous definition can be extended to non-smooth varieties (even non-irreducible ones) and over any field with a set of normalized absolute values which satisfy the product formula, see [68] for further details. From the previous definition one can show the following properties for the height machinery:

Proposition 9.2 ([56, 68]). *With the above notation, the function $h_{X,\_}$ satisfies*

(d) *Let D be an effective divisor in X then, up to bounded functions, $h_{X,D} \geq O(1)$;*

(e) *Northcott's Theorem: Let A be an ample divisor in X with associated height function $h_{X,A}$ then, for all constants C_1, C_2, and every extension k' of k with $[k' : k] \leq C_2$, the following set is finite:*

$$\{P \in X(k') : h_{X,A}(P) \leq C_1\}.$$

The second ingredient we need to introduce to formally state Vojta's conjecture is the notion of local height. Morally we want a function which measures the v-adic distance from a point to a divisor D and such that a linear combination of these functions when v runs over the set of places gives a Weil height for the divisor D. This motivates the following:

Definition 9.3 (Local Height). Let X be a smooth projective variety defined over a number field k. Then there exists a map

$$\lambda_\_ : \mathrm{Pic}(X) \to \{ \text{ functions } \coprod_{v \in M_k} X \setminus \mathrm{supp}\, D(k_v) \to \mathbb{R}\}$$

defined up to M_k-bounded functions, i.e. up to constant maps $O_v(1) : M_k \to \mathbb{R}$ that are nonzero for finitely many places $v \in M_k$, such that:

(a) λ is additive up to M_k bounded functions;

(b) given a rational function f on X with associated divisor $\mathrm{div}(f) = D$. Then

$$\lambda_{D,v}(P) = v(f(P))$$

up to $O_v(1)$, for each $v \in M_k$ where $P \in U \subset X \setminus \mathrm{supp}\, D(k_v)$ with U affine and $\max|P|_v = 0$ for all but finitely many v;

(c) Functoriality: for each $\bar{k}$-morphism $g : X \to Y$ of varieties and for each $D \in \mathrm{Pic}(Y)$ the following holds:

$$\lambda_{g^*D,v} = \lambda_{D,v} \circ g + O_v(1);$$

(d) if D is an effective divisor, then $\lambda_{D,v} \geq O_v(1)$;

(e) if h_D is a Weil height for D, then

$$h_D(P) = \sum_{v \in M_k} d_v \lambda_{D,v}(P) + O(1)$$

for all $P \notin \operatorname{supp} D$, with $d_v = [k_v : \mathbb{Q}_v]/[k : \mathbb{Q}]$.

For the detailed construction and related properties of local height we refer to [68] and [88]. One intuition behind the work of Vojta was the fact that local heights are arithmetic counterparts of proximity functions in value distribution theory: to see this consider a metrized line bundle $\mathscr{L}$ with a section s and metric $|\cdot|_v$: the function $P \mapsto \log|s(P)|_v$ is a local height at v. Following Vojta [96] one can introduce arithmetic proximity and counting functions for algebraic varieties over number fields in the same spirit.

Definition 9.4. Let S be a finite set of places of k, and let (X, D) be a pair defined over k. Then the following functions are well defined:

$$m_{S,D}(P) = \sum_{v \in S} d_v \lambda_{D,v}(P)$$

$$N_{S,D}(P) = \sum_{v \notin S} d_v \lambda_{D,v}(P)$$

called the *arithmetic proximity function* and *arithmetic counting function* respectively. By definition,

$$h_D(P) = N_{S,D}(P) + m_{S,D}(P).$$

With these definitions we can now state the main Vojta conjecture which translates Griffiths' conjectural "Second Main Theorem" in value distribution theory.

Conjecture 9.5 (Vojta). *Let X be a smooth irreducible projective variety defined over a number field k and let S be a finite set of places of k. Let D be a normal crossing divisor and A an ample divisor on X. Then for every $\epsilon > 0$ there exists a proper closed subset Z such that, for all $P \in X(k) \setminus Z$,*

$$m_{S,D}(P) + h_{K_X}(P) \le \epsilon h_A(P) + O(1).$$

Vojta's main conjecture 9.5 is known to imply most of the open conjectures and fundamental theorems of Diophantine Geometry (Masser–Osterlé abc conjecture, Faltings' Theorem, ...).

We end this section by two propositions which show how the above stated conjectured implies the Bombieri–Lang conjecture 5.1 and the Lang–Vojta conjecture 5.2. For other implications and discussions we refer the interested reader to [96] or [83].

Remark 9.6. We recall that by Remark 4.15 (Kodaira's Lemma), a big divisor D has a positive multiple that can be written as the sum of an ample and effective divisor. In the following proofs we will always assume that this multiple is the divisor itself for simplifying the notation, as this can be done without loss of generality.

Proposition 9.7. *Vojta conjecture 9.5 implies Bombieri–Lang conjecture 5.1.*

Proof. If X is of general type, then K_X is big, i.e. there exists a positive integer n such that $nK_X = B + E$ with B ample and E effective, and we will assume $n = 1$. Now Conjecture 9.5 with $D = 0$ and $A = B$ gives

$$(1 - \epsilon)h_B(P) + h_E(P) \le O(1).$$

By Proposition 9.2, $h_E(P) \ge 0$ and hence, by Northcott's Theorem 9.2(e), the set $X(K)$ is not Zariski dense in X. □

In order to prove that Vojta conjecture is stronger than the Lang–Vojta conjecture we need the following reformulation of the property of being S-integral in terms of the functions defined in Definition 9.4: a point P is S-integral if $N_{S,D}(P) = O(1)$ and in particular $m_{S,D}(P) = h_D(P) + O(1)$. Using the characterization of bigness mentioned above (Remark 4.15), we prove the following.

Proposition 9.8. *Vojta's conjecture 9.5 implies the Lang–Vojta conjecture 5.2.*

Proof. For a log general type variety (X, D) one has

$$K_X + D = B + E,$$

for B ample and E effective. Hence Vojta's conjecture with $A = B$ gives, for S-integral points,

$$(1 - \epsilon)h_B(P) + h_E(P) \le O(1).$$

As before, $h_E(P) \ge 0$; thus, using Northcott's Theorem, the set of S-integral points of (X, D) is not Zariski dense. □

10 Function Fields

Function fields in one variable and number fields share several properties. This deep analogy was observed in the second half of the 19th century; one of the first systematic treatments can be found in the famous paper by Dedekind and Weber [40]. Further descriptions, due to Kronecker, Weil, and van der Waerden, settled this profound connection which finally became formally completed with the scheme theory developed by Grothendieck.

Definition 10.1 (Function Field). A *function field* F over an algebraically closed field k is a finitely generated field extension of finite transcendence degree over k. A function field *in one variable*, or equivalently a function field of an algebraic curve, is a function field with transcendence degree equal to one.

Remark 10.2. With the language of schemes the function field of a curve X, or more general of every integral scheme over an algebraic closed field, can be recovered from the structure sheaf $\mathcal{O}_X$ in the following way: given any affine open subset of X, the function field of X is the fraction field of $\mathcal{O}_X(V)$. Moreover, if η is the (unique) generic point of X, then the function field of X is also isomorphic to the stalk $\mathcal{O}_{X,\eta}$.

The analogy between number fields and function fields of curves, also known as algebraic function fields in one variable, comes from the fact that one-dimensional affine integral regular schemes are either smooth affine curves over a field k or an open subset of the spectrum of the ring of integers of a number field. Formally, given a number field k with ring of integers $\mathcal{O}_k$ the scheme $\operatorname{Spec} \mathcal{O}_k$ is one-dimensional affine and integral. From this analogy, several classical properties of number fields find an analogue in the theory of function field. In particular the theory of heights can be defined over function fields.

Definition 10.3. Given a function field F in one variable of a non-singular curve $\mathcal{C}$, each (geometric) point $P \in \mathcal{C}$ determines a non-trivial absolute value by

$$|f|_P := e^{-\operatorname{ord}_P(f)}.$$

Moreover if $Q \neq P$, then the absolute values $|\cdot|_Q$ and $|\cdot|_P$ are not equivalent.

Remark 10.4.

- The definition could have been given more generally for function fields of algebraic varieties regular in codimension one (or rather for regular models of higher dimensional function fields), replacing the point P with prime divisors. Extensions exist also for function fields over non-algebraically closed fields in which one should replace points with orbits under the absolute Galois group.
- From the fact that any rational function f on a projective curve has an associated divisor of degree zero, it follows that the set of absolute values satisfy the product formula.

Table 1 Number Fields
and Function Fields
analogy

Number Field	Function Field
$\mathbb{Z}$	$k[x]$
$\mathbb{Q}$	$k(x)$
$\mathbb{Q}_p$	$k((x))$
k finite extension of $\mathbb{Q}$	F function field of $\mathcal{C}$
Place	Geometric point
Finite set of places	Finite set of points
Ring of S-integers	Ring of regular functions
Spec $\mathcal{O}_{k,S}$	Affine curve $\mathcal{C} \setminus S$
Product formula	deg principal divisor $= 0$
Extension of number fields	Dominant maps
Extension of ideals	Pullback of divisors

Given the set of absolute values M_F for a function field in one variable F, normalized in such a way that they satisfy the product formula, heights can be defined for F in the following way:

Definition 10.5. Let $F = k(\mathcal{C})$ be as before. For any $f \in F$ the *height* of f is

$$h(f) = -\sum_{P \in \mathcal{C}} \min\{0, \mathrm{ord}_P(f)\} = \sum_{P \in \mathcal{C}} \max\{0, \mathrm{ord}_P(f)\}.$$

In the same way for a point $g \in \mathbb{P}^n(F)$, $g = (f_0 : \cdots : f_n)$, its height is defined as

$$h(g) = -\sum_{P \in \mathcal{C}} \min_i\{\mathrm{ord}_P(f_i)\}.$$

From the definition it follows that a rational function on a regular curve has no poles if and only if its height is zero if and only if it is constant on the curve.

We end this subsection with Table 1, which illustrates the interplay and the similarity between number fields and function fields. We stress in particular how each geometric object in the right column, in particular dominant maps and pullbacks, are analogous to purely arithmetic notions like extensions of fields and extensions of ideals. This analogy can be further explored using Arakelov Theory and extending the notion of divisors to number fields by suitably compactifying the affine curve Spec $\mathcal{O}_{k,S}$; in this framework an intersection theory can be defined for such compactified divisors sharing many analogous properties of intersection theory on the geometric side. We refer to [67] for further details on this subject.

10.1   Mordell Conjecture for Function Fields

Over function fields one cannot expect Faltings' Theorem 2.4 to hold as shown by the following examples.

Example 10.6. Let C be a curve with $g(C) > 1$ defined over $\mathbb{C}$ and consider the trivial family $C \times \mathbb{P}^1 \to \mathbb{P}^1$. The family can be viewed as the curve C (trivially) defined over the function field $\mathbb{C}(t)$ of $\mathbb{P}^1$. All fibers of the family, being isomorphic to C have genus greater than one. The Mordell Conjecture over function fields, without any other restriction, should imply that the set of $\mathbb{C}(t)$-rational points of C, i.e. there are finitely many sections $\mathbb{P}^1 \to C \times \mathbb{P}^1$. However this is easily seen to be false by considering constant sections $\mathbb{P}^1 \to \{P\} \times \mathbb{P}^1$ for each point $P \in C(\mathbb{C})$. In particular, the general type curve C defined over $\mathbb{C}(t)$ has infinitely many $\mathbb{C}(t)$-rational points.

From the previous example one could guess that the problem relied on the fact that the family was a product and the curve C was actually defined over the base field $\mathbb{C}$ rather than on the function field $\mathbb{C}(t)$, i.e. the family was trivial. However, as the following example shows, things can go wrong even for non-trivial families.

Example 10.7 (Gasbarri [49]). Consider the curve $C := (x+ty)^4 + y^4 - 1$ defined over $\mathbb{C}(t)$. It has an associated fibration $C \to \mathbb{P}^1$ whose generic fiber $C_{t_0} = (x + t_0 y)^4 + y^4 - 1$ is a smooth projective curve of genus $g(C_{t_0}) = 3$. Again if we consider the same statement of Theorem 2.4 only replacing the number field with the function field $\mathbb{C}(t)$ we would expect that the number of $\mathbb{C}(t)$-rational points of C to be finite. However we claim that $C(\mathbb{C}(t))$ is infinite; to see this consider the equation $\alpha^4 + \beta^4 = 1$ over $\mathbb{C}^2$: it has infinitely many solutions. Each solution gives a $\mathbb{C}$-point of C_{t_0}, namely $(\alpha - t_0\beta, \beta)$ proving the claim. Moreover the family is not trivial in the sense of the previous example, i.e. C is not defined over $\mathbb{C}$. Notice however that each fiber of the family is isomorphic to the curve $x^4 + y^4 = 1$ via $x + ty \mapsto x$ and $y \mapsto y$.

Motivated by the previous examples we give the following:

Definition 10.8. Given a family of irreducible, smooth projective curves $C \to B$ over a smooth base B, we say that the family is *isotrivial* if C_b is isomorphic to a fixed curve C_0 for b in an open dense subset of B. With abuse of notation, we will say that a curve C defined over a function field F is *isotrivial* if the corresponding fibration $C \to B$ is isotrivial, where B is a curve with function field F.

Isotriviality extends the notion of (birational) triviality for family of curves, i.e. a product of curves fibered over one of the factors is immediately isotrivial. At the same time this notion encompasses many other families that are not products, like the one defined in the previous example. However, after a cover of the base of the family, each isotrivial family becomes trivial; in particular the following easy lemma holds:

Lemma 10.9. *Given an isotrivial family $\mathcal{C} \to \mathcal{B}$ of smooth projective irreducible curves, there exists a cover $\mathcal{B}' \to \mathcal{B}$ such that the base changed family $\mathcal{C} \times_{\mathcal{B}} \mathcal{B}' \to \mathcal{B}'$ is a generically trivial family, i.e. is birational to a product $\mathcal{C} \times_{\mathbb{C}} \mathcal{B}_0$.*

Lemma 10.9 implies that rational points for curves defined over function fields will not be finite for isotrivial curves. The analogous form of Mordell Conjecture for function fields thus asks whether this holds only for this class of curves. We can then restate Theorem 2.4 in the following way:

Theorem 10.10. *Let $\mathcal{C}$ be a smooth projective curve defined over a function field F of genus $g(\mathcal{C}) > 1$. If $\mathcal{C}(F)$ is infinite, then $\mathcal{C}$ is isotrivial.*

Theorem 10.10 was proved in the sixties by Manin [73] (although with a gap fixed by Coleman [26]) using analytic arguments, and later by Grauert [50] using algebraic methods. Samuel in [84] gave a proof in characteristic p using ideas of Grauert. A detailed explanation of Grauert methods can be found in Samuel's survey [85]. In Mazur's detailed discussion of Faltings' proof of Mordell Conjecture [74], Mazur stresses the role of Arakelov [7] and Zahrin's [104] results which imply new proofs of the Geometric Mordell Conjecture, using ideas of Parshin: this gives even more importance to the geometric case.

One of the ideas of Grauert's proof, which is central in some of the higher dimensional extensions is the following: suppose $\mathcal{C}$ is a curve defined over a function field F of a curve $\mathcal{B}$, corresponding to a fibration $\pi : X \to \mathcal{B}$. Then one can prove that almost all sections of the fibration, which correspond to rational points, verify a first order differential equation, i.e. almost all sections are tangent to a given horizontal vector field. Formally each section $\sigma : \mathcal{B} \to X$ can be lifted to the projective bundle $\mathcal{B} \to \mathbb{P}(\Omega^1_X) = \mathrm{Proj}(\mathrm{Sym}(\Omega^1_X))$ via the surjective map $\sigma^* \Omega^1_X \to \Omega^1_{\mathcal{B}}$. Grauert proves (in a different language) that there exists a section ϕ of a suitable line bundle over $\mathbb{P}(\Omega^1)$ whose zero section contains all but finitely many images of sections. Grauert then concludes that if infinitely many sections exist, given the fact that they satisfy the differential equation given by $\phi = 0$, a splitting is provided for the relative tangent sequence which implies that the family is isotrivial (via the vanishing of the Kodaira–Spencer class).

In particular, Grauert's construction gives first insights towards the theory of jet spaces which occupy a central role in some degeneracy results in the complex analytic setting. In this direction, recent analogues of Theorem 10.10 in higher dimension have been proved by Mourougane [77] for very general hypersurfaces in the projective space of high enough degree using proper extensions of the ideas briefly described above.

10.2 Vojta Conjecture for Function Fields

Since function fields possess a theory of heights analogous to the theory over number fields, one can translate Vojta's Main Conjecture 9.5 to the function field case. The main conjecture implies the following height bound for varieties of log general type over function fields.

Conjecture 10.11. *Let $(\mathcal{X}, \mathcal{D})$ be a pair over a function field $F = k(B)$ whose generic fiber (X, D) is a pair of log general type. Then, for every $\varepsilon > 0$ there exists a constant C and a proper closed subvariety Z such that for all $P(\in \mathcal{X} \setminus Z)(\overline{F})$ one has*

$$h_{K_{\mathcal{X}}+D}(P) \leq C(\chi(P) + N_D^{(1)}(P)) + O(1) \tag{1}$$

where, given a point $P \in \mathcal{X}(L)$ corresponding to a cover $B_P \to B$ of degree n, corresponding to the field extension $L \supset F$, we have that $\chi(P) = \chi(B_P)/n$. Moreover, the truncated counting function $N_D^{(1)}(P)$ is the cardinality of the support of P^*D.

Note that for varieties of log general type the height in (1) is associated with a big divisor. In this case, if the set of points of bounded height is Zariski dense, then the model is isotrivial. Moreover, if one considers only points defined over F, then the characteristic of the point P reduces to $2(g(B)) - 2$ and one recovers the usual conjecture for (D, S)-integral points where $\#S \geq N_D^{(1)}(P)$.

In this latter case one can relate Conjecture 10.11 to hyperbolicity using the following result of Demailly.

Theorem 10.12 (Demailly [41]). *Let X be a projective complex variety embedded in some projective space for a choice of a very ample line bundle. Then if the associated manifold is Kobayashi hyperbolic the following holds: there exists a constant $A > 0$ such that each irreducible curve $\mathcal{C} \subset X$ satisfies*

$$\deg \mathcal{C} \leq A(2g(\widetilde{\mathcal{C}}) - 2) = A\chi(\widetilde{\mathcal{C}}),$$

where $\widetilde{\mathcal{C}}$ is the normalization of $\mathcal{C}$.

Following this result, Demailly introduced the following notion.

Definition 10.13. *A smooth projective variety X is algebraically hyperbolic if there exists a constant A such that for each irreducible curve $\mathcal{C} \subset X$ the following holds:*

$$\deg \mathcal{C} \leq A\chi(\widetilde{\mathcal{C}}).$$

Using strong analogies between hyperbolicity and degeneracy of rational points Lang conjectured that a general type variety should be hyperbolic outside a proper exceptional set and therefore one could also conjecture that the variety should be algebraically hyperbolic outside that set (for more on algebraically hyperbolic varieties we refer to [58, 59]). This allows one to rephrase Conjecture 10.11 as follows.

Conjecture 10.14 (Lang–Vojta for Function Fields). *Given an affine variety X embedded as $\overline{X} \setminus D$ for a smooth projective variety $\overline{X}$ and a normal crossing divisor D, if X is of log general type, then there exists a proper subvariety Exc (called the* exceptional set*) such that there exists a bound for the degree of images of non-constant morphisms $C \to X$ from affine curves whose image is not entire contained in Exc, in terms of the Euler Characteristic of C.*

By the previous remark it is easy to see that Conjecture 10.11 implies Conjecture 10.14.

We note that most of the known techniques used for the number field case can be used to prove analogous results in the function field setting. However, due to the presence of tools that are not available over number fields, most notably the presence of derivation, one can obtained stronger results that lead to cases of Conjecture 10.14 and Conjecture 10.11 in settings that are currently out of reach in the function field case. We refer to the articles [18, 22, 24, 31, 33, 77, 81, 94, 100, 101] as some examples of results over function fields along these lines.

Remark 10.15. For the sake of completion, we discuss briefly how *algebraic hyperbolicity* fits in with our previous discussions on hyperbolicity (see [70, Example 6.3.24].

- If X is algebraically hyperbolic, then X contains no rational or elliptic curves.
- If X is algebraically hyperbolic, then there are no non-constant maps $f : A \to X$ from an abelian variety A.
- Kobayashi (and thus Brody) hyperbolicity implies algebraic hyperbolicity for projective varieties.

Furthermore, a theorem of Kobayashi (see [70, Theorem 6.3.26]) states that if Ω_X^1 is ample, then X is algebraically hyperbolic.

10.3 Moriwaki for Function Fields

The analogue of Theorem 5.6 over function fields is the following theorem due to Noguchi.

Theorem 10.16 (Noguchi [79]). *Let X be a smooth variety over a function field F. If Ω^1_X is ample, then the conclusion of Conjecture 10.11 holds.*

It is therefore natural to consider the analogous question for pairs. As pointed out several times in these notes, for a pair (X, D), the analogous assumption on the positivity of the log cotangent, is to require that $\Omega^1_X(\log D)$ is almost ample. In this setting the following was suggested to us by Carlo Gasbarri.

Expectation 10.17. Let (X_F, D) be a log smooth non-isotrivial pair over F. If $\Omega^1_{X_F/F}(\log D)$ is almost ample, then there exists a constant A and a proper closed subset $Z \subsetneq X_F$ such that for every $p \in (X_F \setminus Z)(\overline{F})$ we have that

$$h_{K_X+D}(P) \leq A(\chi(P) + N_D^{(1)}(P)) + O(1)$$

where P is a model of p over $\mathcal{C}$.

The intuition is as follows: first one obtains a height bound for lifts of sections over the projectivization of the model of the log cotangent sheaf. Then using the almost ample hypothesis together with the non-isotriviality of the pair, one shows that the base locus of the structure sheaf of the projectivized bundle does not dominate the base.

11 Consequences of Lang's Conjecture

For the sake of completeness, and due to our personal interests, we conclude these notes with a few consequences of Lang's conjecture.

11.1 Consequences of Lang's Conjecture – Uniformity

Caporaso–Harris–Mazur [21] showed that Conjecture 5.1 implies that $\#\mathcal{C}(K)$ in Faltings' Theorem is not only finite, but is also uniformly bounded by a constant $N = N(g, K)$ that does *not* depend on the curve $\mathcal{C}$.

Theorem 11.1 (See [21]). *Let K be a number field and $g \geq 2$ an integer. Assume Lang's conjecture. Then there exists a number $B = B(K, g)$ such that for any smooth curve $\mathcal{C}$ defined over K of genus g the following holds:* $\#\mathcal{C}(K) \leq B(g, K)$

Pacelli [80] (see also [1]), proved that N only depends on g and $[K : \mathbb{Q}]$. Some cases of Theorem 11.1 have been proven unconditionally [60, 82, 92] depending on the Mordell–Weil rank of the Jacobian of the curve and for [82], on an assumption related to the Height conjecture of Lang–Silverman. It has also been shown that families of curves of high genus with a uniformly bounded number of rational points in each fiber exist [38].

Näive translations of uniformity fail in higher dimensions as subvarieties can contain infinitely many rational points. However, one can expect that after removing such subvarieties the number of rational points is bounded. Hassett proved that for surfaces of general type this follows from Conjecture 5.1, and that the set of rational points on surfaces of general type lie in a subscheme of uniformly bounded degree [54].

The main tool used to prove the above uniformity results is the *fibered power theorem* and was shown for curves in [21], for surfaces [54] and in general by Abramovich [2]. In higher dimensions, similar uniformity statements hold conditionally on Lang's conjecture, and follow from the fibered power theorem under some additional hypotheses that take care of the presence of subvarieties that are not of general type [6].

11.1.1   Consequences of the Lang–Vojta Conjecture – Uniformity

We saw above that Lang's conjecture had far-reaching implications for uniformity results on rational points for varieties of general type. One can analogously ask if the Lang–Vojta conjecture implies uniformity results for integral points on varieties of log general type. It turns out that such results are much more subtle in the pairs case, but we review some of the known results here.

This question was first addressed in [3] when Abramovich asked if the Lang–Vojta conjecture implies uniformity statements for integral points. Abramovich showed this cannot hold unless one restricts the possible models used to define integral points (see Example 3.9). Instead, Abramovich defined *stably integral points*, and proved uniformity results (conditional on the Lang–Vojta conjecture), for stably integral points on elliptic curves, and together with Matsuki [5] for principally polarized abelian varieties. While we do not give a precise definition of stably integral points in these notes, we remark that they are roughly integral points which remain integral after stable reduction. We refer the interested reader to our paper [10].

In [10], we prove various generalizations of the work of Abramovich and Abramovich–Matsuki. In particular, we prove that the Lang–Vojta conjecture implies that the set of stably integral points on curves of log general type is uniformly bounded. Additionally, we prove a generalization of Hassett's result, showing that the Lang–Vojta conjecture implies that (stably) integral points on families of log canonically polarized surfaces lie in a subscheme whose degree is uniformly bounded. Finally, we prove, assuming the Lang–Vojta conjecture, and under the assumption that the surfaces have almost ample log cotangent, that the set of stably integral points on polarized surfaces is uniformly bounded.

Finally, we note that results present in [10] have two key ingredients. One is a generalization of the fibered power theorems mentioned in Section 11.1 to the case of pairs [8]. The other, is a generalization of Theorem 7.14, which gives a condition so that subvarieties of a *singular* surface of log general type are curves of log general type. It turns out that proving a result for stably integral points requires the use of the compact moduli space of stable pairs, and as such, we are forced to work with singular surfaces.

11.2   Consequences of Lang's Conjecture – Rational Distance Sets

A *rational distance set* is a subset S of $\mathbb{R}^2$ such that the distance between any two points of S is a rational number. In 1946, Ulam asked if there exists a rational distance set that is dense for the Euclidean topology of $\mathbb{R}^2$. While this problem is still open, Shaffaf [89] and Tao [93] independently showed that Lang's conjecture implies that the answer to the Erdős-Ulam question is "no." In fact, they showed that if Lang's conjecture holds, a rational distance set cannot even be dense for the Zariski topology of $\mathbb{R}^2$, i.e. must be contained in a union of real algebraic curves.

Solymosi and de Zeeuw [91] proved (unconditionally, using Faltings' proof of Mordell's conjecture) that a rational distance contained in a real algebraic curve must be finite, unless the curve has a component which is either a line or a circle. Furthermore, any line (resp. circle) containing infinitely many points of a rational distance set must contain all but at most four (resp. three) points of the set. One can rephrase the result of [91] by saying that almost all points of an infinite rational distance set contained in a union of curves tend to concentrate on a line or circle. It is therefore natural to consider the "generic situation," and so we say that a subset $S \subseteq \mathbb{R}^2$ is in *general position* if no line contains all but at most four points of S, and no circle contains all but at most three points of S. For example, a set of seven points in $\mathbb{R}^2$ is in general position if and only if no line passes through $7 - 4 = 3$ of the points and no circle passes through $7 - 3 = 4$ of the points.

In particular, the aforementioned results show that Lang's conjecture implies that rational distance sets in general position must be finite. With Braune, we proved the following result.

Theorem 11.2 ([9, Theorem 1.1]). *Assume Lang's conjecture. There exists a uniform bound on the cardinality of a rational distance set in general position.*

We can rephrase this theorem as follows.

Corollary 11.3 ([9, Corollary 1.2]). *If there exist rational distance sets in general position of cardinality larger than any fixed constant, then Lang's conjecture does not hold.*

We note that we are unaware of any examples of rational distance sets in general position of cardinality larger than seven (the case of seven answered a question of Erdős, see [64]).

11.3   Acknowledgments

These notes grew out of our minicourse given at the workshop "Geometry and arithmetic of orbifolds" which took place December 11–13, 2018 at the Université du Québec à Montréal. We are grateful to the organizers: Steven Lu, Marc-Hubert Nicole, and Erwan Rousseau for the opportunity to give these lectures, and we thank the audience who were wonderful listeners and provided many useful comments which are reflected in these notes. We are grateful to Marc-Hubert Nicole for organizing this collection. We also thank Erwan Rousseau for pointing us to the paper [20]. We thank Damian Brotbek, Pietro Corvaja, Carlo Gasbarri, De-Qi Zhang and Ariyan Javanpeykar for helpful discussions and comments; we are also grateful to the anonymous referees who greatly helped improving the first version of this notes. Finally, we thank our coauthor Kristin DeVleming whose work is represented here, and for many comments and conversations related to this work, our lectures, and the writing of these notes. Research of K.A. was supported in part by an NSF Postdoctoral Fellowship. Research of A.T. was supported in part by funds from NSF grant DMS-1553459 and in part by Centro di Ricerca Matematica Ennio de Giorgi. Part of these notes were written while K.A. was in residence at the Mathematical Sciences Research Institute in Berkeley, California, during the Spring 2019, supported by the National Science Foundation under Grant No. 1440140.

References

1. Dan Abramovich.   Uniformité des points rationnels des courbes algébriques sur les extensions quadratiques et cubiques. *C. R. Acad. Sci., Paris, Sér. I*, 321(6):755–758, 1995.
2. Dan Abramovich. A high fibered power of a family of varieties of general type dominates a variety of general type. *Invent. Math.*, 128(3):481–494, 1997.
3. Dan Abramovich. Uniformity of stably integral points on elliptic curves. *Invent. Math.*, 127(2):307–317, 1997.

4. Dan Abramovich. Birational geometry for number theorists. In *Arithmetic geometry*, volume 8 of *Clay Math. Proc.*, pages 335–373. Amer. Math. Soc., Providence, RI, 2009.

5. Dan Abramovich and Kenji Matsuki. Uniformity of stably integral points on principally polarized Abelian varieties of dimension ≤ 2. *Isr. J. Math.*, 121:351–380, 2001.

6. Dan Abramovich and José Felipe Voloch. Lang's conjectures, fibered powers, and uniformity. *New York J. Math.*, 2:20–34, electronic, 1996.

7. Suren Ju. Arakelov. Families of algebraic curves with fixed degeneracies. *Izv. Akad. Nauk SSSR Ser. Mat.*, 35:1269–1293, 1971.

8. Kenneth Ascher and Amos Turchet. A fibered power theorem for pairs of log general type. *Algebra Number Theory*, 10(7):1581–1600, 2016.

9. Kenneth Ascher, Lucas Braune, and Amos Turchet. The Erdös-Ulam problem, Lang's conjecture, and uniformity. Bulletin of the London Math. Society, to appear. https://doi.org/10.1112/blms.12381

10. Kenneth Ascher, Kristin DeVleming, and Amos Turchet. Hyperbolicity and uniformity of log general type varieties. *arXiv e-prints*, Jul. 2018.

11. Pascal Autissier. Sur la non-densité des points entiers. *Duke Math. J.*, 158(1):13–27, 2011.

12. Enrico Bombieri. The Mordell conjecture revisited. *Ann. Scuola Norm. Sup. Pisa Cl. Sci. (4)*, 17(4):615–640, 1990.

13. Enrico Bombieri and Walter Gubler. *Heights in Diophantine geometry*, volume 4 of *New Mathematical Monographs*. Cambridge University Press, Cambridge, 2006.

14. Sébastien Boucksom, Salvatore Cacciola, and Angelo Felice Lopez. Augmented base loci and restricted volumes on normal varieties. *Math. Z.*, 278(3–4):979–985, 2014.

15. Robert Brody. Compact manifolds and hyperbolicity. *Trans. Amer. Math. Soc.*, 235:213–219, 1978.

16. Damian Brotbek and Lionel Darondeau. Complete intersection varieties with ample cotangent bundles. *Invent. Math.*, 212(3):913–940, 2018.

17. Damian Brotbek and Ya Deng. On the positivity of the logarithmic cotangent bundle. *Ann. Inst. Fourier (Grenoble)*, 68(7):3001–3051, 2018.

18. W. D. Brownawell and D. W. Masser. Vanishing sums in function fields. *Math. Proc. Cambridge Philos. Soc.*, 100(3):427–434, 1986.

19. Frédéric Campana. Arithmetic aspects of orbifold pairs. In *Arithmetic Geometry of Logarithmic Pairs and Hyperbolicity of Moduli Spaces*, CRM Short Courses. Springer-Verlag, 2020.

20. Frédéric Campana and Mihai Păun. Orbifold generic semi-positivity: an application to families of canonically polarized manifolds. *Ann. Inst. Fourier (Grenoble)*, 65(2):835–861, 2015.

21. Lucia Caporaso, Joe Harris, and Barry Mazur. Uniformity of rational points. *J. Am. Math. Soc.*, 10(1):1–35, 1997.

22. Laura Capuano and Amos Turchet. Lang-Vojta Conjecture over function fields for surfaces dominating $\mathbb{G}_m^2$. *arXiv e-prints*, Nov 2019.

23. Claude Chabauty. Sur les points rationnels des courbes algébriques de genre supérieur à l'unité. *C. R. Acad. Sci. Paris*, 212:882–885, 1941.

24. Xi Chen. On algebraic hyperbolicity of log varieties. *Commun. Contemp. Math.*, 6(4):513–559, 2004.

25. Robert F. Coleman. Effective Chabauty. *Duke Math. J.*, 52(3):765–770, 1985.

26. Robert F. Coleman. Manin's proof of the Mordell conjecture over function fields. *Enseign. Math. (2)*, 36(3–4):393–427, 1990.

27. Pietro Corvaja. *Integral points on algebraic varieties*, volume 3 of *Institute of Mathematical Sciences Lecture Notes*. Hindustan Book Agency, New Delhi, 2016. An introduction to Diophantine geometry.

28. Pietro Corvaja and Umberto Zannier. A subspace theorem approach to integral points on curves. *C. R. Math. Acad. Sci. Paris*, 334(4):267–271, 2002.
29. Pietro Corvaja and Umberto Zannier. On integral points on surfaces. *Ann. of Math. (2)*, 160(2):705–726, 2004.
30. Pietro Corvaja and Umberto Zannier. On the integral points on certain surfaces. *Int. Math. Res. Not.*, pages Art. ID 98623, 20, 2006.
31. Pietro Corvaja and Umberto Zannier. Some cases of Vojta's Conjecture on integral points over function fields. *J. Algebraic Geometry*, 17:195–333, 2008.
32. Pietro Corvaja and Umberto Zannier. Integral points, divisibility between values of polynomials and entire curves on surfaces. *Adv. Math.*, 225(2):1095–1118, 2010.
33. Pietro Corvaja and Umberto Zannier. Algebraic hyperbolicity of ramified covers of $\mathbb{G}_m^2$ (and integral points on affine subsets of $\mathbb{P}_2$). *J. Differential Geom.*, 93(3):355–377, 2013.
34. Pietro Corvaja and Umberto Zannier. *Applications of Diophantine Approximation to Integral Points and Transcendence.* Cambridge Tracts in Mathematics. Cambridge University Press, 2018.
35. Pietro Corvaja, Aaron Levin, and Umberto Zannier. Integral points on threefolds and other varieties. *Tohoku Math. J. (2)*, 61(4):589–601, 2009.
36. Izzet Coskun and Eric Riedl. Effective bounds on ampleness of cotangent bundles. *arXiv e-prints*, Oct 2018.
37. Pranabesh Das and Amos Turchet. Invitation to integral and rational points on curves and surfaces. In *Rational points, rational curves, and entire holomorphic curves on projective varieties*, volume 654 of *Contemp. Math.*, pages 53–73. Amer. Math. Soc., Providence, RI, 2015.
38. Sinnou David, Michael Nakamaye, and Patrice Philippon. Bornes uniformes pour le nombre de points rationnels de certaines courbes. In *Diophantine geometry*, volume 4 of *CRM Series*, pages 143–164. Ed. Norm., Pisa, 2007.
39. Olivier Debarre. Varieties with ample cotangent bundle. *Compos. Math.*, 141(6):1445–1459, 2005.
40. Richard Dedekind and Heinrich Weber. *Theory of algebraic functions of one variable*, volume 39 of *History of Mathematics*. American Mathematical Society, Providence, RI, 2012. Translated from the 1882 German original and with an introduction, bibliography and index by John Stillwell.
41. Jean-Pierre Demailly. *Algebraic criteria for Kobayashi hyperbolic projective varieties and jet differentials*, volume 62 of *Proc. Sympos. Pure Math.* Amer. Math. Soc., Providence, RI, 1997.
42. Ya Deng, On the Diversio-Trapani Conjecture, ASÉNS, 4e série, t. 53, 2020, 787–814.
43. Lawrence Ein, Robert Lazarsfeld, Mircea Mustaţă, Michael Nakamaye, and Mihnea Popa. Asymptotic invariants of base loci. *Ann. Inst. Fourier (Grenoble)*, 56(6):1701–1734, 2006.
44. Hélène Esnault and Eckart Viehweg. *Lectures on vanishing theorems*, volume 20 of *DMV Seminar*. Birkhäuser Verlag, Basel, 1992.
45. Gerd Faltings. Endlichkeitssätze für abelsche Varietäten über Zahlkörpern. *Invent. Math.*, 73(3):349–366, 1983.
46. Gerd Faltings. Diophantine approximation on abelian varieties. *Ann. of Math. (2)*, 133(3):549–576, 1991.
47. Gerd Faltings. The general case of S. Lang's conjecture. In *Barsotti Symposium in Algebraic Geometry (Abano Terme, 1991)*, volume 15 of *Perspect. Math.*, pages 175–182. Academic Press, San Diego, CA, 1994.
48. Osamo Fujino. On quasi-Albanese maps. http://www4.math.sci.osaka-u.ac.jp/~fujino/quasi-albanese2.pdf, 2015.

49. Carlo Gasbarri. Lectures on the abc conjecture over function fields. T$_E$Xed notes for a course at *Doctoral program on Diophantine Geometry*, Rennes 2009.
50. Hans Grauert. Mordells Vermutung über rationale Punkte auf algebraischen Kurven und Funktionenkörper. *Publications Mathématiques de l'Institut des Hautes Études Scientifiques*, 25(1):131–149, 1965.
51. Daniel Greb, Stefan Kebekus, Sándor J. Kovács, and Thomas Peternell. Differential forms on log canonical spaces. *Publ. Math. Inst. Hautes Études Sci.*, (114):87–169, 2011.
52. Robin Hartshorne. Ample vector bundles. *Inst. Hautes Études Sci. Publ. Math.*, (29):63–94, 1966.
53. Robin Hartshorne. *Algebraic geometry*. Springer-Verlag, New York, 1977. GTM.
54. Brendan Hassett. Correlation for surfaces of general type. *Duke Math. J.*, 85(1):95–107, 1996.
55. Milena Hering, Mircea Mustaţă, and Sam Payne. Positivity properties of toric vector bundles. *Ann. Inst. Fourier (Grenoble)*, 60(2):607–640, 2010.
56. Marc Hindry and Joseph H. Silverman. *Diophantine geometry*, volume 201 of *GTM*. Springer-Verlag, New York, 2000.
57. Ariyan Javanpeykar. The Lang-Vojta conjectures on projective pseudo-hyperbolic varieties. In *Arithmetic Geometry of Logarithmic Pairs and Hyperbolicity of Moduli Spaces*, CRM Short Courses. Springer-Verlag, 2020.
58. Ariyan Javanpeykar and Ljudmila Kamenova. Demailly's notion of algebraic hyperbolicity: geometricity, boundedness, moduli of maps. *Math. Z.* 296(3–4):1645–1672, 2020.
    Ariyan Javanpeykar and Ljudmila Kamenova. Demailly's notion of algebraic hyperbolicity: geometricity, boundedness, moduli of maps, Jul 2018. arXiv:1807.03665.
59. Ariyan Javanpeykar and Junyi Xie. Finiteness properties of pseudo-hyperbolic varieties, 2019. arXiv:1909.12187.
60. Eric Katz, Joseph Rabinoff, and David Zureick-Brown. Uniform bounds for the number of rational points on curves of small Mordell-Weil rank. *Duke Math. J.*, 165(16):3189–3240, 2016.
61. Minhyong Kim. The non-abelian (or non-linear) method of Chabauty. In *Noncommutative geometry and number theory*, Aspects Math., E37, pages 179–185. Friedr. Vieweg, Wiesbaden, 2006.
62. Minhyong Kim. Fundamental groups and Diophantine geometry. *Cent. Eur. J. Math.*, 8(4):633–645, 2010.
63. Shoshichi Kobayashi. *Hyperbolic manifolds and holomorphic mappings*. World Scientific Publishing Co. Pte. Ltd., Hackensack, NJ, second edition, 2005. An introduction.
64. Tobias Kreisel and Sascha Kurz. There are integral heptagons, no three points on a line, on four on a circle. *Discrete Comput. Geom.*, 39(4):786–790, 2008.
65. Serge Lang. Integral points on curves. *Inst. Hautes Études Sci. Publ. Math.*, 6:27–43, 1960.
66. Serge Lang. Higher dimensional diophantine problems. *Bull. Amer. Math. Soc.*, 80:779–787, 1974.
67. Serge Lang. *Introduction to Arakelov theory*. Springer-Verlag, New York, 1988.
68. Serge Lang. *Survey on Diophantine Geometry*. Encyclopaedia of mathematical sciences. Springer, 1997.
69. Robert Lazarsfeld. *Positivity in algebraic geometry. I. Classical setting: line bundles and linear series. Ergebnisse der Mathematik und ihrer Grenzgebiete. 3. Folge. A Series of Modern Surveys in Mathematics*, 48. Springer-Verlag, Berlin, 2004.

70. Robert Lazarsfeld. Positivity in algebraic geometry. II. *Positivity for vector bundles, and multiplier ideals. Ergebnisse der Mathematik und ihrer Grenzgebiete. 3. Folge. A Series of Modern Surveys in Mathematics*, 49. Springer-Verlag, Berlin, 2004.

71. Aaron Levin. Generalizations of Siegel's and Picard's theorems. *Ann. of Math. (2)*, 170(2):609–655, 2009.

72. Kurt Mahler. Über die rationalen Punkte auf Kurven vom Geschlecht Eins. *Journal für die reine und angewandte Mathematik*, 170:168–178, 1934.

73. Juri I. Manin. Rational points on algebraic curves over function fields. *Izv. Akad. Nauk SSSR Ser. Mat.*, 27:1395–1440, 1963.

74. Barry Mazur. On some of the mathematical contributions of Gerd Faltings. In *Proceedings of the International Congress of Mathematicians, Vol. 1, 2 (Berkeley, Calif., 1986)*, pages 7–12, Providence, RI, 1987. Amer. Math. Soc.

75. William McCallum and Bjorn Poonen. The method of Chabauty and Coleman. In *Explicit methods in number theory*, volume 36 of *Panor. Synthèses*, pages 99–117. Soc. Math. France, Paris, 2012.

76. Atsushi Moriwaki. Remarks on rational points of varieties whose cotangent bundles are generated by global sections. *Math. Res. Lett.*, 2(1):113–118, 1995.

77. Christophe Mourougane. Families of hypersurfaces of large degree. *Journal of the European Mathematical Society*, pages 911–936, 2012.

78. David Mumford. A remark on Mordell's conjecture. *Amer. J. Math.*, 87:1007–1016, 1965.

79. Junjiro Noguchi. A higher-dimensional analogue of Mordell's conjecture over function fields. *Math. Ann.*, 258(2):207–212, 1981/82.

80. Patricia L. Pacelli. Uniform boundedness for rational points. *Duke Math. J.*, 88(1):77–102, 1997.

81. Gianluca Pacienza and Erwan Rousseau. On the logarithmic Kobayashi conjecture. *J. Reine Angew. Math.*, 611:221–235, 2007.

82. Fabien Pazuki. Bornes sur le nombre de points rationnels des courbes - en quête d'uniformitè, arxiv 1512.04907. *ArXiv e-prints*, Dec. 2015.

83. Min Ru. *Nevanlinna theory and its relation to Diophantine approximation.* World Scientific Publishing Co., Inc., River Edge, NJ, 2001.

84. Pierre Samuel. Compléments à un article de Hans Grauert sur la conjecture de Mordell. *Publications Mathématiques de l'IHÉS*, 29:55–62, 1966.

85. Pierre Samuel. *Lectures on old and new results on algebraic curves.* Notes by S. Anantharaman. Tata Institute of Fundamental Research Lectures on Mathematics, No. 36. Tata Institute of Fundamental Research, Bombay, 1966.

86. Christian Schnell. On a theorem of Campana and Păun. *Épijournal Geom. Algébrique*, 1:Art. 8, 9, 2017.

87. Jean-Pierre Serre. Morphismes universels et variétés d'Albanese. Variétés de Picard. Sém. C. Chevalley 3 (1958/59), No. 10, 22 p., 1960.

88. Jean-Pierre Serre. *Lectures on the Mordell-Weil theorem.* Aspects of Mathematics, E15. Friedr. Vieweg & Sohn, Braunschweig, 1989. Translated from the French and edited by Martin Brown from notes by Michel Waldschmidt.

89. Jafar Shaffaf. A solution of the Erdős-Ulam problem on rational distance sets assuming the Bombieri-Lang conjecture. *Discrete Comput. Geom.*, 60(2):283–293, 2018.

90. Carl Ludwig Siegel. Über einege Anwendungen diophantischer Approximationen. *Preuss. Akad. Wiss. Phys. Math. KL.*, 1:1–70, 1929.

91. Jozsef Solymosi and Frank de Zeeuw. On a question of Erdős and Ulam. *Discrete Comput. Geom.*, 43(2):393–401, 2010.

92. Michael Stoll. Uniform bounds for the number of rational points on hyperelliptic curves of small Mordell-Weil rank. *J. Eur. Math. Soc. (JEMS)*, 21(3):923–956, 2019.

93. Terence Tao. The Erdős-Ulam problem, varieties of general type, and the Bombieri-Lang conjecture, 2014. https://terrytao.wordpress.com/2014/12/20/the-erdos-ulam-problem-varieties-of-general-type-and-the-bombieri-lang-conjecture/.

94. Amos Turchet. Fibered threefolds and Lang-Vojta's conjecture over function fields. *Trans. Amer. Math. Soc.*, 369(12):8537–8558, 2017.

95. Eckart Viehweg. *Quasi-projective moduli for polarized manifolds*, volume 30 of *Ergebnisse der Mathematik und ihrer Grenzgebiete (3)*. Springer-Verlag, Berlin, 1995.

96. Paul Vojta. *Diophantine approximations and value distribution theory*, volume 1239 of *Lecture Notes in Mathematics*. Springer-Verlag, Berlin, 1987.

97. Paul Vojta. Siegel's theorem in the compact case. *Ann. of Math. (2)*, 133(3):509–548, 1991.

98. Paul Vojta. Integral points on subvarieties of semiabelian varieties. I. *Invent. Math.*, 126(1):133–181, 1996.

99. Paul Vojta. Integral points on subvarieties of semiabelian varieties. II. *Amer. J. Math.*, 121(2):283–313, 1999.

100. José Felipe Voloch. Diagonal equations over function fields. *Bol. Soc. Brasil. Mat.*, 16(2):29–39, 1985.

101. Julie Tzu-Yueh Wang. An effective Schmidt's subspace theorem over function fields. *Math. Z.*, 246(4):811–844, 2004.

102. Song-Yan Xie. On the ampleness of the cotangent bundles of complete intersections. *Invent. Math.*, 212(3):941–996, 2018.

103. Umberto Zannier. *Some Applications of Diophantine Approximation to Diophantine Equations*. Editrice Universitaria Udinese, 2003.

104. Ju. G. Zarhin. A finiteness theorem for isogenies of abelian varieties over function fields of finite characteristic. *Funkcional. Anal. i Priložen.*, 8(4):31–34, 1974.

Printed in the United States
by Baker & Taylor Publisher Services